NOISE-INDUCED PHENOMENA IN THE ENVIRONMENTAL SCIENCES

Randomness is ubiquitous in nature. Random drivers are generally considered a source of disorder in environmental systems. However, the interaction between noise and nonlinear dynamics may lead to the emergence of a number of ordered behaviors (in time and space) that would not exist in the absence of noise. This counterintuitive effect of randomness may play a crucial role in environmental processes. For example, seemingly "random" background events in the atmosphere can grow into larger instabilities that have great effects on weather patterns. This book presents the basics of the theory of stochastic calculus and its application to the study of noise-induced phenomena in environmental systems. It will be an invaluable reference text for ecologists, geoscientists, and environmental engineers interested in the study of stochastic environmental dynamics.

LUCA RIDOLFI is Professor in the Department of Water Engineering at the Polytechnic of Turin. His research focuses on ecohydrology, fluvial morphodynamics, biogeography, river water quality, and hydrodynamic instabilities.

PAOLO D'ODORICO is Professor in the Department of Environmental Sciences at the University of Virginia. His research focuses on ecohydrology, soil-moisture dynamics, global environmental change, nonlinear ecosystem dynamics, and desertification.

FRANCESCO LAIO is Associate Professor in the Department of Water Engineering at the Polytechnic of Turin. His research focuses on ecohydrology, stochastic hydrology, surface-water hydrology, and extreme value statistics.

NOISE-INDUCED PHENOMENA IN THE ENVIRONMENTAL SCIENCES

LUCA RIDOLFI
Politecnico di Torino

PAOLO D'ODORICO
University of Virginia

FRANCESCO LAIO
Politecnico di Torino

CAMBRIDGE
UNIVERSITY PRESS

University Printing House, Cambridge CB2 8BS, United Kingdom

One Liberty Plaza, 20th Floor, New York, NY 10006, USA

477 Williamstown Road, Port Melbourne, VIC 3207, Australia

4843/24, 2nd Floor, Ansari Road, Daryaganj, Delhi - 110002, India

79 Anson Road, #06-04/06, Singapore 079906

Cambridge University Press is part of the University of Cambridge.

It furthers the University's mission by disseminating knowledge in the pursuit of education, learning and research at the highest international levels of excellence.

www.cambridge.org
Information on this title: www.cambridge.org/9781108446785

© Luca Ridolfi, Paolo D'Odorico, and Francesco Laio 2011

First published 2011
First paperback edition 2017

A catalogue record for this publication is available from the British Library

Library of Congress Cataloging in Publication data
Ridolfi, Luca, 1963–
Noise-induced phenomena in the environmental sciences / Luca Ridolfi,
Paolo D'Odorico, Francesco Laio.
p. cm.
Includes bibliographical references and index.
ISBN 978-0-521-19818-9 (hardback)
1. Geophysical prediction – Mathematics.   2. Random noise theory.
3. Environmental sciences – Mathematics.   I. D'Odorico, Paolo. 1969–   II. Laio,
Francesco, 1973–   III. Title.
QC809.M37R53   2011
550–dc22   2011002162

ISBN 978-0-521-19818-9 Hardback
ISBN 978-1-108-44678-5 Paperback

A Gisella e Lorenzo
L.R.

Contents

Preface

Noise-induced phenomena are characterized by the ability of noise to induce order (either in space or in time) in dynamical systems. These phenomena are caused by the randomness of external drivers, and they would not exist in the absence of noise. The ability of noise to create order is counterintuitive. In fact, until recently, noise was generally associated with disordered random fluctuations around the steady states of the underlying deterministic dynamics. However, in the past few years the scientific community has become aware that noise can also have a more fundamental effect, in that it can determine new states and new dynamical patterns.

The speculative "beauty" of these dynamical behaviors, as well as the ubiquitous occurrence of random drivers in a number of natural and engineered systems, explains the great attention that has been recently paid to the study of noise-induced phenomena. A number of recent contributions have shown that the emergence of order and patterns in nature may result as an effect of the noise inherent in environmental variability. A typical example is climate fluctuations and their ability to induce dynamical behaviors that would not exist in the absence of random climate variability.

The main reason for writing this book is that there is a rich body of literature on noise-induced phenomena in the environmental sciences, and it has become difficult to keep track of the main theories, methods, and findings that have been presented in a number of research articles spread throughout the physics, mathematics, geoscience, and ecology journals. After working for a few years in this research field, we have become aware of the need for a book that (1) describes the main mechanisms of noise-induced order in space and in time; (2) presents rigorous mathematical tools addressing a relatively broad readership of environmental scientists, who are not necessarily familiar with the theory of stochastic processes; (3) focuses on applications to the environmental sciences; and (4) reviews a number of recent studies on noise-induced phenomena in environmental dynamics.

The goal of this book is to provide a synthesis of theories and methods for the study of noise-induced phenomena in the environment and to draw the attention of the earth and environmental science communities toward this fascinating and challenging research area. Through a number of examples of noise-induced phenomena we stress how in the natural environment random fluctuations are the rule and interesting behaviors may emerge from the interactions between the deterministic and stochastic components of environmental dynamics.

This book is not intended to be a comprehensive treatise on noise-induced phenomena. This relatively vast and fast-moving research field is enriched every day with new studies appearing in the literature. It would not be possible to contain in this volume an exhaustive review of all the existing theories of noise-induced order and their application to the environmental sciences. This book tries to provide an organized synthesis of the main contributions to this subject, drawing from material that is currently spread through a number of journals and other publications.

The completion of this book would have not been possible without the help, motivation, and support of a few collaborators and colleagues. We thank Stefania Scarsoglio and Fabio Borgogno (Politecnico di Torino) for providing invaluable help in performing the numerical simulations and contributing to the analysis of the results on noise-induced pattern formation (Chapters 5 and 6). We are grateful to Ignacio Rodriguez-Iturbe (Princeton University), Amilcare Porporato (Duke University), and Andrea Rinaldo (Ecóle Polytechnique Federale de Lausanne) for their unfailing encouragement and support through years of continued collaboration and companionship. We also acknowledge René Lefever (Université Libre de Bruxelles), whose work has inspired our research on noise-induced phenomena. We are also indebted to our institutions, the Polytechnic of Turin (Dipartimento di Idraulica, Trasporti e Infrastrutture Civili) and the University of Virginia (Department of Environmental Sciences) for providing high-quality academic environments that constantly stimulate our work.

Luca Ridolfi   Paolo D'Odorico   Francesco Laio

1

Introduction

1.1 Noise-induced phenomena

Most environmental dynamics are affected by a number of random drivers. This randomness typically results from the uncertainty inherent to the temporal or spatial variability of the driving processes. For example, if we consider the temperature record measured at a certain meteorological station, we can easily notice some obvious deterministic components of climate variability associated with the daily rotation of the Earth or with the annual seasonal cycle. At longer time scales we might recognize some patterns of interannual climate variability (e.g., the El Niño Southern Oscillation or the North Atlantic Oscillation) associated with temporally and spatially coherent anomalies in the atmospheric and oceanic circulations. These anomalies exhibit a certain degree of regularity in addition to unpredictable random fluctuations. However, besides these daily, annual, and interannual oscillations (and other deterministic signals), the temperature record will also exhibit some disorganized fluctuations that are typically ascribed to *environmental randomness*. In stochastic models of environmental dynamics this randomness is commonly expressed as *noise*.

Random environmental drivers are ubiquitous in nature. The occurrence of rainfall, sea storms, droughts, fires, or insect outbreaks are typical examples of random environmental processes. The noise underlying these processes is an important cause of environmental variability. What is the effect of this noise on the dynamics of environmental systems? Systems forced by random drivers are commonly expected to exhibit random fluctuations in their state variables. Thus the effect of noise is typically associated with the emergence of disorganized random fluctuations in the state of the system about its stable state(s). However, this trivial effect of noise is not the only possible way in which random drivers can affect a dynamical system. In the physics literature it was reported that noise can have a more fundamental role (e.g., Horsthemke and Lefever, 1984; Cross and Hohenberg, 1993; Garcia-Ojalvo and Sancho, 1999; Sagues et al., 2007). In fact it can induce new ordered states and new

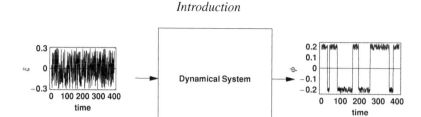

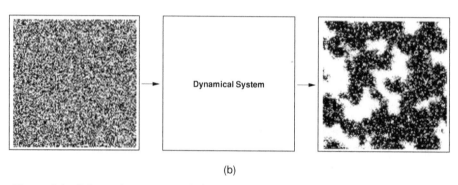

Figure 1.1. Schematic representation of noise-induced phenomena in (a) time and (b) space. Nonlinear systems forced by random drivers (left-hand panels) may lead to the emergence of ordered states in both time and space (right-hand panels).

bifurcations that would not exist in the deterministic counterpart of these systems. Noise can also modify the stability and resilience of deterministic states and induce coherence in the spatial and temporal variabilities of the state variables, including the emergence of periodic oscillations or the formation of spatial patterns. The ability of noise to induce order and organization (the so-called *constructive effect of noise*) is a quite counterintuitive effect that has seldom been investigated in the environmental science literature (e.g., May, 1972; Benzi et al., 1982a; Rodriguez-Iturbe et al., 1991; Katul et al., 2007). This book concentrates on this constructive effect of noise and on its ability to induce dynamical behaviors [i.e., states, bifurcations, spatial or temporal coherence (or both)] that do not exist in the underlying deterministic dynamics. We generically refer to these behaviors as *noise-induced phenomena*.

Figure 1.1 shows a schematic representation of some of these noise-induced behaviors: A nonlinear dynamical system forced by disordered random fluctuations either in space or in time may lead to the emergence of different forms of coherence, including for example noise-induced bistable dynamics [e.g., Fig. 1.1(a)] or morphogenesis [e.g., Fig. 1.1(b)]. In all of these cases, noise-induced behaviors appear when the noise intensity exceeds a critical level, whereas they disappear when it tends to zero.

The purpose of this book is to provide conceptual and mathematical tools that allow environmental scientists to familiarize themselves with the notion of noise-induced phenomena and with the idea that environmental noise (e.g., random climate fluctuations) is not necessarily a mere source of disturbance in environmental systems. The existence of a more fundamental role of noise should also be recognized in that it could have a crucial role in the way these systems respond to changes in environmental variability.

An example of the relevance of these constructive effects of noise can be found in the study of ecosystems' response to climate variability. Research in the field of ecosystem and population ecology has been investigating the effect of climate and land-use changes on ecosystem structure and function. In most cases the focus has been on how ecosystems respond to changes in the mean values of environmental parameters (e.g., mean annual precipitation or temperature) whereas the impact of changes in the variance has seldom been studied. However, recent climate-change studies indicate that, in addition to trends in the mean values of climate variables, interannual variability is also increasing (Katz and Brown, 1992; Easterling et al., 2000a, 2000b). It becomes therefore important to understand how this increase in the variance of environmental parameters will affect the dynamics of natural systems.

1.2 Time scales and noise models

The dynamics investigated in this book include four major components, namely, (i) a dynamical system, (ii) the external environment, (iii) a stochastic forcing, and (iv) some possible feedbacks between the state of the system and its environment or stochastic drivers. The first element is the deterministic *dynamical system* of interest. This system is a conceptually separate "entity" from a much more complex dynamical system, called *the environment*. The dynamical system generally involves a limited number of physical variables. We model it with a minimalist approach, which captures only the fundamental features of the dynamics. The physical variables representing the state of the system (e.g., plant biomass, soil moisture, soil thickness) are usually referred to as *state variables*. In this book we focus on systems that we can investigate by considering only one state variable, though we also consider the case of noise-induced phenomena that can emerge only in multivariate systems. The focus on univariate dynamics is motivated by their conceptual simplicity, the possibility of investigating them with analytical mathematical models, and the fact that their study allows us to show how noise-induced behaviors may emerge even without invoking complex interactions among a number of environmental variables. We express these univariate dynamics by using a first-order differential model,

$$\frac{d\phi}{dt} = f(\phi), \tag{1.1}$$

where $\phi(t)$ is the state variable characterizing the state of the system, t is time, and $f(\phi)$ is a deterministic algebraic function of the state variable. A spatially extended version of Eq. (1.1) would include also a term representing the effects on the dynamics of the values of ϕ in the neighboring sites; spatially extended systems are considered in the second half of this book, but the spatial coupling is neglected here to speculate more easily on the role of noise in the dynamics of ϕ.

The environment from which dynamical system (1.1) is extracted is generally much bigger than the dynamical system itself and is also called *external environment* to stress the fact that it is external to the dynamical system. Because the external environment is often too "large," complex, and also partially unknown to be modeled deterministically, its action on the dynamical system is represented as a stochastic process. Therefore we account for the randomness of environmental conditions through a stochastic forcing, which is modeled as noise $\xi(t)$. Thus the dynamics of the state variable read

$$\frac{\mathrm{d}\phi}{\mathrm{d}t} = f(\phi) + g(\phi)\xi(t), \tag{1.2}$$

where the (linear or nonlinear) algebraic function $g(\phi)$ accounts for the possibility that the effect of the random forcing on the system is modulated by the state of the system itself. The noise is *additive* when $g(\phi) = $ const, whereas it is *multiplicative* otherwise.

Because the dynamical system is much smaller than the external environment, it is generally unable to affect its environmental drivers. However, in some dynamics the impact of the system on its environment can be important. In these cases a *feedback* exists between the state of the system and environmental conditions. This book discusses some examples of feedbacks relevant to the biogeosciences. Feedbacks with random environmental drivers are typically expressed either through the multiplicative function $g(\phi)$ or through a state dependency in the stochastic forcing [i.e., $\xi(t, \phi)$].

One of the most crucial issues in the representation of these stochastic dynamics arises from the modeling of the random forcing. To address this point, we need to consider two time scales underlying the dynamics, namely, the time scale τ_s of the deterministic dynamical system and the time scale τ_n of the random forcing. The former describes the response time of the deterministic system after a displacement from its steady state(s) ϕ_s. In other words, τ_s expresses how slowly or quickly the system will converge to its stable state(s). For example, τ_s can be expressed as inversely proportional to the first-order derivative of $f(\phi)$ calculated in ϕ_s, $\tau_s \simeq 1/|f'(\phi_s)|$.

The time scale τ_n of the random forcing is a function of its autocorrelation, which expresses the interrelations existing within the noise signal, i.e., how the values of random forcing at different times depend on their temporal separation (a formal definition is provided in Chapter 2). The time scale τ_n can be expressed as the integral of the autocorrelation function, which represents the (linear) temporal memory of the noise. In other words, values of $\xi(t)$ calculated at two different times, t_1 and t_2, are significantly interrelated if the temporal separation, $|t_2 - t_1|$, is less than τ_n. The stochastic forcing is modeled in different ways, depending on the ratio, τ_s/τ_n.

- **Case with $\tau_s/\tau_n \gg 1$:** In this case the dynamical system is very slow with respect to the temporal variability of its random drivers. Thus, because the overall dynamics are not able to "perceive" the autocorrelation of the random forcing, this autocorrelation can be reasonably neglected. The random forcing can be therefore modeled as *white noise* (i.e., uncorrelated noise). Despite its being an idealization and a mathematical singularity, white noise is a cornerstone of the theory of stochastic processes in that it lends itself to analytical mathematical solutions. Therefore, even though it suffers from physically unrealistic behaviors (e.g., noncontinuous-noise realizations), white noise is very often used to simulate stochastic forcing in environmental systems. In this case the dynamics of the state variable $\phi(t)$, driven by a white noise, do not need to be analyzed at the τ_n scale – at which the nonphysical behaviors of the noise would emerge [e.g., nondifferentiable realizations of $\phi(t)$] – but should be investigated at the time scale τ_s of the deterministic system. At this scale the predictions of theories based on the use of white noise are indistinguishable – for all practical purposes – from the behavior of the system forced by autocorrelated noise with $\tau_n \ll \tau_s$. The combination of analytical tractability and success in providing a realistic description of stochastic dynamics explains the widespread usage of the white-noise approximation, when $\tau_s/\tau_n \gg 1$.

  It is worth noticing that our ability to obtain analytical results for the dynamics of $\phi(t)$ depends also on the Markovian character of the $\phi(t)$ process forced by the white noise. We recall that a stochastic process is called Markovian when its future evolution depends on only the present state of the process. Most of the exact analytical results in the theory of stochastic processes were obtained in the case of Markovian processes.

  In the modeling of the stochastic forcing, the high dimension of the phase space of the external environment and the absence of significant correlations are very often invoked in order to apply the central-limit theorem and assume that the noise is Gaussian. Thus Gaussian white noise is commonly adopted. However, it is important to understand that whiteness and Gaussianity are two distinct noise properties, and white non-Gaussian noises are often important drivers of dynamical systems. For example, in the following chapters we show how in the biogeosciences a number of non-Gaussian, intermittent stochastic processes can be conveniently modeled as white shot noise.

- **Case with $\tau_s/\tau_n \simeq 1$:** In this case the dynamics are slow enough to be sensitive to the autocorrelation of the random forcing. Thus the white-noise approximation would not provide an appropriate representation of the stochastic driver, and therefore autocorrelated (or *colored*) noises should be used.

  It should be stressed that the process $\phi(t)$ driven by a colored noise is not Markovian. In fact, at any time the noise component depends on the past through the autocorrelation of the noise term. This non-Markovianity introduces great complications that limit our ability to obtain exact analytical results. To make the representation of the dynamics mathematically more tractable, we can assume that at least the colored noise is Markovian, i.e., it is generated by a Markovian process as in the case of dichotomous Markov noise and Gaussian colored noise (i.e., the so-called Ornstein–Uhlenbeck process) presented in the next chapter. With this assumption the bivariate system composed of the state variable ϕ and the noise ξ becomes Markovian, and some analytical representations of the stochastic dynamics can be obtained.

- **Case with $\tau_s/\tau_n \ll 1$:** In this case the system responds very quickly to the noise forcing, thereby adjusting (almost) instantaneously to the random forcing. In other words, the state variable ϕ is always in equilibrium with the noise term [i.e., $d\phi/dt \simeq 0$]. In these conditions we can use the so-called *adiabatic elimination* of the ϕ variable, whereby the dynamics of ϕ are described as $f(\phi) + g(\phi)\xi = 0$ and the probabilistic properties of ϕ are derived from those of the noise.

In the following chapters we consider different types of noise, including the case of both white and colored noise as well as continuous and intermittent noises. We review the major properties of each type of noise as well as their possible use in the development of stochastic models of environmental systems. To this end, we use a number of examples and case studies to show the possible impact of both additive and multiplicative noise on environmental dynamics.

2

Noise-driven dynamical systems

2.1 Introduction

We consider dynamical systems that can be represented through a stochastic differential equation in the form

$$\frac{d\phi}{dt} = f(\phi) + g(\phi)\xi(t), \tag{2.1}$$

where ϕ is the state variable, $f(\phi)$ and $g(\phi)$ are deterministic functions of ϕ, and $\xi(t)$ is a noise term accounting for the random external fluctuations forcing the dynamics of ϕ.

The solution of Eq. (2.1) requires that the noise term $\xi(t)$ be suitably specified. The scope of this chapter is to describe the main features of the noise term and of the resulting dynamics of ϕ in four cases, which are particularly interesting in the environmental sciences: We model $\xi(t)$ as (i) dichotomous Markov noise (DMN), (ii) white shot noise (WSN), (iii) white Gaussian noise, and (iv) Markovian colored Gaussian noise. These representations of $\xi(t)$ are very well suited for investigating the role of the random drivers typically found in the biogeosciences, and they are simple enough to allow for the analytical (probabilistic) solution of Eq. (2.1).

We first consider (in Section 2.2) the case of dichotomous noise because it is more general in that both WSN and white Gaussian noise can be obtained as limit cases of the dichotomous noise. For this reason, these two white noises are described in detail right after the case of DMN noise (i.e., in Sections 2.3 and 2.4); colored Gaussian noise is presented in Section 2.5.

2.2 Dichotomous noise

2.2.1 Definition and properties

The dichotomous Markov process is a stochastic process described by a state variable $\xi_{dn}(t)$ that can take only two values, namely $\xi_{dn} = \Delta_1$ and $\xi_{dn} = \Delta_2$, with transition

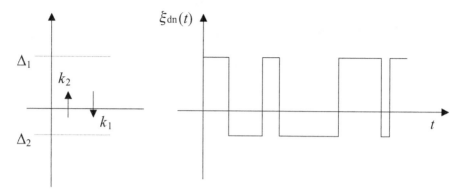

Figure 2.1. Parameters of the dichotomous noise and representation of a typical realization.

rate k_1 for the transition $\Delta_1 \rightarrow \Delta_2$ and k_2 for $\Delta_2 \rightarrow \Delta_1$. A realization of the process is shown in Fig. 2.1. The path of the noise is a step function with instantaneous jumps between Δ_1 and Δ_2 and random permanence times, t_1 and t_2, in these two states. The mean permanence times in the two states are $\langle t_1 \rangle = \tau_1 = 1/k_1$ and $\langle t_2 \rangle = \tau_2 = 1/k_2$. Moreover, when the transition rates k_1 and k_2 are constant in time, the permanence times are exponentially distributed random variables (e.g., Bena, 2006). If $\Delta_1 = |\Delta_2|$, the noise is called *symmetric* DMN; otherwise it is called *asymmetric* DMN. This type of noise was first introduced in information theory under the name of *random telegraph noise* or *Poisson square wave* (e.g., McFadden, 1959; Pawula, 1967); this process, studied in detail by physicists (Hongler, 1979; Kitahara et al., 1980), is called a two-state Markov process or DMN.

The probability $P_1(t)$ that the process is in the state Δ_1 at time t obeys the kinetic equation

$$\frac{dP_1(t)}{dt} = k_2 P_2(t) - k_1 P_1(t), \tag{2.2}$$

which includes a gain term $k_2 P_2(t)$ accounting for the probability of being in $\xi_{dn} = \Delta_2$ and jumping to $\xi_{dn} = \Delta_1$ and a loss term $-k_1 P_1(t)$ that accounts for the probability of escaping from the state $\xi_{dn} = \Delta_1$. Analogously, for the probability $P_2(t)$ that the process is in the state $\xi_{dn} = \Delta_2$ at time t, we have

$$\frac{dP_2(t)}{dt} = k_1 P_1(t) - k_2 P_2(t). \tag{2.3}$$

We obtain the steady solutions by neglecting the temporal derivatives on the left-hand side of Eqs. (2.2) and (2.3):

$$P_1 = \frac{k_2}{k_1 + k_2}, \qquad P_2 = \frac{k_1}{k_1 + k_2}. \tag{2.4}$$

The steady-state probability distribution of the state variable ξ_{dn} is then a discrete-valued distribution that can assume only two values, Δ_1 and Δ_2, with probability P_1 and P_2, respectively. The steady-state moment-generating function (see, for example, van Kampen, 1992, for a definition) is then

$$M_{dn}(v) = \sum_{i=1}^{2} e^{v\Delta_i} P_i = \frac{k_2 e^{v\Delta_1} + k_1 e^{v\Delta_2}}{k_1 + k_2}, \tag{2.5}$$

and the corresponding cumulant-generating function (see van Kampen, 1992, for a definition) is

$$K_{dn}(v) = \log[M_{dn}(v)] = \log[k_2 e^{v\Delta_1} + k_1 e^{v\Delta_2}] - \log[k_1 + k_2]. \tag{2.6}$$

By definition of the cumulant-generating function, the steady-state cumulant of the order of m is obtained as the mth derivative of $K_{dn}(v)$ with respect to v, calculated in $v = 0$. Therefore the mean of the process, κ_{1dn}, is

$$\langle \xi_{dn} \rangle = \kappa_{1dn} = \left. \frac{dK_{dn}(v)}{dv} \right|_{v=0} = \frac{k_2 \Delta_1 + k_1 \Delta_2}{k_1 + k_2}. \tag{2.7}$$

Because the DMN is used as a noise term in Eq. (2.1), it can be useful to consider a zero-average process. If this is the case, using Eq. (2.7) we have

$$\Delta_1 k_2 + \Delta_2 k_1 = \frac{\Delta_1}{\tau_2} + \frac{\Delta_2}{\tau_1} = 0. \tag{2.8}$$

In this case the (stationary) dichotomous Markov process is characterized by three independent parameters. For example, we can choose (i) the two transition rates k_1 and k_2 (or the mean durations τ_1 and τ_2) and (ii) assign the value of one of the states of ξ_{dn}, say Δ_1, and obtain the other value (i.e., Δ_2) by using Eq. (2.8). Unless explicitly stated otherwise, in what follows we refer to the case of zero-mean [Eq. (2.8)] DMN.

The variance of the dichotomous process is

$$\langle (\xi_{dn} - \kappa_{1dn})^2 \rangle = \kappa_{2dn} = \left. \frac{d^2 K_{dn}(v)}{dv^2} \right|_{v=0} = \frac{k_1 k_2 (\Delta_2 - \Delta_1)^2}{(k_1 + k_2)^2} = -\Delta_1 \Delta_2, \tag{2.9}$$

and the autocovariance function is

$$\langle \xi_{dn}(t) \xi_{dn}(t') \rangle = \frac{k_1 k_2 (\Delta_2 - \Delta_1)^2}{(k_1 + k_2)^2} e^{-|t-t'|(k_1+k_2)} = -\Delta_1 \Delta_2 e^{-|t-t'|(k_1+k_2)}, \tag{2.10}$$

as demonstrated in Box 2.1, Eq. (B2.1-5). The structure of the autocovariance function shows that the dichotomous noise is a colored noise, i.e., it is autocorrelated. This is an important characteristic that explains why this type of noise is commonly used to mimic natural processes in the biogeosciences (see Subsection 2.2.2). A typical temporal scale of a correlated process is the integral scale $\mathcal{I}$, defined as the ratio between the area subtended by the autocovariace function (i.e., the integral of the autocovariance function with respect to the lag) and the variance of the process. The

---

Box 2.1: Transient dynamics of the dichotomous Markov process

Some considerations of the transient dynamics of the dichotomous Markov process can be of interest. Equations (2.2) and (2.3) can be solved to give

$$P_1(t) = \tau_c k_2 \left(1 - e^{-t/\tau_c}\right) + P_1(0)e^{-t/\tau_c}, \tag{B2.1-1}$$

$$P_2(t) = \tau_c k_1 \left(1 - e^{-t/\tau_c}\right) + P_2(0)e^{-t/\tau_c}, \tag{B2.1-2}$$

where $P_1(0)$ and $P_2(0) = 1 - P_1(0)$ are the initial conditions and

$$\tau_c = \frac{1}{k_1 + k_2} = \frac{\tau_1 \tau_2}{\tau_1 + \tau_2} \tag{B2.1-3}$$

is the characteristic relaxation time of the process.

It can be argued from Eqs. (B2.1-1) and (B2.1-2) that the joint probability for ξ_{dn} at time t and t' is

$$p_{i,j} = \text{prob}\left[\xi_{dn}(t) = \Delta_i, \xi_{dn}(t') = \Delta_j\right]$$

$$= (1 - k_i \tau_c)(1 - k_j \tau_c)\tau_c^2 \left(1 - e^{-\frac{|t-t'|}{\tau_c}}\right)$$

$$+ \delta_{i,j}(1 - k_i \tau_c)e^{-\frac{|t-t'|}{\tau_c}}, \tag{B2.1-4}$$

where $\delta_{i,j}$ is the Kronecker delta function and $i, j = 1, 2$.

The steady-state autocorrelation function is then

$$\langle \xi_{dn}(t)\xi_{dn}(t') \rangle = \sum_{i,j=1}^{2} p_{i,j}\Delta_i \Delta_j = \tau_c(\Delta_1^2 k_2 + \Delta_2^2 k_1)e^{-\frac{|t-t'|}{\tau_c}}$$

$$= -\Delta_1 \Delta_2 e^{-\frac{|t-t'|}{\tau_c}} = \frac{s_{dn}}{\tau_c}e^{-\frac{|t-t'|}{\tau_c}}, \tag{B2.1-5}$$

where Eq. (2.8) has been repeatedly used. The term

$$s_{dn} = k_1 k_2 \tau_c^3 (\Delta_2 - \Delta_1)^2 = -\Delta_1 \Delta_2 \tau_c \tag{B2.1-6}$$

in Eq. (B2.1-5) represents the noise *amplitude* or *intensity*.

---

integral scale is generally interpreted as a measure of the memory of the process, and in the case of dichotomous noise it is

$$\mathcal{I} = \frac{1}{k_1 + k_2} = \tau_c. \tag{2.11}$$

Some generalization of dichotomous noise were proposed in the literature. Notable examples include the so-called *trichotomous* noise (Mankin et al., 1999), characterized by a three-valued state space and its further generalization, multivalued noise (Weiss et al., 1987); compound dichotomous noise (van den Broeck, 1983), in which the value assumed in one of the two states is a random variable; *complicated* DMN (Li, 2007) in which both states are (Gaussian) random variables; the gamma and McFadden dichotomous noise (Pawula et al., 1993), in which the distribution of the permanence times in the two states follows a gamma or a McFadden probability distribution rather than an exponential distribution, as in the classical DMN.

2.2.2 Dichotomous noise in the environmental sciences

Dichotomous noise can be encountered in a wide variety of physical and mathematical models for two main reasons. First, dichotomous noise is a simple and analytically tractable form of colored noise; in fact, it is possible to obtain exact analytical solutions for a stochastic differential equation driven by DMN in steady-state conditions. Thus DMN can be used to investigate the effect of an autocorrelated random driver on a dynamical system. We define this approach as the *functional* usage of the DMN because of its function as a tool to conveniently represent a correlated (i.e., colored) random forcing. In this case (functional usage) the starting point is a given deterministic system, say $d\phi/dt = f(\phi)$, and DMN is typically used to investigate the effect of a zero-mean correlated random driver in this system. There are several examples of processes in which the autocorrelation is one of the key characteristics of the external forcing. For example, consider the variety of biogeochemical processes that are affected by (random) daily temperature or the case of fluvial processes forced by river flow. In these processes the autocorrelation of the random forcing is relevant, and it cannot be neglected. Dichotomous noise is one of the two main mathematical tools available for the study of the effects of colored noise on dynamical systems. Colored Gaussian noise, described in Section 2.5, is another type of autocorrelated noise, which is often used in dynamical models with analytical solutions. The functional usage of DMN can be also motivated by the fact that both white Poisson noise and white Gaussian noise can be recovered from the dichotomous noise by taking suitable limits, as shown in Subsections 2.3.2 and 2.4.2.

Dichotomous noise is commonly used also for its ability to model a broad class of systems that randomly switch between two dynamical states. This approach is called the *mechanistic* usage of DMN, in which DMN is used to represent a dynamical behavior, i.e., the mechanism of random switching between two states.

The distinction between the functional and the mechanistic use of DMN is crucial in the stochastic modeling of a process. The mechanistic approach is frequently used for a class of processes characterized by the following three components: (i) the dynamical system, whose state is expressed by one state variable, $\phi(t)$; (ii) a random driver $q(t)$; (iii) a threshold value θ of $q(t)$, marking the transition between conditions favorable to growth or to decay of ϕ. For example, the variable ϕ could represent vegetation biomass in semiarid environments (D'Odorico et al., 2005) or riparian vegetation along a river (Camporeale and Ridolfi, 2006); correspondingly, q could represent random rainfall fluctuations that determine the occurrence of water-limited conditions or of flooded or unflooded states, respectively. Thus the stochastic driver determines the random alternation between stressed and unstressed conditions for the ecosystem.

The two alternating dynamics of ϕ involve growth and decay and can be modeled by two functions, $f_1(\phi)$ and $f_2(\phi)$, respectively,

$$\frac{d\phi}{dt} = \begin{cases} f_1(\phi) & \text{if} \quad q(t) \geq \theta \\ f_2(\phi) & \text{if} \quad q(t) < \theta \end{cases},$$

(2.12a)

(2.12b)

whre $f_1(\phi) > 0$ and $f_2(\phi) < 0$. Equations (2.12a) and (2.12b) are written assuming that q is a resource, in that values of q exceeding the threshold are associated with unstressed conditions (in the sense that ϕ grows). However, the general results do not change when the random driver is a stressor. In this case the conditions in (2.12a) and (2.12b) are reversed, i.e., growth or decay occurs when q is below or above the threshold, respectively.

The class of processes defined by (2.12a) and (2.12b) can be conveniently represented through a suitable dichotomous Markov process, thereby leading to a mechanistic usage of DMN. Thus the process is random and switches between two possible states: "success" (or "no stress") when q is above the threshold or "failure" (or "stress") when q is below the threshold [see Fig. 2.2(a)]. This is by definition a dichotomous process. If we further suppose that q is uncorrelated, the driving noise is the outcome of a Bernoulli trial with probability of success $k_2 = 1 - P_Q(\theta)$, where $P_Q(\theta)$ is the cumulative probability distribution of q, evaluated in $q = \theta$. The residence time in the "above-threshold" state is then an integer number n_1 with a geometric probability distribution of $p_{N_1}(n_1) = k_2^{n_1-1}(1 - k_2), n_1 = 1, \ldots, \infty$, with average $\langle n_1 \rangle = 1/(1 - k_2)$. Analogously, the residence time n_2 in the "below-threshold" state is distributed as $p_{N_2}(n_2) = (1 - k_2)^{n_1-1}k_2, n_1 = 1, \ldots, \infty$, with average $\langle n_1 \rangle = 1/k_2$. The DMN (in its mechanistic interpretation) is obtained as the continuous-time approximation of this driving process [see Fig. 2.2(b)]. In fact, in continuous time the residence time in each state becomes exponentially distributed (the exponential distribution is the continuous counterpart of the geometric distribution: e.g. Kendall and Stuart, 1977), which is a basic property of DMN (see Subsection 2.2.1).

The overall dynamics of the variable ϕ can then be expressed by a stochastic differential equation forced by DMN $\xi_{dn}(t)$, assuming (constant) values Δ_1 and Δ_2 (see Fig. 2.1):

$$\frac{d\phi}{dt} = f(\phi) + g(\phi)\xi_{dn}(t),$$

(2.13)

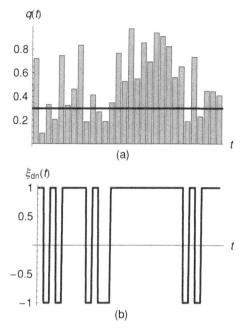

Figure 2.2. (a) The behavior of an uncorrelated random variable $q(t)$ around a threshold value θ (continuous horizontal line), (b) the path of the dichotomous noise that mimics the threshold effect on the dynamics of $q(t)$.

with

$$f(\phi) = -\frac{\Delta_2 f_1(\phi) - \Delta_1 f_2(\phi)}{\Delta_1 - \Delta_2}, \qquad g(\phi) = \frac{f_1(\phi) - f_2(\phi)}{\Delta_1 - \Delta_2}. \tag{2.14}$$

The transition rates are defined by $k_1 = P_Q(\theta)$ and $k_2 = 1 - k_1 = 1 - P_Q(\theta)$. As for the values of Δ_1 and Δ_2, in the mechanistic approach, DMN is used as a tool to randomly switch between $f_1(\phi)$ and $f_2(\phi)$. The only mechanistically relevant characteristics of DMN are in this case the switching rates k_1 and k_2, whereas the other noise characteristics, including its mean $\Delta_1 k_2 + \Delta_2 k_1$ and variance $-\Delta_1 \Delta_2$, are not relevant to the representation of the dynamics of ϕ. In fact, in this case ϕ switches between two dynamics [$f_1(\phi)$ and $f_2(\phi)$] that are independent of Δ_1 and Δ_2. As a consequence, Δ_1 and Δ_2 may assume arbitrary values, and it is important to assign values of the switching rates k_1 and k_2 that are consistent with the fluctuations of $q(t)$ across the threshold.

The functional interpretation of the DMN, in contrast, is commonly used to simply investigate how an autocorrelated random forcing would affect the dynamics of a system. Thus the dynamical model has two components, namely (i) the deterministic dynamics $d\phi/dt = f(\phi)$ and (ii) an autocorrelated random forcing $\xi(t)$. The effect of $\xi(t)$ on the dynamics can be in general modulated by a function $g(\phi)$ of the state variable. The temporal dynamics are therefore modeled by the stochastic differential equation $d\phi/dt = f(\phi) + g(\phi)\xi(t)$. The functional usage of the DMN consists in approximating the colored noise, $\xi(t)$, as a DMN, i.e., $\xi(t) = \xi_{dn}(t)$. In this case none of the parameters k_1, k_2, Δ_1, and Δ_2 has an arbitrary value. In fact, these parameters need to be determined by adapting the DMN to the characteristics of the driving noise (i.e., for example, by matching the mean, variance, skewness, and correlation scale). Moreover, the functions $f(\phi)$ and $g(\phi)$ are in this case assigned a priori, whereas $f_1(\phi)$ and $f_2(\phi)$ are obtained from (2.14) and depend on the noise characteristics

$$f_1(\phi) = f(\phi) + g(\phi)\Delta_1, \qquad f_2(\phi) = f(\phi) + g(\phi)\Delta_2. \qquad (2.15)$$

To summarize, the functional or mechanistic usage of the dichotomous noise corresponds to two distinct approaches to the stochastic modeling of processes driven by DMN. The differences may be relevant, in particular when dealing with noise-induced transitions (see Chapter 3). Once the approach that is suitable for the study of a specific problem is selected, the dichotomous noise provides a useful modeling framework with a number of applications to the environmental sciences, as shown in Chapter 4. Thus in the following subsection we present some probabilistic methods to solve stochastic equations driven by dichotomous noise.

2.2.3 Stochastic processes driven by dichotomous noise

2.2.3.1 General framework

Consider the stochastic process $\phi(t)$ driven by multiplicative dichotomous noise,

$$\frac{d\phi}{dt} = f(\phi) + g(\phi)\xi_{dn}(t), \qquad (2.16)$$

where $f(\phi)$ and $g(\phi)$ are two deterministic functions of the state variable ϕ. We assume that both $f(\phi)$ and $g(\phi)$ are continuous for any value of ϕ. Depending on the approach or interpretation used for the DMN, the functions $f(\phi)$ and $g(\phi)$ are assigned with a direct physical meaning (functional approach) or are obtainable from functions $f_{1,2}(\phi)$ by use of Eqs. (2.14) (mechanistic approach).

Some examples can help us understand the role of the driving force in the dynamics of $\phi(t)$. We consider four simple cases. The first three examples refer to the mechanistic usage of DMN, and the fourth case considers the functional usage:

- **Example 2.1:** $\phi(t)$ exponentially increases (decreases) when the noise is in the Δ_1 (Δ_2) state:

$$f_1(\phi) = 1 - \phi, \qquad f_2(\phi) = -\phi. \qquad (2.17)$$

  An example is shown in Fig. 2.3(a).
- **Example 2.2:** $\phi(t)$ linearly increases (decreases) when the noise is in the Δ_1 (Δ_2) state:

$$f_1(\phi) = 1, \qquad f_2(\phi) = -1. \qquad (2.18)$$

  An example of the resulting dynamics of $\phi(t)$ is shown in Fig. 2.3(b).

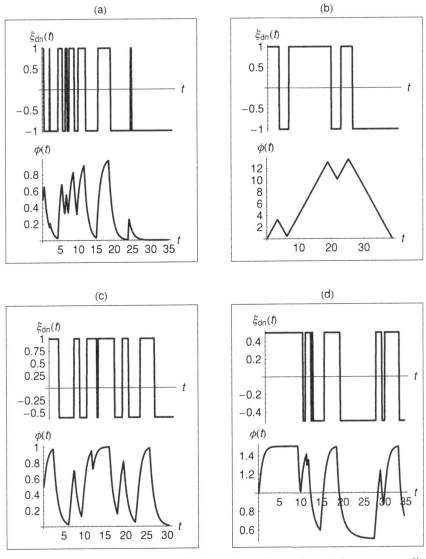

Figure 2.3. The four panels (a)–(d) show the noise path and the corresponding evolution of the $\phi(t)$ variable from Eq. (2.16): (a) Example 2.1, Example 2.2, (c) Example 2.3, (d) Example 2.4.

- **Example 2.3:** $\phi(t)$ increases following a (shifted) logistic law (see Subsection 3.2.1.1) when the noise is in the Δ_1 state, whereas it decreases exponentially in the Δ_2 state:

$$f_1(\phi) = (\phi - a)(1 - \phi), \qquad f_2(\phi) = -\phi, \tag{2.19}$$

where a is a constant. An example for $a = -0.5$ is shown in Fig. 2.3(c).

- **Example 2.4:** $\phi(t)$ follows logistic-type deterministic dynamics $f(\phi)$, perturbed by a dichotomous noise modulated by a linear $g(\phi)$:

$$f(\phi) = \phi(\beta - \phi), \qquad g(\phi) = \phi, \tag{2.20}$$

where $\beta > 0$ is a parameter. An example, calculated with $\beta = 1$, is shown in Fig. 2.3(d).

2.2.3.2 Derivation of the steady-state probability density function

This subsection is devoted to obtaining the steady-state probability density function (pdf) for the process described by Langevin equation (2.16). The standard procedure typically followed to address this task involves (i) deriving the master equations for the process, i.e., the forward differential equations that relate the state probabilities at different points in time; (ii) taking the limit as $t \to \infty$ in the master equation to attain statistically steady-state conditions; and (iii), solving the resulting forward differential equation to find the steady-state pdf. The detailed derivation of the steady-state probability distribution of ϕ following this approach is described in Box 2.2. In this subsection, we describe a simpler approach in a way that the nonexpert reader can more easily follow how the solution of Langevin equation (2.16) is determined.

Consider the probability that, at time $t + \Delta t$, the state variable takes a value contained within the interval $[\phi, \phi + d\phi]$ and the noise is in state $\xi_{\rm dn} = \Delta_1$. These conditions may be attained either when $\xi_{\rm dn} = \Delta_1$ at time t, no jumps of $\xi_{\rm dn}$ occur in $[t, t + \Delta t]$, and the value of ϕ at time t is $\phi - f_1(\phi)\Delta t$, or when at time t we have $\xi_{\rm dn} = \Delta_2$, the random variable $\xi_{\rm dn}$ shifts from Δ_2 to Δ_1 in the interval $[t, t + \Delta t]$, and the state variable at time t is $\phi - f^*(\phi)\Delta t$, where $f^*(\phi)$ is a suitable combination of $f_1(\phi)$ and $f_2(\phi)$ to account for the fact that in the interval $[t, t + \Delta t]$ both functions contribute to determine the trajectory of ϕ. Thus the joint probability that $\xi_{\rm dn} = \Delta_1$ and ϕ is comprised within $[\phi, \phi + \Delta\phi]$ can be expressed as

$$\mathcal{P}[\phi, \Delta_1; t + \Delta t]d\phi = (1 - k_1\Delta t)\mathcal{P}[\phi - f_1(\phi)\Delta t, \Delta_1; t]d[\phi - f_1(\phi)\Delta t]$$

$$+ k_2\Delta t\mathcal{P}[\phi - f^*(\phi)\Delta t, \Delta_2; t]d[\phi - f^*(\phi)\Delta t]. \tag{2.21}$$

The first term on the right-hand-side of Eq. (2.21) is the product of three factors: (i) the probability that the noise $\xi_{\rm dn}$ remains in the state $\xi_{\rm dn} = \Delta_1$ in the interval $(t, t + \Delta t)$. This probability is 1 minus the probability $k_1\Delta t$ that a jump occurs from Δ_1 to Δ_2 in the same interval; (ii) the joint probability $\mathcal{P}$ that at time t noise is equal to Δ_1 and the state variable is at $\phi - \Delta\phi$, where $\Delta\phi = f_1(\phi)\Delta t$ from Eqs. (2.15); and (iii) the infinitesimal amplitude of the interval $d[\phi - f_1(\phi)\Delta t]$. Similarly, the second term represents the probability that $\xi_{\rm dn} = \Delta_2$, the state variable is equal to $\phi - \Delta\phi$ at time t, and a jump from Δ_2 to Δ_1 occurs during the interval $(t, t + \Delta t)$. This jump in $\xi_{\rm dn}$ occurs with probability $k_2\Delta t$. Note that, because the jump may occur at any time during $(t, t + \Delta t)$, in this case $\Delta\phi$ is expressed as $\Delta\phi = f^*(\phi)\Delta t$, where $f^*(\phi)$ is a combination of $f_1(\phi)$ and $f_2(\phi)$ (see Horsthemke and Lefever, 1984, Eq. 9.22).

The probability of occurrence of two or more jumps can be neglected in Eq. (2.21) because it is supposed that Δt is small.

Using a Taylor's expansion truncated to the first order we have[1]

$$\mathcal{P}[\phi, \Delta_1; t + \Delta t] d\phi$$

$$= k_2 \Delta t \mathcal{P}[\phi, \Delta_2; t] d\phi$$

$$+ (1 - k_1 \Delta t) \left(\mathcal{P}[\phi, \Delta_1; t] - \frac{\partial \mathcal{P}[\phi, \Delta_1; t]}{\partial \phi} f_1(\phi) \Delta t \right) \left(1 - \frac{\partial f_1(\phi)}{\partial \phi} \Delta t \right) d\phi.$$

$$(2.22)$$

Notice how Eq. (2.22) is independent of the form of the function $f^*(\phi)$ describing the trajectory of ϕ in correspondence to jump occurrences. Rearranging the terms, dividing by $d\phi$ and Δt, and taking the limit for $\Delta t \to 0$, we finally obtain the forward Kolmogorov equation:

$$\frac{\partial \mathcal{P}(\phi, \Delta_1, t)}{\partial t} = -\frac{\partial}{\partial \phi}[\mathcal{P}(\phi, \Delta_1, t) f_1(\phi)] - \mathcal{P}(\phi, \Delta_1, t) k_1 + \mathcal{P}(\phi, \Delta_2, t) k_2. \quad (2.23)$$

Analogously, we can write for the probability that at time $t + \Delta t$ the state variable is within $(\phi, \phi + d\phi)$ and $\xi_{dn} = \Delta_2$,

$$\mathcal{P}[\phi, \Delta_2; t + \Delta t] d\phi = k_1 \Delta t \mathcal{P}[\phi - f^{**}(\phi) \Delta t, \Delta_1; t] d[\phi - f^{**}(\phi) \Delta t]$$

$$+ (1 - k_2 \Delta t) \mathcal{P}[\phi - f_2(\phi) \Delta t, \Delta_2; t] d[\phi - f_2(\phi) \Delta t], \quad (2.24)$$

where $f^{**}(\phi)$ describes the trajectory of ϕ in the interval $(t, t + \Delta t)$ in the case in which ξ_{dn} switches from Δ_1 to Δ_2 in that interval. After a Taylor expansion for $\Delta t \to 0$ we obtain the second forward Kolmogorov equation:

$$\frac{\partial \mathcal{P}(\phi, \Delta_2, t)}{\partial t} = -\frac{\partial}{\partial \phi}[\mathcal{P}(\phi, \Delta_2, t) f_2(\phi)] - \mathcal{P}(\phi, \Delta_2, t) k_2 + \mathcal{P}(\phi, \Delta_1, t) k_1. \quad (2.25)$$

We refer the interested reader to Box 2.2 for the derivation of the full master equation in the time-dependent case. Here we concentrate on steady-state solutions of (2.16).

---

[1] $\mathcal{P}[\phi - f_1(x)\Delta t, \Delta_1; t]$ can be expanded in a Taylor's series around $\Delta t = 0$:

$$\mathcal{P}[\phi - f_1(\phi)\Delta t, \Delta_1; t] = \mathcal{P}[\phi, \Delta_1; t] + \frac{\partial \mathcal{P}[\phi - f_1(\phi)\Delta t, \Delta_1; t]}{\partial \Delta t} \bigg|_{\Delta t=0} \Delta t$$

$$= \mathcal{P}[\phi, \Delta_1; t] + \frac{\partial \mathcal{P}[z, \Delta_1; t]}{\partial z} \bigg|_{z=\phi} \frac{\partial z}{\partial \Delta t} \bigg|_{\Delta t=0} \Delta t$$

$$= \mathcal{P}[\phi, \Delta_1; t] - \frac{\partial \mathcal{P}[\phi, \Delta_1; t]}{\partial \phi} f_1(\phi)\Delta t,$$

where the series has been truncated to the first order and $z = \phi - f_1(\phi)\Delta t$.

Box 2.2: Master equation of a stochastic process driven by dichotomous Markov noise

In this box, we show the key steps to determine the nonsteady-state master equation of the stochastic process described by (2.16); further details can be found in Horsthemke and Lefever (1984).

We introduce the two quantities

$$\mathcal{P}(\phi, t) = \mathcal{P}(\phi, \Delta_1, t) + \mathcal{P}(\phi, \Delta_2, t), \tag{B2.2-1}$$

$$q(\phi, t) = k_2 \mathcal{P}(\phi, \Delta_2, t) - k_1 \mathcal{P}(\phi, \Delta_1, t), \tag{B2.2-2}$$

where $q(\phi, t)$ is an auxiliary function, and $\mathcal{P}(\phi, t)$ expresses the time-dependent probability distribution of ϕ independently of the state of noise.

Adding Eq. (2.23) to Eq. (2.25) and using zero-average condition (2.8), we obtain

$$\frac{\partial \mathcal{P}}{\partial t} = -\frac{\partial}{\partial \phi}[\mathcal{P} f(\phi)] - \frac{\Delta_2 - \Delta_1}{k_1 + k_2} \frac{\partial}{\partial \phi}[q g(\phi)], \tag{B2.2-3}$$

and, if Eqs. (2.23) and (2.25) are multiplied by k_1 and k_2, respectively, and then (2.23) is subtracted from (2.25), we obtain

$$\frac{\partial q}{\partial t} = \frac{\partial}{\partial \phi} \left\{ \left[f(\phi) + \frac{k_2 \Delta_2 - k_1 \Delta_1}{k_1 + k_2} + (k_1 + k_2) \right] q \right\}$$
$$- \frac{(\Delta_2 - \Delta_1) k_1 k_2}{k_1 + k_2} \frac{\partial}{\partial x}[g(\phi) \mathcal{P}]. \tag{B2.2-4}$$

Using the independent variable

$$\eta = \int \frac{d\phi}{f(\phi) + \frac{k_2 \Delta_2 - k_1 \Delta_1}{k_1 + k_2} g(\phi)}, \tag{B2.2-5}$$

we can reduce differential equation (B2.2-4) to a form that can be analytically integrated (Polyanin et al., 2002), leading to

$$q(\phi, t) = \int_{-\infty}^{t} e^{-\left\{ \frac{\partial}{\partial x}\left[f(\phi) + \frac{k_2 \Delta_2 - k_1 \Delta_1}{k_1 + k_2} g(\phi) \right] + k_1 + k_2 \right\}(t - t')}$$
$$\times \frac{(\Delta_2 - \Delta_1) k_1 k_2}{k_1 + k_2} \frac{\partial}{\partial \phi}[g(\phi) \mathcal{P}(\phi, t')] dt', \tag{B2.2-6}$$

where the statistical independence between the noise ξ_{dn} and the process $\phi(t)$ at $t \to -\infty$ has been assumed as the initial condition.

Equation (B2.2-6) can be substituted into (B2.2-3) to obtain the master equation:

$$\frac{\partial \mathcal{P}(\phi, t)}{\partial t} = -\frac{\partial}{\partial \phi} \left[f(\phi) + \frac{k_2 \Delta_2 - k_1 \Delta_1}{k_1 + k_2} g(\phi) \right] \mathcal{P}(\phi, t)$$

$$+ \frac{k_1 k_2 (\Delta_2 - \Delta_1)^2}{(k_1 + k_2)^2} \frac{\partial g(\phi)}{\partial \phi}$$

$$\times \int_{-\infty}^{t} e^{-\left\{ \frac{\partial}{\partial \phi}\left[f(x) + \frac{k_2 \Delta_2 - k_1 \Delta_1}{k_1 + k_2} g(\phi) \right] + k_1 + k_2 \right\}(t - t')}$$

$$\times \frac{(\Delta_2 - \Delta_1) k_1 k_2}{k_1 + k_2} \frac{\partial}{\partial \phi}[g(\phi) \mathcal{P}(\phi, t')] dt'. \tag{B2.2-7}$$

This rather intricate integrodifferential equation shows that, in general, $\phi(t)$ is not a Markovian process. In fact, the probability distribution of ϕ at time t depends on the integral between $-\infty$ and t of a function of ϕ. Equation (B2.2-7) can be analytically solved in only very simple cases (Bena, 2006). An important case is the so-called persistent diffusion on a line ($f(\phi) = 0$ and $g(\phi) = 1$) that has interesting applications in chemistry and physics (van den Broeck, 1990; Bena, 2006).

In steady-state conditions the temporal derivatives in Eqs. (2.23) and (2.25) are equal to zero and the forward Kolmogorov equations become

$$\frac{\partial}{\partial \phi}[\mathcal{P}(\phi, \Delta_1) f_1(\phi)] + \mathcal{P}(\phi, \Delta_1) k_1 - \mathcal{P}(\phi, \Delta_2) k_2 = 0,$$

$$\frac{\partial}{\partial \phi}[\mathcal{P}(\phi, \Delta_2) f_2(\phi)] + \mathcal{P}(\phi, \Delta_2) k_2 - \mathcal{P}(\phi, \Delta_1) k_1 = 0. \qquad (2.26)$$

By summing up Eqs. (2.26) and integrating with respect to ϕ, we obtain

$$\mathcal{P}(\phi, \Delta_2) = -\mathcal{P}(\phi, \Delta_1) \frac{f_1(\phi)}{f_2(\phi)}, \qquad (2.27)$$

where the integration constant is set to zero. Equation (2.27) inserted into the first of Eqs. (2.26) leads to

$$\frac{\partial}{\partial \phi}[\mathcal{P}(\phi, \Delta_1) f_1(\phi)] + \mathcal{P}(\phi, \Delta_1) k_1 + \mathcal{P}(\phi, \Delta_1) \frac{f_1(\phi)}{f_2(\phi)} k_2 = 0. \qquad (2.28)$$

The integration of (2.28) provides the probability distribution

$$\mathcal{P}(\phi, \Delta_1) = \frac{C}{f_1(\phi)} \exp\left\{ -\int_\phi \left[\frac{k_1}{f_1(\phi')} + \frac{k_2}{f_2(\phi')} \right] d\phi' \right\}, \qquad (2.29)$$

where C is an integration constant. Equation (2.29) can be set in (2.27) to obtain

$$\mathcal{P}(\phi, \Delta_2) = -\frac{C}{f_2(x)} \exp\left\{ -\int_\phi \left[\frac{k_1}{f_1(\phi')} + \frac{k_2}{f_2(\phi')} \right] d\phi' \right\}. \qquad (2.30)$$

We now use these two joint distributions to determine the marginal steady-state pdf $p_\Phi(\phi)$ for the state variable ϕ, as $p_\Phi(\phi) = \mathcal{P}(\phi, \Delta_1) + \mathcal{P}(\phi, \Delta_2)$ (Pawula, 1977; Kitahara et al., 1980; van den Broeck, 1983):

$$p_\Phi(\phi) = C \left[\frac{1}{f_1(\phi)} - \frac{1}{f_2(\phi)} \right] \exp\left\{ -\int_\phi \left[\frac{k_1}{f_1(\phi')} + \frac{k_2}{f_2(\phi')} \right] d\phi' \right\}. \qquad (2.31)$$

We can determine the integration constant C as a normalization constant by imposing the condition that the integral[2] of $p(\phi)$ over its domain (see Subsection 2.2.3.3) be equal to one.

Using the definitions of $f_1(\phi)$ and $f_2(\phi)$ given in Eqs. (2.15) and zero-mean condition (2.8), we also obtain

$$p(\phi) = C(\Delta_2 - \Delta_1)\frac{g(\phi)}{\Xi(\phi)}\exp\left[-\frac{1}{\tau_c}\int_\phi \frac{f(\phi')}{\Xi(\phi')}d\phi'\right], \tag{2.32}$$

where

$$\Xi(\phi) = [f(\phi) + \Delta_1 g(\phi)][f(\phi) + \Delta_2 g(\phi)]. \tag{2.33}$$

We refer again to the four examples introduced in Subsection 2.2.3.1:

- **Example 2.1:** $f_{1,2}(\phi)$ are defined as in Eqs. (2.17). The resulting steady-state pdf,

$$p(\phi) \propto (1 - \phi)^{k_1-1}\phi^{k_2-1}, \tag{2.34}$$

  is a standard beta distribution (Johnson et al., 1994) with parameters k_1 and k_2.
- **Example 2.2:** $f_{1,2}(\phi)$ are defined as in Eqs. (2.18). The resulting steady-state pdf,

$$p(\phi) \propto \exp[-(k_1 - k_2)\phi], \tag{2.35}$$

  is an exponential distribution with parameter $k_1 - k_2$.
- **Example 2.3:** $f_{1,2}(\phi)$ are defined as in Eqs. (2.19). The resulting steady-state pdf is

$$p(\phi) \propto \frac{\phi^2 - a\phi + a}{(\phi - a)(1 - \phi)}\left(\frac{\phi - a}{1 - \phi}\right)^{-\frac{k_1}{1-a}}\phi^{k_2-1}. \tag{2.36}$$

- **Example 2.4:** $f(\phi)$ and $g(\phi)$ are defined as in Eqs. (2.20), and a symmetric noise (i.e., $\Delta_1 = -\Delta_2 = \Delta$ and $k_1 = k_2 = k$) is assumed. The resulting steady-state pdf is

$$p(\phi) \propto -\frac{\phi^{\frac{2k\beta}{\Delta^2-\beta^2}}(\Delta + \beta - \phi)^{\frac{k}{\Delta+\beta}}(\phi + \Delta - \beta)^{\frac{k}{\beta-\Delta}}}{\phi[(\phi - \beta)^2 - \Delta^2]}. \tag{2.37}$$

The plots of these pdfs are provided in the following subsection, after the methods for determining the domain of the steady-state pdf are described. In the next two subsections we discuss the domain of the pdf and its behavior at the boundaries; an analysis of the modes of the pdf is made in Chapter 3 within the context of the theory of noise-induced transitions.

2.2.3.3 *Domain of the steady-state probability distribution*

The domain of the steady-state pdf, $p(\phi)$, i.e., the range of values within which the asymptotic dynamics of ϕ are confined, depends on the stationary points of the functions $f_{1,2}(\phi)$ and on their stability. We recall that a stationary point ϕ_{st} of dynamics

---

[2] The pdf should in general be denoted with a capitalized variable as a subscript [in our case $p_\Phi(\phi)$]. However, for the sake of simplicity, we omit this subscript whenever it is not essential.

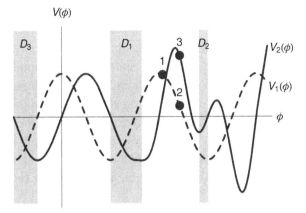

Figure 2.4. Example of two potentials and resulting domains of the steady-state stochastic dynamics of $\phi(t)$.

1 is stable if $f_1(\phi_{st}) = 0$ and $\frac{\mathrm{d}f_1(\phi)}{\mathrm{d}\phi}\big|_{\phi=\phi_{st}} < 0$, whereas it is unstable if $f_1(\phi_{st}) = 0$ and $\frac{\mathrm{d}f_1(\phi)}{\mathrm{d}\phi}\big|_{\phi=\phi_{st}} > 0$.

The characteristics of the stationary points can be easily represented by the potentials $V_1(\phi)$ and $V_2(\phi)$, defined as

$$f_1(\phi) = -\frac{\mathrm{d}V_1(\phi)}{\mathrm{d}\phi}, \qquad f_2(\phi) = -\frac{\mathrm{d}V_2(\phi)}{\mathrm{d}\phi}; \qquad (2.38)$$

the stable (unstable) stationary points correspond to the minima (maxima) of the potentials. The dynamics of ϕ can then be represented as those of a particle moving along the ϕ axis driven by the switching between the two potentials. Thus the particle remains trapped between any pair of stable points [minima of the potentials $V_1(\phi)$ and $V_2(\phi)$] that are not separated by an unstable point [i.e., a maximum of either $V_1(\phi)$ or $V_2(\phi)$]. These pairs of minima define the domain of the steady-state pdf.

To demonstrate this rule for the determination of the extremes of the domains in which ϕ fluctuates at steady state, we can consider the example shown in Fig. 2.4 and assume that the initial condition (at time $t = 0$) of the process corresponds to position 1 on potential $V_1(\phi)$. At time $t > 0$ the particle starts moving downhill to its right, along potential $V_1(\phi)$ (dashed line). When the noise randomly switches to Δ_2, the particle is at a certain position 2 (see Fig. 2.4). At this point the particle will start moving following potential $V_2(\phi)$ (solid line), i.e., like a particle released in position 3. From position 3 the particle can move only to its right, regardless of whether the noise switches to Δ_1. Thus it will necessarily enter (and remain confined within) the domain indicated by D_2. In fact, once the particle enters D_2, it necessarily remains trapped inside it, and D_2 becomes the domain of the steady-state pdf. Notice that the jump from position 2 to 3 could have occurred before the maximum of the $V_2(\phi)$ potential was reached. In this case the particle would have been attracted to the D_1

domain. The possible boundaries of the domain, $[\phi_-, \phi_+]$, are then all pairs of stable stationary points that are not separated by an unstable point.

As noted, the stable points are the boundaries of the domain only if neither one of the two dynamics has an unstable point between them. For example, to the right of the domain D_2 (see Fig. 2.4) there are two minima (one for each potential function), but the maximum of potential $V_2(\phi)$ existing between these minima prevents the long-term persistence of the particle in this interval. Note that the same criteria for the determination of the extremes of the steady-state domain apply when the minima of the potential are at $\pm\infty$. Finally, it should be noted that if the stable points were coincident the pdf would reduce to a Dirac delta function centered in the two overlapping stable points.

The previous discussion focused on the emergence of boundaries intrinsic to the dynamics. However, boundaries can also be externally imposed. For example, a frequent case in the biogeosciences is when the variable ϕ is positive valued or it has a boundary at a certain threshold value ϕ_{th}. This corresponds to changing the potential of the deterministic dynamics by setting $V_1^*(\phi) = V_1(\phi)$ and $V_2^*(\phi) = V_2(\phi)$ for $\phi \le \phi_{\text{th}}$, and $V_1^*(\phi) \to \infty$ and $V_2^*(\phi) \to \infty$ for $\phi > \phi_{\text{th}}$ (if ϕ_{th} is assumed to be an upper bound). The general rule presented in this section to determine the boundaries of the domain can now be applied to the modified potentials $V_1^*(\phi)$ and $V_2^*(\phi)$. The presence of the external bound may create a new minimum in the potential and affect the original boundaries of the domain if the sign of the derivative of $V_1(\phi)$ and $V_2(\phi)$ is negative at ϕ_{th}.

An important property of the functions $f_1(\phi)$ and $f_2(\phi)$ emerges from these analyses of the boundaries of the domain: In all the cases discussed before (Fig. 2.4), one of the two potentials monotonically decreases whereas the other monotonically increases inside each of the possible domains. As a consequence, inside the domain we have $f_1(\phi) \ge 0$ and $f_2(\phi) \le 0$ [or $f_1(\phi) \le 0$ and $f_2(\phi) \ge 0$], with the equal sign occurring at the boundaries. Thus ϕ always increases when the noise is in one state, whereas it always decreases when it is in the other.

Finally, a very particular situation is represented in Fig. 2.5. In this case two stable points are always separated by an unstable point. The particle therefore cannot remain in the long term inside any of these domains, and it will necessarily continue to move from the left to the right of the ϕ axis. A noise-induced drift is therefore generated, a situation that is known in the literature as the *ratchet effect*; see Bena et al. (2003) and Chapter 3.

To clarify the conditions determining the boundaries of the domain, we use examples similar to those presented in the previous subsections:

- **Example 2.1:** Both $f_1(\phi) = 1 - \phi$ and $f_2(\phi) = -\phi$ have a single stable fixed point. The boundaries of the domain correspond to the minima of the two potentials, $V_1(\phi) = -\phi +$

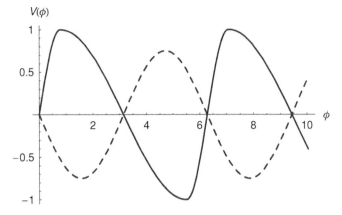

Figure 2.5. Example of potentials inducing a stochastic drift.

$\phi^2/2$ and $V_2(\phi) = \phi^2/2$, $\phi_- = 0$ and $\phi_+ = 1$. The expression for the steady-state pdf in relation (2.34) is therefore valid for $\phi \in [0, 1]$ [see Fig. 2.6(a)].

• **Example 2.2:** Neither of the two functions $f_1(\phi) = 1$ and $f_2(\phi) = -1$ has stationary points. The potentials $V_1(\phi) = -\phi$ and $V_2(\phi) = \phi$ are straight lines whose minima (i.e., boundaries of the domain) are $\phi_- \to -\infty$ and $\phi_+ \to \infty$. For any choice of k_1 and k_2, the pdf in relation (2.35) diverges, i.e., the process does not have a steady state. However, when an external (lower) bound is set, for example, at $\phi = 0$ (i.e., $\phi > 0$), both potentials can be assumed to diverge to infinity for $\phi < 0$. As a consequence, a new minimum of the modified

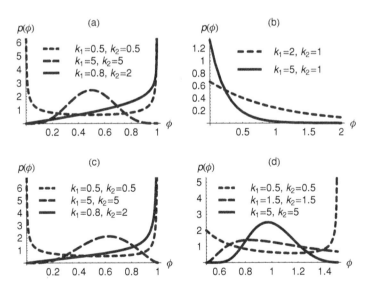

Figure 2.6. Steady-state pdf's for (a) Example 2.1, (b) Example 2.2, (c) Example 2.3 with $a = -1$, (d) Example 2.4 with $\beta = 1$ and $\Delta = 0.5$.

potential $V_2^*(\phi)$ is generated in $\phi = 0$, and the boundaries of the domain become $\phi_- = 0$ and $\phi_+ \to \infty$. In this case the pdf does not diverge if $k_1 > k_2$, and an *atom* of probability (i.e., a local *spike* in the pdf) appears in $\phi = 0$: The pdf assumes the form[3]

$$p(\phi) = \frac{k_1 - k_2}{k_1 + k_2}\delta(\phi) + \frac{2k_2(k_1 - k_2)}{k_1 + k_2}e^{-(k_1-k_2)\phi}, \quad (\phi \geq 0), \tag{2.39}$$

which is plotted in Fig. 2.6(b) (the atom of probability is not plotted in the figure).

- **Example 2.3:** One of the two functions, $f_2(\phi) = -\phi$, has only one stable point whereas the other, $f_1(\phi) = (\phi - a)(1 - \phi)$, has both a stable and an unstable point. We assume $a < 1$, which implies that a is the unstable fixed point and 1 is the stable one. We have to distinguish two subcases depending on the values of a. If $a < 0$, the two stable points 0 and 1 are not separated by the unstable point; as a consequence, the boundaries of the domain of the asymptotic dynamics are $\phi_- = 0$ and $\phi_+ = 1$. This is the case of Fig. 2.6(c). Conversely, if $0 < a < 1$, the unstable point is between the stable points 0 and 1. The potential $V_1(\phi) = \phi^3/3 - (a + 1)\phi^2/2 + a\phi$ tends to $-\infty$ for $\phi \to -\infty$; therefore the boundaries of the domain are $\phi_- \to -\infty$ and $\phi_+ = 0$. In the special case $a = 0$ the boundaries of the domain can be either $] - \infty, 0]$, if the initial condition ϕ_0 is negative or $[0, 1]$ if $\phi_0 > 0$.

- **Example 2.4:** If we consider a symmetric dichotomic noise $(\Delta_1 = \Delta_2 = \Delta)$ we have $f_{1,2}(\phi) = -\phi(\phi - \beta \pm \Delta)$ and $V_{1,2}(\phi) = \phi^3/3 - (\beta \pm \Delta)\phi^2/2$. If $\beta > \Delta$ the domain is therefore $[\beta - \Delta, \beta + \Delta]$, whereas the domain is $[0, \beta + \Delta]$ in the reverse case. An example of a pdf with $\beta = 1$ and $\Delta = 0.5$ is reported in Fig. 2.6(d).

2.2.3.4 Behavior of the steady-state pdf at the boundaries

It may be interesting to investigate the behavior of the steady-state pdf near the boundaries of the domain. Suppose the boundary ϕ_i (with $\phi_i = \phi_-$ or $\phi_i = \phi_+$) is a stable point of the $f_1(\phi)$ dynamics, i.e., $f_1(\phi_i) = 0$. If $f_2(\phi_i) \neq 0$, the steady-state pdf in the vicinity of ϕ_i is determined as a limit of Eq. (2.31) for $f_1(\phi) \to 0$:

$$p(\phi) \sim \frac{1}{f_1(\phi)}\exp\left[-\int_\phi \frac{k_1}{f_1(\phi')}d\phi'\right]. \tag{2.40}$$

Using in approximation (2.40) the Taylor expansion of $f_1(\phi)$ around ϕ_i truncated to the first order [i.e., $f_1(\phi) = (\phi - \phi_i)\frac{df_1(\phi)}{d\phi}\big|_{\phi=\phi_i}$], we can represent the pdf as

$$p(\phi) \sim \left(\frac{1}{|\phi - \phi_i|}\right)^{1+\frac{k_1}{\frac{df_1(\phi)}{d\phi}\big|_{\phi_i}}}. \tag{2.41}$$

This limit behavior reflects the competition between two time scales: the time scale characteristic of the switching between the two deterministic dynamics k_1^{-1} and the time scale of the deterministic dynamics $f_1(\phi)$ near the attractor ϕ_i.

---

[3] The first term on the right-hand side of (2.39) represents the probability atom and is calculated as the integral between ϕ_- and 0 of the function expressing the pdf of ϕ in the absence of the externally imposed bound at $\phi = 0$ (i.e., in the case $\phi \to -\infty$).

- In fact, when the random switching (i.e., the transition rate k_1) is relatively slow with respect to the deterministic dynamics for $\phi \to \phi_i$,

$$-\left.\frac{\mathrm{d}f_1(\phi)}{\mathrm{d}\phi}\right|_{\phi_i} > k_1 \qquad (2.42)$$

(recall that $\left.\frac{\mathrm{d}f_1(\phi)}{\mathrm{d}\phi}\right|_{\phi_i}$ is negative by definition of stable fixed point), the particle tends "to spend much time" near the boundary and the pdf diverges at the boundary (i.e., as $\phi \to \phi_i$). This is the case, for example, of the continuous line in Fig. 2.6(a) for $\phi \to 1$.

- And vice versa when the switching between the two dynamics is sufficiently fast to prevent ϕ from "spending much time" near the attractors, i.e.,

$$-\left.\frac{\mathrm{d}f_1(\phi)}{\mathrm{d}\phi}\right|_{\phi_i} < k_1, \qquad (2.43)$$

the pdf becomes null at the boundary. This is the case, for example, of the dashed curve in Fig. 2.6(a). In particular, the slope of the pdf is also zero when $-2\left.\frac{\mathrm{d}f_1(\phi)}{\mathrm{d}\phi}\right|_{\phi_i} < k_1$.

These results are valid only when $f_1(\phi_i) = 0$ and $f_2(\phi_i) \neq 0$, which excludes the cases when the bound is externally imposed. Moreover relations (2.40)–(2.43) are not valid for the cases in which ϕ_i is also an unstable stationary point of $f_2(\phi)$. Similar results are obtained for the other boundary. In this case the behavior of the system close to the boundary is expressed again by relations (2.40)–(2.43) but with $f_2(\phi)$ in place of $f_1(\phi)$ and k_2 in place of k_1.

2.2.3.5 *State-dependent DMN*

An interesting generalization of the process described by Eq. (2.16) is the state-dependent DMN (Laio et al., 2008). In the state-dependent DMN the transition rates k_1 and k_2 depend on the state variable ϕ. This dependency may arise when, in the dynamics expressed by Eq. (2.12), a feedback exists between the state ϕ of the system and the random driver $q(t)$ [see Subsection 2.2.2 for a definition of $q(t)$]. Under the mechanistic interpretation this feedback translates into a dependency of q on ϕ or of the threshold value θ (i.e., the transition point between the two states) on ϕ [continuous line in Fig. 2.7(a)]. We introduce the feedback by assuming that either the cumulative distribution of q, $p_Q(q)$, or the threshold value θ (or both) depends on the state system, namely $p_Q(q) = p_Q(q|\phi)$ or $\theta = \theta(\phi)$. This implies that the rates of the DMN also depend on ϕ, $k_1(\phi) = \int_0^\theta p_Q(q|\phi)\mathrm{d}q$ or $k_1(\phi) = \int_0^{\theta(\phi)} p_Q(q)\mathrm{d}q$, and $k_2(\phi) = 1 - k_1(\phi)$. Under the functional interpretation, the feedback may produce a state dependency in any of the parameters (k_1, k_2, Δ_1, and Δ_2) of the DMN. However, a possible ϕ dependency of Δ_1 or Δ_2 (or both) can be accounted for through a suitable modification of the $g(\phi)$ function in Eq. (2.16), whereas the ϕ dependency of k_1 and k_2 intrinsically modifies the dynamical system. In this case the multiplicative noise cannot be factorized, i.e., it cannot be expressed as the by-product of ξ_{dn} with a suitable function of ϕ. Even though it has been seldom considered in the literature (special

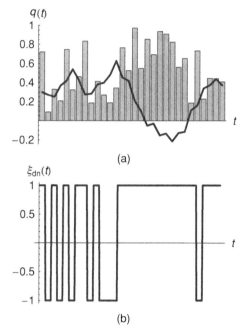

Figure 2.7. Example of the relations among (a) the external forcing $q(t)$ (gray bars), the threshold θ (continuous line), and (b) the corresponding DMN.

cases are considered by Masoliver and Weiss, 1994; Christophorov, 1996; and Julicher et al., 1997), the presence of the state dependency in k_1 and k_2 profoundly affects the dynamics of ϕ because of the modification of the distribution of the residence times in states Δ_1 and Δ_2 [Fig. 2.7(b)]. The fact that this distribution is not exponential as in the case of the standard DMN is consistent with a general property of processes with state-dependent rates (Daly and Porporato, 2006, 2007).

When transition rates k_1 and k_2 depend on the state of the system, the process remains analytically solvable. In fact, the steps leading to steady-state solution (2.31) are not affected by a possible state dependency of the transition rates. As a consequence, we simply obtain the solution in the state-dependent case from Eq. (2.31) by setting $k_1 = k_1(\phi)$ and $k_2 = k_2(\phi)$:

$$p(\phi) = C\left[\frac{1}{f_1(\phi)} - \frac{1}{f_2(\phi)}\right]\exp\left\{-\int_\phi\left[\frac{k_1(\phi')}{f_1(\phi')} + \frac{k_2(\phi')}{f_2(\phi')}\right]d\phi'\right\}, \qquad (2.44)$$

where C is a normalization constant we calculate by imposing the condition that the integral of $p(\phi)$ in the domain of the definition of ϕ be equal to 1. The zeros of $f_1(\phi)$ and $f_2(\phi)$ are the natural boundaries for the dynamics and represent the limits of the domain of ϕ (see also Bena, 2006). An alternative representation of the pdf, which is of interest under the functional interpretation of the DMN, is obtained as a function of $f(\phi)$ and $g(\phi)$ by use of Eqs. (2.14):

$$p(\phi) = \frac{Cg(\phi)\exp\left\{-\int_\phi \left[\frac{k_1(\phi')}{f(\phi')+\Delta_1 g(\phi')} + \frac{k_2(\phi')}{f(\phi')+\Delta_2 g(\phi')}\right]d\phi'\right\}}{[f(\phi)+\Delta_1 g(\phi)][f(\phi)+\Delta_2 g(\phi)]}. \qquad (2.45)$$

To demonstrate the possible impact of the feedback between noise and the dynamical system, we consider the simple case in which the alternating processes of growth and decay are expressed by two linear functions:

$$f_1(\phi) = \alpha(1-\phi), \qquad f_2(\phi) = -\alpha\phi, \qquad (2.46)$$

where $\alpha > 0$ determines the rates of growth and decay. The stationary states, $\phi_{st,1} = 1$ and $\phi_{st,2} = 0$, are also the boundaries of the dynamics. We also assume a linear dependence of θ on ϕ, $\theta(\phi) = \theta_0 + b\phi$, and a logistic distribution to represent the variability of the resource q, with cumulative density function

$$P_Q(q) = \left(1 + e^{-\frac{q-q_*}{\sigma}}\right)^{-1}, \qquad (2.47)$$

where σ is a scale parameter. The mean and the standard deviation of the distribution are q_* and $\frac{\pi}{\sqrt{3}}\sigma$, respectively. The choice of a linear dependence of θ on ϕ and the use of the logistic distribution are aimed at simplifying the mathematical treatment of the problem, but other choices [i.e., other monotonic forms of the $\theta(\phi)$ function or other probability distributions] could be equally adopted. Under the preceding assumptions, the transition rates are found as

$$k_1(\phi) = 1 - k_2(\phi) = \left(1 + e^{-\frac{\theta_0 - q_* + b\cdot\phi}{\sigma}}\right)^{-1}, \qquad (2.48)$$

and the corresponding steady-state pdf from Eq. (2.44) is

$$p(\phi) = C\phi^{\frac{1}{\alpha}-1}(1-\phi)^{-1}\exp\left[-\frac{1}{\alpha^2}\int_\phi \frac{dy}{y(1-y)\left(1+e^{-\frac{\theta_0-q_*+b\cdot y}{\sigma}}\right)}\right], \qquad (2.49)$$

where C is the normalization constant we calculate by imposing the condition that the integral of $p(\phi)$ in the domain [0,1] be equal to 1.

Figure 2.8 shows an example of pdf (2.49) along with the one corresponding to no feedback (i.e., $b = 0$). In particular, the feedback is chosen to be positive ($b < 0$). It is obvious that the state dependency of the transition rates induces a remarkable change in $p(\phi)$. This change is not only quantitative but also qualitative. In Chapter 3 we discuss thoroughly the ability of feedbacks to induce structural changes to the shape of the pdf's.

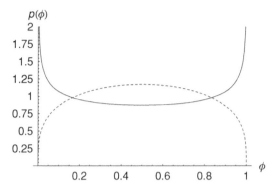

Figure 2.8. The pdfs corresponding to state-dependent (dashed curve, $b = -0.5$, $\theta_0 = 1.25$) and constant (continuous curve, $b = 0$, $\theta_0 = 1.1$) transition rates. The common parameters are $q_* = 1$, $\alpha = 0.05$, and $\sigma = 0.1$.

2.3 White shot noise

2.3.1 Definition and properties

WSN, or white Poisson noise, is a stochastic process described by a state variable $\xi_{sn}(t)$ that is defined by a sequence of pulses at random times τ_i, with each pulse having an infinitesimal duration and an infinite random height $h_i \delta(0)$, where $\delta(\cdot)$ is the Dirac delta function (see Fig. 2.9). The random times $\{\tau_i\}$ form a Poisson sequence with rate λ, which implies that the probability distribution of the interarrival times $t_i = \tau_i - \tau_{i-1}$ is $p_T(t) = \lambda e^{-\lambda t}$ (e.g., Ross, 1996). The probability distribution of the random heights [divided by $\delta(0)$] is exponential with mean α, $p_H(h) = \frac{1}{\alpha} e^{-h/\alpha}$. A mathematical formulation for the process is

$$\xi_{sn}(t) = \sum_i h_i \delta(t - \tau_i). \qquad (2.50)$$

Shot noise has been extensively investigated by scientists since the beginning of the 20th century. The first documented works on shot noise were written by Campbell (1909a, 1909b). A comprehensive analysis of shot noise was conducted by Rice (1944, 1945).

WSN is a singular mathematical object, just as the Dirac delta function is a singular function. We can better understand these singularities by considering WSN as the formal derivative of the homogeneous compound Poisson process

$$Z(t) = \int_0^t \xi_{sn}(t')dt' = \sum_i h_i \Theta(t - \tau_i), \qquad (2.51)$$

where $\Theta(\cdot)$ is the unit step function. A realization of the stochastic process $Z(t)$ is shown in Fig. 2.9 and resembles a staircase with random steps.

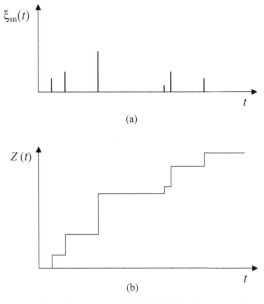

Figure 2.9. Examples of realizations of (a) WSN $\xi_{sn}(t)$, and (b) the homogeneous compound Poisson process $Z(t)$. The spikes in (a) are represented as bars of unit base, instead of delta functions as in Eq. (2.50).

The average and the covariance function of the compound Poisson process are (Parzen, 1967, p. 130)

$$\langle Z(t) \rangle = \lambda \alpha t, \tag{2.52}$$

$$\langle Z(t)Z(t') \rangle = 2\lambda \alpha^2 \min(t, t'), \tag{2.53}$$

respectively.

We can now obtain the covariance function and the moments of WSN by considering that WSN is the formal derivative of the compound Poisson process; as a consequence, the mean is (VanMarcke, 1983, pp. 110–111)

$$\langle \xi_{sn}(t) \rangle = \frac{\partial \langle Z(t) \rangle}{\partial t} = \lambda \alpha, \tag{2.54}$$

and the covariance is

$$\langle \xi_{sn}(t)\xi_{sn}(t') \rangle = -\frac{\partial^2 \langle Z(t)Z(t') \rangle}{\partial (t - t')^2} = 2\lambda \alpha^2 \delta(t - t'). \tag{2.55}$$

The variance reads $\kappa_{2sn} = 2\lambda \alpha^2 \delta(0) = 2s_{sn}\delta(0)$, where s_{sn} is the strength of the noise, i.e., $s_{sn} = \lambda \alpha^2$ [see Eq. (B2.1-6)]. Only the mean is finite, whereas all higher-order moments and cumulants diverge.

2.3.2 *White shot noise as a limiting case of the DMN*

WSN can be obtained from DMN by taking the following parameter values:

$$\Delta_1 = \alpha k_1, \quad \Delta_2 = 0, \quad k_1 \to \infty, \quad k_2 = \lambda. \qquad (2.56)$$

It is easily verified that the average of WSN is recovered from Eq. (2.7) when we use the values in (2.56). Analogously, the higher-order moments and cumulants are found in the limiting case to diverge with rate ∞^{m-1}, consistent with results obtained in the previous subsection. Also, autocorrelation function (2.55) can be obtained from Eq. (B2.1-5) by taking the limit for $\tau_c \to 0$ [note that $\delta(t) = \lim_{a \to 0} e^{-|t/a|}/(2a)$].

In the following subsection, a zero-average shot-noise process, $\xi'_{sn} = \xi_{sn} - \lambda\alpha$, is considered. We obtain the zero-average shot noise from the DMN by taking the same parameter values as in (2.56), except $\Delta_2 = -\lambda\alpha$.

2.3.3 *Relevance of white shot noise in the biogeosciences*

The infinitesimal duration of the spikes and the uncorrelation could suggest that WSN is not well suited to realistically represent an external forcing in "real-world" processes. However, this is not necessarily true. In fact, to determine whether WSN is suitable for the modeling of a random driver, we should look at the ratio between the duration of a single random event and the typical time scale of the deterministic process forced by this random event. If this ratio is low, then the temporal structure of the forcing in the course of each event has no influence on the overall dynamics, and – at the time scale of the deterministic process – the external forcing can then be modeled as a sequence of episodic instantaneous impulses.

For example, consider the plots shown in Fig. 2.10. They report the same rainfall sequence visualized at different time scales. When the rainfall is described at a 10-min temporal resolution, several complex features of the rainfall events can be observed: For example, the rainfall intensity is irregular and shows strong temporal gradients. But as the process is observed at coarser time scales, the temporal resolution decreases until the rainfall sequence appears as a sequence of spikes, whose height is equal to the total rainfall in the corresponding event. In the study of a process driven by this random forcing, the time scale of the process will obviously dictate the scale that has to be adopted to model the noise. For example, if we are interested in the modeling of the flood process in a relatively small basin (e.g., contributing area ≈ 100 km^2), we need to account for the hourly structure of precipitation; in contrast, if we are investigating seasonal dynamics of vegetation, the daily time scale will be sufficient and rainfall can be modeled as WSN (Rodriguez-Iturbe et al., 1999b).

The previous example clarifies that when the time scale of the deterministic process is much greater than the duration of single random events, WSN is a suitable

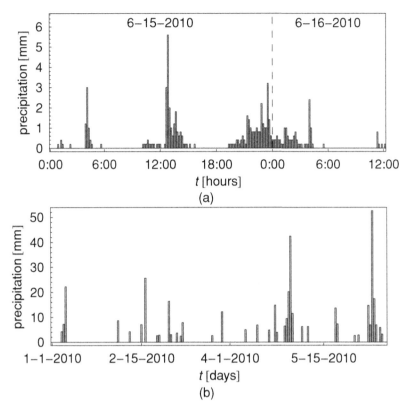

Figure 2.10. Rainfall time sequence recorded in Torino (Italy) with (a) 10-min and (b) daily aggregation windows.

representation of the random forcing. Apart from rainfall, several other intermittent stochastic processes in the biogeosciences can be modeled as WSN processes. For example, the effect of fires in vegetation dynamics, the dynamics of epidemics in ecological models, the occurrence of landslides in models of landform evolution, or the dynamics of earthquakes can be represented as WSN, as discussed in Chapter 4.

2.3.4 Stochastic process driven by white shot noise

Consider the stochastic equation

$$\frac{\mathrm{d}\phi}{\mathrm{d}t} = f(\phi) + g(\phi)\xi'_{\mathrm{sn}}(t), \tag{2.57}$$

driven by a WSN with zero average. Because the noise is white (i.e., nonautocorrelated), a problem of interpretation of Eq. (2.57) arises: If $g(\phi)$ does depend on ϕ (i.e., if the noise is multiplicative), Eq. (2.57) is meaningless (van Kampen, 1981) unless an interpretation rule is specified.

In fact, in Eq. (2.57), each pulse in $\xi'_{\text{sn}}(t)$ causes a pulse in $\mathrm{d}\phi/\mathrm{d}t$ and hence a jump in ϕ. This fact has the important consequence that the value of ϕ that needs to be used in $g(\phi)$ is undetermined. *Ito's convention* assigns the rule that $g(\phi)$ should be calculated with the value of ϕ just before the jump (van Kampen, 1981). This is a very simple convention that, however, has the consequence that the standard rules of calculus do not apply anymore to stochastic equations with multiplicative WSN such as Eq. (2.57). For example, if we define a transformed variable $z = \int_{\phi} 1/g(\phi')\mathrm{d}\phi'$, the resulting stochastic equation under Ito's convention is (see van Kampen, 1981)

$$\frac{\mathrm{d}z}{\mathrm{d}t} = \frac{f[\phi(z)]}{g[\phi(z)]} - \frac{1}{2}\frac{g'[\phi(z)]}{g^2[\phi(z)]} + \xi'_{\text{sn}}(t), \tag{2.58}$$

rather than

$$\frac{\mathrm{d}z}{\mathrm{d}t} = \frac{f[\phi(z)]}{g[\phi(z)]} + \xi'_{\text{sn}}(t), \tag{2.59}$$

as we would expect from the ordinary rules of calculus.

For a physicist, this mathematical peculiarity of Ito's convention may cause some interpretation problems: Consider an equation similar to (2.57) but with a noise term with short, though not null, autocorrelation time τ_c. For example, we can consider the case of DMN (see Section 2.2). In this case no interpretation rule would be required for Eq. (2.57) (see Subsection 2.2.3 and van Kampen, 1981) and the standard rules of calculus would apply, leading to Eq. (2.59) in the limit for $\tau_c \to 0$. The physicist's interpretation (van den Broeck, 1983) of the white noise as the limit of a colored noise with $\tau_c \to 0$ would therefore be undermined by use of Ito's convention. The rule that directly arises from the use of the ordinary rules of calculus is called the *Stratonovich* rule (e.g., van Kampen, 1981). In this book we adopt in most cases the Stratonovich interpretation of Eq. (2.57) because, as noted, it is physically more plausible and it allows us to establish a more direct relationship between processes forced by white and colored noises. Examples of the implications of adopting Ito's interpretation are also given for some specific cases in the following chapters.

The Stratonovich rule is often defined as the rule that calculates $g(\phi)$ by using a value of ϕ obtained as half of the sum of the values of ϕ before and after the jump (e.g. van Kampen, 1981; van den Broeck, 1997). However, this definition is valid only when the driving force is a white Gaussian noise (see Section 2.4), whereas it would lead to some inconsistent results in the case of WSN. This is not just a matter of definition: The knowledge of the correct size of the jump in the ϕ trajectory is essential to numerically simulate a process driven by WSN. We therefore need to clarify this incongruence. If we want the ordinary rules of calculus to apply to Eq. (2.57), we have from (2.57) that the relation between the value ϕ_0 before the jump and the value ϕ_1 after the jump needs to be

$$\int_{\phi_0}^{\phi_1} \frac{d\phi'}{g(\phi')} = h, \tag{2.60}$$

where h is the size of the jump [the contribution of the first term on the right-hand side of Eq. (2.57) is negligible in the time frame in which a jump occurs]. If the left-hand side of Eq. (2.60) can be analytically integrated, the size of the jump in the trajectory of ϕ, $\phi_1 - \phi_0$, is directly found; otherwise the solution should be found numerically.

We can obtain an interesting representation of $\phi_1 - \phi_0$ by considering the transformed variable $z(\phi) = \int_\phi 1/g(\phi')d\phi'$, using (2.60) to express ϕ_1 as $\phi_1 = z^{-1}[z(\phi_0) + h]$ (where the superscript -1 represents the inverse function) and expanding ϕ_1 in a Taylor series around $h = 0$. We find

$$\phi_1 = \phi_0 + \sum_{i=1}^{\infty} \frac{1}{i!} g_{(i)}(\phi_0) h^i, \tag{2.61}$$

where $g_{(1)}(\phi_0) = g(\phi_0)$ and $g_{(i+1)}(\phi_0) = g(\phi_0) \frac{dg_{(i)}(\phi)}{d\phi}\big|_{\phi=\phi_0}$. An analogous representation is obtained by Caddemi and Di Paola (1996) and Pirrotta (2005) following a different reasoning. The first three terms of the expansion are

$$\phi_1 - \phi_0 = g(\phi_0)h + \frac{1}{2}g'(\phi_0)g(\phi_0)h^2 + \frac{1}{6}\left[g''(\phi_0)g^2(\phi_0) + [g'(\phi_0)]^2 g(\phi_0)\right]h^3. \tag{2.62}$$

We recognize that the first term corresponds to Ito's rule and the higher-order terms in general do not vanish.

Consider now the rule frequently used for the Stratonovich convention to calculate ϕ in $g(\phi)$ as half the sum of the values before and after the jump:

$$\phi_1 - \phi_0 = g\left[\frac{\phi_0 + \phi_1}{2}\right]h. \tag{2.63}$$

Setting $\phi_1 = \phi_0 + g\left[(\phi_0 + \phi_1)/2\right]h$ on the right-hand side of Eq. (2.63) and expanding $g[\cdot]$ in a Taylor series in h until $O(h^3)$, we obtain

$$\phi_1 - \phi_0 = g[\phi_0]h + \frac{1}{2}g'[\phi_0]g\left[\frac{\phi_0 + \phi_1}{2}\right]h^2 + \frac{1}{8}g''[\phi_0]g^2\left[\frac{\phi_0 + \phi_1}{2}\right]h^3. \tag{2.64}$$

We repeat the substitution and the expansion with the second and third terms on the right-hand side of Eq. (2.64) by using (2.61); we neglect the terms of orders higher than 3 to obtain

$$\phi_1 - \phi_0 = g(x_0)h + \frac{1}{2}g'(\phi_0)g(\phi_0)h^2 + \left[\frac{1}{8}g''(\phi_0)g^2(\phi_0) + \frac{1}{4}[g'(\phi_0)]^2 g(\phi_0)\right]h^3, \tag{2.65}$$

which is clearly different from Eq. (2.62) for the terms of orders higher than 2 in h. As a consequence, convention (2.63) is in general inconsistent with relation (2.61);

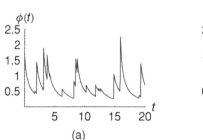
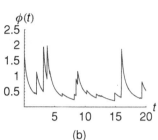

Figure 2.11. Examples of trajectories of (a) process (2.66) and (b) process (2.67). In this simulation $\alpha = 0.5$ and $\lambda = 1$.

hence it is incompatible with the ordinary rules of calculus. We show in Section 2.4 that convention (2.63) is instead appropriate in the case of Gaussian white noise.

2.3.4.1 Some examples of processes driven by WSN

To facilitate the understanding of the properties presented in the following sections, it is useful to introduce some simple examples of processes driven by WSN:

- **Example 2.5:** $\phi(t)$ decreases hyperbolically during the interarrival times and the noise is additive:

$$\frac{d\phi}{dt} = -\phi^2 + \xi_{sn} = -\phi^2 + \alpha\lambda + \xi'_{sn}. \tag{2.66}$$

- **Example 2.6:** The deterministic component is the same as in the previous example, but the noise is multiplicative with $g(\phi) = \phi$:

$$\frac{d\phi}{dt} = -\phi^2 + \phi\xi_{sn} = -\phi(\phi - \alpha\lambda) + \phi\xi'_{sn}. \tag{2.67}$$

Figure 2.11 shows examples of the trajectories of $\phi(t)$ for these two processes. In the second example [Fig. 2.11(b)], the size of the jumps Δ in the trajectories of ϕ is evaluated with Eq. (2.60), which in this simple case [with $g(\phi) = \phi$] can be analytically integrated. We obtain

$$\Delta = \phi_1 - \phi_0 = \phi_0(e^h - 1), \tag{2.68}$$

with a clear dependence of Δ on the spike intensity h and the starting point ϕ_0. Thus the function $g(\phi)$ modulates the magnitude of the jumps. In particular, the multiplicative character of the noise implies that if ϕ_0 is greater (lower) than $h/(e^h - 1)$ the jump size in the ϕ trajectory is greater (lower) than h. The differences of the jump sizes in the two examples can be observed in Fig. 2.11, for which the same realization of the shot noise ξ'_{sn} is used.

2.3.4.2 Steady-state pdf and its properties

Under the Stratonovich interpretation, we obtain the steady-state pdf for a process driven by WSN from (2.31) by taking the limits in (2.56) (but with $\Delta_2 = -\lambda\alpha$ to have zero-mean WSN). These limits lead to

$$p(\phi) = C \frac{1}{f(\phi) - \lambda \alpha g(\phi)} \exp \left\{ -\int_{\phi} \frac{f(\phi')}{[f(\phi') - \lambda \alpha g(\phi')] \alpha g(\phi')} d\phi' \right\}. \qquad (2.69)$$

We can obtain the same solution by determining and solving the master equation for WSN (see Box 2.3).

The derivation of the master equation and the steady-state probability distribution in Box 2.3 holds under the Stratonovich interpretation of Langevin equation (2.57) because the ordinary rules of calculus are used in the transformation to an additive equation. The master equation corresponding to Langevin equation (2.57), interpreted with Ito's convention is (Denisov et al., 2009)

$$\frac{\partial p(\phi, t)}{\partial t} = -\frac{\partial}{\partial \phi} [p(\phi, t) f(\phi)] - \lambda p(\phi, t) + \lambda \int_0^{\phi} \frac{p(\phi', t)}{|g(\phi')|} p_H \left(\frac{\phi - \phi'}{g(\phi')} \right) d\phi'. \qquad (2.70)$$

As expected, when the noise is additive [$g(\phi) = 1$] this master equation becomes equivalent to Eq. (B2.3-7) because there is no difference between Ito's and Stratonivich's interpretation when the noise is additive (van Kampen, 1981). Unfortunately, no analytical solution is known of Eq. (2.70) when $g(\phi) \neq 1$, not even under steady-state conditions.

The domain of the steady-state pdf (under the Stratonovich interpretation) can be inferred from the general rule presented for the case of DMN, i.e., that the boundaries of the domain are defined by the stable points of the dynamics $d\phi/dt = f_1(\phi)$ and $d\phi/dt = f_2(\phi)$. In the case of WSN the stable points correspond to the zeros of $f(\phi) - \lambda \alpha g(\phi)$ and $g(\phi)$. The same conditions already described for DMN apply also to the case of WSN.

In the case of "externally" imposed bounds [i.e., bounds not emerging from dynamics (2.57) of the state variable ϕ], Rodriguez-Iturbe et al. (1999b) showed that the Markovianity of the process is preserved. In this case the pdf of ϕ may have a spike (or "atom" of probability) at the bound that can be calculated as explained in the case of dichotomous noise.

We can now refer to the examples presented in Subsection 2.3.4.1 to show some applications of Eq. (2.69).

- **Example 2.5:** In this case the pdf is

$$p(\phi) = \frac{2}{\sqrt{\alpha \lambda}} B_1 \left[2\frac{\sqrt{\lambda}}{\sqrt{\alpha}} \right] \frac{e^{-\frac{\phi}{\alpha} - \frac{\lambda}{\phi}}}{\phi^2}, \qquad (2.71)$$

where $B_1[\cdot]$ is the first-order modified Bessel function. A plot of $p(\phi)$ is shown in Fig. 2.12.
- **Example 2.6:** in this case the pdf of ϕ is

$$p(\phi) = \left(\frac{1}{\lambda} \right)^{1+\frac{1}{\lambda}} \Gamma \left[1 + \frac{1}{\alpha} \right] e^{-\frac{\lambda}{\phi}} \phi^{-2-\frac{1}{\alpha}}, \qquad (2.72)$$

where $\Gamma[\cdot]$ is the gamma function. An example of $p(\phi)$ is shown in Fig. 2.12.

Box 2.3: Direct derivation of the master equation for the processes driven by WSN

In this box, we determine the master equation for process (2.57) driven by WSN, using the approach by Cox and Miller (1965). We first transform generic multiplicative Langevin equation (2.57) into an additive equation by using the transformation $z = \int_\phi 1/g(\phi')\mathrm{d}\phi'$, $\rho(z) = f[\phi(z)]/g[\phi(z)]$. The Langevin equation for z is

$$\frac{\mathrm{d}z}{\mathrm{d}t} = \rho(z) + \xi_{sn}. \tag{B2.3-1}$$

In the infinitesimal interval $\mathrm{d}t$ the probability of having no jumps is $(1 - \lambda\mathrm{d}t) + \mathrm{o}(\mathrm{d}t)$; in this case, at time $z + \mathrm{d}t$, the variable z will take the value

$$z(t + \mathrm{d}t) = z(t) - \Delta z, \tag{B2.3-2}$$

where

$$\Delta z = -\int_t^{t+\mathrm{d}t} \rho[z(\tau)]\mathrm{d}\tau = \rho[z(\tau)]\mathrm{d}t + \mathrm{o}(\mathrm{d}t), \tag{B2.3-3}$$

where $\mathrm{o}(\cdot)$ is a high-order infinitesimal. The probability that a jump occurs in the same interval $\mathrm{d}t$ is $\lambda\mathrm{d}t + \mathrm{o}(\mathrm{d}t)$. In this case,

$$z(t + \mathrm{d}t) = z(t) + h + \Delta z, \tag{B2.3-4}$$

where h is the size of the jump, with distribution $p_H(h)$. As a consequence, the probability that the process takes a value in $[z, z + \mathrm{d}z]$ at time $t + \mathrm{d}t$ can be expressed as

$$p(z, t + \mathrm{d}t)\mathrm{d}z = (1 - \lambda\mathrm{d}t)p(z + \Delta z, t)\mathrm{d}(z + \Delta z)$$
$$+ \lambda\mathrm{d}t \int_0^z p(z' + \Delta z', t)p_H(z - z')\mathrm{d}(z' + \Delta z')\mathrm{d}z, \tag{B2.3-5}$$

where the second term on the right-hand side accounts for the case in which the process reaches the z level because of a jump induced by the noise. Now, if expression (B2.3-3) for Δz is substituted into Eq. (B2.3-5) and all terms of the order $\mathrm{o}(\mathrm{d}t)$ are neglected, we obtain

$$p(z, t + \mathrm{d}t)\mathrm{d}z) = (1 - \lambda\mathrm{d}t)p[z - \rho(z)\mathrm{d}t, t]\mathrm{d}[z - \rho(z)\mathrm{d}t]$$
$$+ \lambda\mathrm{d}t \int_0^z p[z' - \rho(z')\mathrm{d}t, t]p_H(z - z')\mathrm{d}[z' - \rho(z')\mathrm{d}t]\mathrm{d}z$$
$$= (1 - \lambda\mathrm{d}t)\left[p(z, t) - \rho(z)\mathrm{d}t\frac{\partial}{\partial z}p(z, t)\right]\left[1 - \frac{\partial}{\partial z}\rho(z)\mathrm{d}t\right]$$
$$+ \lambda\mathrm{d}t\mathrm{d}z \int_0^z p(z', t)p_H(z - z')\mathrm{d}z'. \tag{B2.3-6}$$

Finally, dividing by $\mathrm{d}z$, subtracting $p(z, t)$ from both sides, dividing by $\mathrm{d}t$, and taking the limit as $\mathrm{d}t \to 0$, we obtain (see Rodriguez-Iturbe et al., 1999b) the master equation:

$$\frac{\partial p(z, t)}{\partial t} = -\frac{\partial}{\partial z}[p(z, t)\rho(z)] - \lambda p(z, t) + \lambda \int_0^z p(z', t)p_H(z - z')\mathrm{d}z'. \tag{B2.3-7}$$

When the distribution $p_H(h)$ of the jump sizes is exponential with mean α, we obtain the following steady-state solution of Eq. (B2.3-7):

$$p(z) = C \frac{1}{\rho(z)} \exp\left[-\frac{z}{\alpha} - \lambda \int_z \frac{1}{\rho(z')} dz'\right], \qquad (B2.3\text{-}8)$$

where C is an integration constant. Backtransforming Eq. (B2.3-8) from the z to the ϕ domain, we recover Eq. (2.69). To this end, we recall that if $z(\phi)$ is a monotonic function the derived distribution of ϕ is $p_\Phi(\phi) = p_Z(z) |dz/d\phi|$. Moreover, the expressions of master equation (B2.3-7) and probability distribution (B2.3-8) of z determined in this box are for the case of WSN, ξ_{sn} with mean $\langle\xi_{sn}\rangle = \alpha\lambda$, whereas (2.69) refers to the case of zero-mean WSN ξ'_{sn}. Thus, to compare with (2.69) the pdf of ϕ obtained as a derived distribution of $p_Z(\phi)$ using (B2.3-8), we need to use the transformation $f(\phi) \rightarrow f(\phi) - \alpha\lambda g(\phi)$.

The results obtained in this box can be also generalized to the case of processes driven by WSN with state-dependent rate $\lambda(\phi)$. Following the same steps as in (B2.3-3)–(B2.3-7) [with an additional Taylor expansion of $\lambda(\phi)$ about ϕ] and neglecting the higher-order terms, we obtain the pdf of ϕ for the case of state-dependent WSN (Porporato and D'Odorico, 2004):

$$p(z) = C \frac{1}{\rho(z)} \exp\left[-\frac{z}{\alpha} - \int_z \frac{\lambda(z')}{\rho(z')} dz'\right]. \qquad (B2.3\text{-}9)$$

The effect of the state dependency of the rate λ on the properties of the probability distribution of ϕ are discussed in Chapter 4; in Subsection 2.3.4.3 we show that the same result can be obtained as a limit of the pdf of ϕ in the case of state-dependent dichotomous noise [Eq. (2.44)].

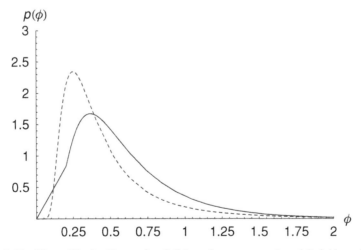

Figure 2.12. The pdf's for Examples 2.5 (continuous curve) and 2.6 (dotted curve) with $\alpha = 0.5$ and $\lambda = 1$.

2.3.4.3 State-dependent WSN

In Box 2.3 we obtained the pdf of ϕ in the case of a state-dependent Poisson process with rate $\lambda(\phi)$. We can now determine the same expression as the limiting behavior of the steady-state pdf of ϕ when the correlation time of DMN tends to zero. This applies only to the case in which DMN is used as a simple form of colored noise to perturb the deterministic dynamics $d\phi/dt = f(\phi)$ (functional usage), i.e., when the functions $f(\phi)$ and $g(\phi)$ are assigned a priori and are independent of the noise parameters. In fact, in the case of mechanistic use of DMN, the underlying dynamics are intrinsically dichotomous [i.e., the process randomly switches between two alternative dynamics, $d\phi/dt = f_{1,2}(\phi)$], and it would be unclear what the limiting properties of these dynamics represent.

We can obtain a state-dependent form of WSN from state-dependent dichotomous noise (with functional usage) discussed in Subsection 2.2.3.5 by taking the limits in (2.56). It is interesting to observe that when such limits are taken in the case of state-dependent dichotomous noise (see also Laio et al., 2008), the ϕ dependency of k_1 is transferred to $\alpha = \alpha(\phi)$ [e.g., if $k_1(\phi) = k_1 h(\phi)$, with $h(\phi)$ being a generic function of ϕ, then $k_1 \to \infty$ implies $\alpha(\phi) = \alpha/h(\phi)$], whereas the state dependency of $k_2(\phi)$ translates into a state dependency of $\lambda = \lambda(\phi) = k_2(\phi)$.

We then obtain the steady-state distribution of ϕ by using the Stratonovich integration rule (which naturally arises when taking the limit from correlated to white noise) as a limit of the dichotomous noise by using (state-dependent) parameter values (2.56) in (2.45):

$$p(\phi) = C \frac{1}{f(\phi)} \exp\left[-\int_\phi \frac{f(\phi') + \lambda(\phi')\alpha(\phi')g(\phi')}{f(\phi')\alpha(\phi')g(\phi')} d\phi' \right] \qquad (2.73)$$

(see also Porporato and D'Odorico, 2004). Note that $\alpha(\phi)$ always appears, in Eq. (2.73), multiplied by $g(\phi)$: This implies that the state dependency in $\alpha(\phi)$ is simply translated into a modification of the $g(\phi)$ function, which becomes $\alpha\overline{g}(\phi) = \alpha(\phi)g(\phi)$. This component of the state-dependent noise therefore reduces to a standard multiplicative noise while a state dependency remains in $\lambda(\phi)$. In other words, by taking the limits in Eq. (2.56) from Eq. (2.13), we obtain the Langevin equation $d\phi/dt = f(\phi) + \overline{g}(\phi)\xi_{sn}$, where ξ_{sn} is a shot-noise process with state-dependent rate $\lambda(\phi)$.

2.4 White Gaussian noise

2.4.1 Definition and properties

Gaussian white noise $\xi_{gn}(t)$ is a stochastic process characterized by the following properties: (i) its average vanishes, $\langle \xi_{gn}(t) \rangle = 0$; (ii) its autocorrelation function has a sharp peak in zero and then instantaneously drops to zero for time lags $|t - t'| > 0$,

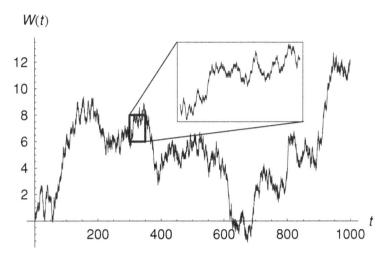

Figure 2.13. Example of a typical realization of the Wiener process. An enlargement of a small portion of the path is shown in the inset to underline the fractal nature of Brownian motion.

i.e., it can be mathematically represented as

$$\langle \xi_{gn}(t)\xi_{gn}(t') \rangle = 2s_{gn}\delta(t - t'), \tag{2.74}$$

where s_{gn} is the strength of the noise[4]; (iii) all cumulants of $\xi_{gn}(t)$ of orders higher than the second vanish. The first two properties are valid for any white-noise process; the third one is peculiar to white Gaussian noise. As for WSN, the problem with the Gaussian white noise is that, even though the notion of white noise is commonly used, it refers to a singular mathematical object. Thus Gaussian white noise is commonly treated as the formal derivative of the Wiener process,

$$W(t) = \int_0^t \xi_{gn}(t')dt', \tag{2.75}$$

a mathematical model for the displacement of the Brownian particle from some arbitrary starting point at $t = 0$ (Brownian motion). A typical path of the Wiener process is represented in Fig. 2.13.

The average and the covariance function of the Wiener process are (Parzen, 1967, p. 68)

$$\langle W(t) \rangle = 0, \tag{2.76}$$

$$\langle W(t)W(t') \rangle = \sigma^2 \min(t, t'), \tag{2.77}$$

respectively, where σ^2 is the parameter of the Wiener process.

---

[4] The autocorrelation function of white Gaussian noise is often denoted as $\sigma^2\delta(t - t')$. However, we believe this notation may induce some confusion in the reader, because the symbol σ^2 is commonly used to represent the variance of a process, and the variance of white Gaussian noise diverges. Moreover, s_{gn} is obtained as the limiting value of the intensity of DMN, $s_{dn} = -\Delta_1\Delta_2/(k_1 + k_2)$, as explained in Subsection 2.4.2.

We obtain the covariance function and first moments of the Gaussian white noise as found in VanMarcke (1983, pp. 110–111) by considering that Gaussian white noise is the formal derivative of the Wiener process; as a consequence the mean $\langle \xi_{gn}(t) \rangle = \partial \langle W(t) \rangle / \partial t$ is zero and the covariance $\langle \xi_{gn}(t) \xi_{gn}(t') \rangle = -\partial^2 \langle W(t) W(t') \rangle / \partial (t - t')^2$ is reported in Eq. (2.74), where noise intensity $s_{gn} = \sigma^2 / 2$. The variance is $\kappa_{2gn} = \sigma^2 \delta(0) = 2 s_{gn} \delta(0)$, i.e., it diverges.

2.4.2 White Gaussian noise as a limiting case of DMN and WSN

We can obtain white Gaussian noise from DMN by taking the following parameter values:

$$\Delta_1 = -\Delta_2 \to \infty, \qquad k_1 = k_2 = \frac{\Delta_1^2}{2 s_{gn}}. \qquad (2.78)$$

It is easily verified that we can obtain moments and autocorrelation function (2.74) from Eq. (B2.1-5) by taking the limit for $\tau_c \to 0$ [note that $\delta(t) = \lim_{a \to 0} e^{-|t/a|} / (2a)$].

We can also obtain Gaussian white noise from WSN by letting the frequency of the shots go to infinity and the average intensity go to zero. This corresponds to taking, for the zero-average WSN process ξ'_{sn},

$$\lambda \to \infty, \qquad \alpha = \sqrt{\frac{s_{gn}}{\lambda}}. \qquad (2.79)$$

2.4.3 Relevance of Gaussian noise in the biogeosciences

In a large class of environmental processes, external noise varies at a much faster time scale than the deterministic component of the system. This justifies the use of a white-noise (memoryless) representation of the external forcing in a number of environmental systems. Moreover, in many situations, fluctuations in the external forcing are the result of several factors simultaneously acting on the environmental system. When the number of these factors is large enough, the central-limit theorem ensures that the fluctuations in the external parameter have approximatively a Gaussian distribution. The quality of the approximation depends on the number of concurring factors, the shape of their probability distributions, and the cross correlation among the factors. The frequent simultaneous presence of these two conditions (memoryless external forcing and numerous environmental factors) explains the common use of Gaussian white noise to represent the external forcing in models of biological and geophysical processes.

We refer again, as in Subsection 2.3.3, to an example in which rainfall plays the role of the external random force; we showed in Figure 2.10 that, considering a daily resolution, rainfall can be approximatively modeled as a WSN process, in which rainfall events occur at random times and are separated by relatively long dry periods. The intermittent nature of a rainfall time series is lost when the temporal resolution (or aggregation scale) of rainfall is much larger than the diurnal one. For example, if we consider rainfall at an annual resolution, we typically find a sequence of random uncorrelated precipitation amounts, approximatively Gaussian distributed. If the response time of the dynamical system subjected to precipitation is larger than the year (e.g., slowly growing vegetation or geomorphological systems), the forcing can therefore be modeled as white Gaussian noise. Other examples of environmental processes in which the forcing can be represented as white Gaussian noise are given in Chapters 4 and 6.

2.4.4 Stochastic process driven by Gaussian noise

In this case the stochastic equation regulating the dynamics of the state variable ϕ is

$$\frac{d\phi}{dt} = f(\phi) + g(\phi)\xi_{gn}. \tag{2.80}$$

As in the case of a system driven by WSN, a problem of interpretation arises when the noise is multiplicative [i.e., $g(\phi)$ is a function of ϕ]. We follow a similar reasoning as in Subsection 2.3.4 and adopt a Stratonovich convention to interpret Eq. (2.80). In the case of systems forced by Gaussian white noise, the Stratonovich convention corresponds to calculating ϕ in $g(\phi)$ as half the sum of the values of ϕ before and after the jump; see Eq. (2.63). In fact, in this case only the terms up to h^2 remain in Eqs. (2.62) and (2.65), because h is infinitesimal, of the order of $\infty^{-0.5}$ [see relations (2.79)]. As a consequence, Eqs. (2.62) and (2.65) are identical for systems driven by Gaussian white noise, i.e., the Stratonovich rule (which we have broadly defined as the rule that arises from the use of the ordinary rules of calculus) corresponds to calculating $g(\phi)$ in $(\phi_0 + \phi_1)/2$, where ϕ_0 and ϕ_1 are the values immediately before and after the jump. We remark again that this is valid only for systems driven by Gaussian white noise.

2.4.4.1 Some examples of processes driven by Gaussian white noise

To facilitate the understanding of the properties presented in following subsections, it is useful to introduce some simple examples of processes driven by Gaussian white noise.

- **Example 2.7:** $f(\phi) = -\phi^3 + 2$ and the noise is additive:

$$\frac{d\phi}{dt} = -\phi^3 + 2 + \xi_{gn}. \tag{2.81}$$

- **Example 2.8:** The deterministic component is the same as in the previous example, but the noise is multiplicative with $g(\phi) = \phi$:

$$\frac{d\phi}{dt} = -\phi^3 + 2 + \phi\xi_{gn}. \tag{2.82}$$

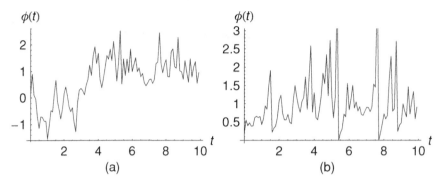

Figure 2.14. Example of a realizations of processes driven by Gaussian white noise: (a) Eq. (2.81), (b) Eq. (2.82). The noise intensity is $s_{gn} = 1$, and the realization of the noise is the same in the two panels.

Figure 2.14 shows the path of $\phi(t)$ for the two processes. From the comparison of the two panels it is clear that in this case the presence of the multiplicative term has an even stronger influence than in the case of the shot-noise process (see also Fig. 2.11). In fact, the presence of the multiplicative term changes the domain of the process and the shape of the corresponding probability distribution, as explained in the following subsections.

2.4.4.2 Steady-state pdf and its properties

We obtain the steady-state pdf for a process driven by Gaussian white noise either from Eq. (2.32) by taking limiting values (2.78) or from (2.69) by using (2.79). We obtain (e.g., Horsthemke and Lefever, 1984)

$$p(\phi) = \frac{C}{g(\phi)} \exp\left[\int_\phi \frac{f(\phi')}{s_{gn} g(\phi')^2} d\phi'\right], \qquad (2.83)$$

where C is an integration constant that ensures that the pdf has unit area. We obtain the same solution by considering the master equation for the evolution of the probability density in time, which is reported in Box 2.4.

The domain of steady-state pdf (2.83) is bounded by the zeros of $g(\phi)$, i.e., by the minima of the potentials $V_1(\phi)$ and $V_2(\phi)$ (defined in Subsection 2.2.3.3), that in this case are defined by

$$g(\phi) = -\frac{dV_1(\phi)}{d\phi}, \qquad g(\phi) = \frac{dV_2(\phi)}{d\phi}, \qquad (2.84)$$

which can be obtained from Eqs. (2.38) by use of limit values (2.78). For example, if the noise is additive, i.e., $g(\phi) = \text{const}$, the domain covers the whole real axis. This is the case for Example 2.7. For Example 2.8, $g(\phi) = \phi$, which implies that the dynamics are constrained within the interval $[0, \infty]$ or $[-\infty, 0]$, depending on whether the initial condition ϕ_0 is positive or negative valued.

Box 2.4: The master equation for processes driven by Gaussian white noise

The master equation for a process driven by Gaussian white noise is known as the Fokker–Planck equation; the derivation of the Fokker–Planck equation is a standard problem treated in several textbooks on stochastic processes, including those of Parzen (1967), Gardiner (1983), and van Kampen (1992), among others. The interested reader is referred to these books for details. The Fokker–Planck equation corresponding to Langevin equation (2.80) under the Stratonovich interpretation is

$$\frac{\partial p(\phi, t)}{\partial t} = -\frac{\partial [f(\phi) p(\phi, t)]}{\partial \phi} + s_{gn} \frac{\partial}{\partial \phi} \left\{ g(\phi) \frac{\partial}{\partial \phi} [g(\phi) p(\phi, t)] \right\}. \tag{B2.4-1}$$

This equation has the standard form of a convection–diffusion equation, with diffusion coefficient s_{gn}. The steady-state solution of Eq. (B2.4-1) is (2.83). Transient solutions in the form of time-dependent probability distributions exist in particular cases, including the Ornstein–Uhlenbeck (O-U) process [represented by Eq. (2.80) with $f(\phi) = -\gamma\phi$ and $g(\phi) = 1$]: In this case the transient solution is a Gaussian distribution with average $\phi_0 e^{-\gamma t}$ and variance $2s_{gn}/\gamma \left(1 - e^{-2\gamma t}\right)$, where ϕ_0 is the initial state of the system.

Under Ito's interpretation, the Fokker–Planck equation corresponding to Langevin equation (2.80) is

$$\frac{\partial p(\phi, t)}{\partial t} = -\frac{\partial [f(\phi) p(\phi, t)]}{\partial \phi} + s_{gn} \frac{\partial^2}{\partial \phi^2} [g(\phi)^2 p(\phi, t)], \tag{B2.4-2}$$

with the steady-state solution

$$p(\phi) = \frac{C}{g(\phi)^2} \exp \left[\int_\phi \frac{f(\phi')}{s_{gn} g(\phi')^2} d\phi' \right], \tag{B2.4-3}$$

where C is an integration constant that ensures that the pdf has unit area. Equation (B2.4-3) differs from (2.83) only for the presence of the exponent 2 in the function $g(\phi)$ in the denominator.

To exemplify the application of Eq. (2.83), we report the solutions for the two examples previously defined.

- **Example 2.7:** The steady-state pdf is

$$p(\phi) = C \exp \left[\frac{8\phi - \phi^4}{4s_{gn}} \right]. \tag{2.85}$$

- **Example 2.8:** Under the Stratonovich interpretation, the steady-state pdf is

$$p(\phi) = \frac{C}{\phi} \exp \left[-\frac{4 + \phi^3}{2s_{gn}\phi} \right]. \tag{2.86}$$

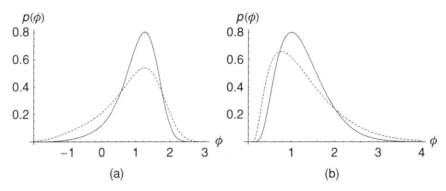

Figure 2.15. Examples of steady-state pdfs for processes driven by Gaussian white noise, corresponding to (a) Eq. (2.81) and (b) Eq. (2.82). The continuous curves correspond to a noise intensity $s_{\mathrm{gn}} = 1$, the dashed curves to $s_{\mathrm{gn}} = 2$.

The pdfs in Eqs. (2.85) and (2.86) are shown in Figs. 2.15(a) and 2.15(b), respectively.

2.5 Colored Gaussian noise

In the introduction we discussed the fact that when the integral scale τ_n of the noise and the typical time scale τ_s of the modeled dynamical system are comparable, the noise correlation plays a fundamental role in the stochastic dynamics of the state variable ϕ and cannot be neglected. In spite of the existence of a plethora of different types of correlated noises, exact analytical results are known only for some classes of systems driven by non-Markovian Gaussian colored noises and for two types of Markovian colored noise, namely the dichotomous noise and the Markovian Gaussian colored noise, i.e., the Ornstein–Uhlenbeck (O-U) process. The case of dichotomous noise was presented at the beginning of this chapter (Section 2.2); in this section we describe some basic properties of Gaussian colored noise.

We consider the stochastic model

$$\frac{\mathrm{d}\phi}{\mathrm{d}t} = f(\phi) + g(\phi)\xi_{\mathrm{cn}}(t), \tag{2.87}$$

driven by generic (Markovian or non-Markovian) Gaussian colored noise $\xi_{\mathrm{cn}}(t)$, with zero mean [i.e., $\langle \xi_{\mathrm{cn}}(t) \rangle = 0$] and autocovariance function

$$\langle \xi_{\mathrm{cn}}(t)\xi_{\mathrm{cn}}(t') \rangle = C(t - t'), \qquad t \geq t', \tag{2.88}$$

where $C(t - t')$ is a generic function of $t - t'$. The pdf $p(\phi, t)$ can be formally written as

$$p(\phi, t) = \langle \delta(\phi(t) - \phi) \rangle, \tag{2.89}$$

where $\delta(\cdot)$ is the Dirac delta function, $\phi(t)$ is the solution of stochastic equation (2.87) for a given noise realization $\xi_{cn}(t)$ and a given initial condition, and ϕ is a value of the state variable. In Eq. (2.89) the ensemble average is taken over the noise realizations and the initial conditions.

The master equation for the evolution in time of the pdf, corresponding to Langevin equation (2.87) is (Hanggi, 1989; Hanggi and Jung, 1995)

$$\frac{\partial p(\phi, t)}{\partial t} = -\frac{\partial p(\phi, t) f(\phi)}{\partial \phi} - \frac{\partial}{\partial \phi} g(\phi) \int_0^t C(t - t') \left\langle \frac{\varrho[\delta(\phi(t) - \phi)]}{\varrho \xi_{cn}(t')} \right\rangle dt', \quad (2.90)$$

where the notation $\varrho F(z)/\varrho z$ represents the functional derivative of the generic functional $F(z)$ with respect to the function z.

2.5.1 Solution for linear Langevin equations

Master equation (2.90) can be solved (Hanggi and Jung, 1995) only if $\varrho \phi(t)/\varrho \xi_{cn}(t)$ does not depend on the process $\phi(t)$, so that the ensemble average in Eq. (2.90) reduces to

$$\left\langle \frac{\varrho[\delta(\phi(t) - \phi)]}{\varrho \xi_{cn}(t')} \right\rangle = \left\langle \left[-\frac{\partial}{\partial \phi} \delta(\phi(t) - \phi) \right] \frac{\varrho \phi(t)}{\varrho \xi_{cn}(t')} \right\rangle = -\frac{\partial p(\phi, t)}{\partial \phi} \frac{\varrho \phi(t)}{\varrho \xi_{cn}(t')}. \quad (2.91)$$

In particular, transformation (2.91) is valid for all linear processes (or processes that can be linearized through changes in variables) that can be written in the form

$$\frac{d\phi}{dt} = a + b\phi + c\xi_{cn}(t). \quad (2.92)$$

In this case we obtain

$$\frac{\varrho \phi(t)}{\varrho \xi_{cn}(t')} = \theta(t - s) c \, e^{b(t - t')}. \quad (2.93)$$

The master equation then becomes a Fokker–Planck-like equation,

$$\frac{\partial p(\phi, t)}{\partial t} = -a \frac{\partial p(\phi, t)}{\partial \phi} - b \frac{\partial \phi p(\phi, t)}{\partial \phi} + \left[\int_0^t C(t - t') e^{b(t - t')} dt' \right] \frac{\partial^2 p(\phi, t)}{\partial \phi^2}, \quad (2.94)$$

whose solution is

$$p(\phi, t) = \frac{\exp\left\{ -\frac{[\phi - \beta(t, \phi_0)]^2}{2\alpha(t)} \right\}}{\sqrt{2\pi \alpha(t)}}, \quad (2.95)$$

where ϕ_0 is the initial condition (at $t = 0$),

$$\alpha(t) = \int_0^t e^{2b(t - t')} D(t') dt', \qquad D(t') = 2c \int_0^{t'} C(t' - t'') e^{b(t' - t'')} dt'', \quad (2.96)$$

and

$$\beta(t, \phi_0) = \phi_0 e^{bt} + a \int_0^t e^{b(t-t')}\mathrm{d}t'. \tag{2.97}$$

2.5.2 Dynamics driven by the Ornstein–Uhlenbeck process

In the case of nonlinear Langevin equations, the solution of master equation (2.90) remains a rather intractable problem, unless the noise is assumed to be Markovian. In this case some important approximated results can be obtained. It can be shown (Doob's theorem; Doob, 1942) that there is only one form of Markovian colored Gaussian noise. This noise is known as the Ornstein-Uhlenbeck (O-U) process (the O-U process is therefore the only existing Markovian, Gaussian colored process) and is described by the equation

$$\frac{\mathrm{d}\xi_{\mathrm{ou}}}{\mathrm{d}t} = -\frac{1}{\tau_{\mathrm{ou}}}\xi_{\mathrm{ou}} + \frac{\sqrt{s_{\mathrm{ou}}}}{\tau_{\mathrm{ou}}}\xi_{\mathrm{gn}}(t), \tag{2.98}$$

where τ_{ou} is the correlation scale of the O-U process, s_{ou} is the intensity of ξ_{ou}, and ξ_{gn} is Gaussian white noise with strength $s_{\mathrm{gn}} = 1$ [i.e., $\langle \xi_{\mathrm{gn}}(t)\xi_{\mathrm{gn}}(t') \rangle = 2\delta(t - t')$]. The autocorrelation function of the O-U process is exponential:

$$C(t - t') = \frac{s_{\mathrm{ou}}}{\tau_{\mathrm{ou}}}e^{-\frac{|t-t'|}{\tau_{\mathrm{ou}}}}. \tag{2.99}$$

Notice that, although the O-U process is Markovian, a process driven by O-U noise is not Markovian. Unfortunately, no exact mathematical expressions exist for the steady-state pdf $p(\phi)$ associated with stochastic dynamical system (2.87) driven by an O-U process [i.e., with driving colored noise $\xi_{\mathrm{cn}}(t)$ corresponding to $\xi_{\mathrm{ou}}(t)$]. However, some approximated solutions were proposed in the literature. Essentially, two approaches were followed. The first one is to enlarge the state space, including $\xi_{\mathrm{ou}}(t)$ as an auxiliary variable driven by Gaussian white noise. In this way, the $(\phi, \xi_{\mathrm{ou}})$ dynamics are a bivariate Markovian process and an exact two-dimensional (2D) Fokker–Planck equation can be written for the density function $p(\phi, \xi_{\mathrm{ou}}, t)$. From this starting point, several techniques were proposed to investigate the process $\phi(t)$, including a perturbative expansion in τ_{ou} (Horsthemke and Lefever, 1984) or continued fraction expansions (Risken, 1984). The other approach treats the process as driven by non-Markovian Gaussian noise and tries to obtain from exact master equation (2.90) a Fokker–Planck-like master equation. This equation is assumed to be able to capture some characteristics of the non-Markovian process $\phi(t)$, at least in the steady-state condition. The differences between these two approaches translate into different ranges of validity of the approximated expressions, particularly with respect to the correlation scale of the noise. Here we summarize the main results along with their range of validity.

- **Small-correlation expansion.** An often-adopted approach consists of dealing with colored noises close to the white-noise limit, i.e., for small values of τ_{ou}. Thus, starting from exact master equation (2.90), the term $\varrho\phi(t)/\varrho\xi_{cn}(t)$ is expanded into a power series in $(t - t')$ truncated to the first order. We obtain (Hanggi, 1989; Hanggi and Jung, 1995)

$$\frac{\partial p(\phi, t)}{\partial t} = -\frac{\partial}{\partial \phi}[f(\phi)p(\phi, t)] + s_{ou}\frac{\partial}{\partial \phi}g(\phi)\frac{\partial}{\partial \phi}g(\phi)\left[1 + \tau_{ou}g(\phi)\left(\frac{f}{g}\right)'\right]p(\phi, t),$$
(2.100)

which is the standard small-τ_{ou} approximation of the master equation; therefore the non-Markovian process $\phi(t)$ has been approximated by a Markovian process characterized by Fokker–Planck equation (2.100), whose solution is

$$p(\phi) = \frac{C}{|g(\phi)\{1 + \tau_{ou}g(\phi)[f(\phi)/g(\phi)]'\}|}$$
$$\times \exp\left[\int_\phi \frac{f(\phi')d\phi'}{s_{ou}g^2(\phi')\{1 + \tau_{ou}g(\phi')[f(\phi')/g(\phi')]'\}}\right],$$
(2.101)

where C is a normalization constant. This expression of $p(\phi)$ [Eq. (2.101)] can also be obtained with other methods based on a truncated τ_{ou} expansion. The main weakness of this approach is that the τ_{ou} expansion is a singular perturbation expansion, so that the relative importance of the neglected terms is not easy to estimate. Moreover, the approximation is not homogeneous for all values of ϕ, and this introduces artificial boundaries at the point where $\tau_{ou}g(\phi)[f(\phi)/g(\phi)]' = 1$. Thus, if the pdf [Eq. (2.101)] tends to zero at these boundaries, the approximation can be considered reasonable, whereas if it diverges other approximation schemes need to be adopted. To overcome these shortcomings some authors refined the approach based on the small-τ_{ou} approximation. In particular, we recall the steady-state pdf determined by Fox (1986),

$$p(\phi) = C\frac{|1 - \tau_{ou}g(\phi)[f(\phi)/g(\phi)]'|}{|g(\phi)|}$$
$$\times \exp\left[\int_\phi \frac{f(\phi')\{1 - \tau_{ou}g(\phi')[f(\phi')/g(\phi')]'\}}{s_{ou}g^2(\phi')}d\phi'\right],$$
(2.102)

and the pdf obtained by Sancho and San Miguel (1989),

$$p(\phi) = Cp_0(\phi)e^{-\tau_{ou}G(\phi)}, \qquad G(\phi) = g(\phi)\left[\frac{f(\phi)}{g(\phi)}\right]' + \frac{1}{2s_{ou}}\left[\frac{f(\phi)}{g(\phi)}\right]^2,$$
(2.103)

where C is the normalization factor and $p_0(\phi)$ is the pdf in the case of white noise (i.e., $\tau_{ou} = 0$).

Equations (2.102) and (2.103) give in general good results, though their validity remains limited to (i) small-correlation times and (ii) small values of τ_{ou}/s_{ou}. Moreover we can observe that in all the approximated expressions the white-noise case (according to the Stratonovich interpretation of the Langevin equation) is recovered when $\tau_{ou} \to 0$.

- **Decoupling approximation.** This approach was proposed by Hanggi et al. (1985) to provide an expression of $p(\phi)$ whose validity is not limited to the case of small-autocorrelation scales for the noise term. The starting point is to decouple the correlation present in master equation

(2.90). In this way the main mathematical difficulty is removed. Thus, a Fokker–Planck-like equation is formally obtained, and at steady state the pdf of ϕ is obtained as

$$p(\phi) = \frac{C}{|g(\phi)|}\exp\left[\frac{1 - \tau_{\mathrm{ou}}(\langle f'\rangle - \langle fg'/g\rangle)}{s_{\mathrm{ou}}}\int_\phi \frac{f(\phi')}{g^2(\phi')}\mathrm{d}\phi'\right], \qquad (2.104)$$

where the ensemble averages are taken over the probability distribution $p(\phi)$ [i.e., $\langle\cdot\rangle = \int p(\phi)\mathrm{d}\phi$]. It follows that it is generally impossible to obtain an explicit expression of $p(\phi)$. To avoid this difficulty, we usually approximate the ensemble averages by taking them over the probability distribution $p_{\mathrm{gn}}(\phi)$ corresponding to the white-noise approximation (i.e., $\tau_{\mathrm{ou}} \to 0$).

Approximated form (2.104) corresponds to the usual steady-state pdf for Gaussian white noise with an effective noise strength

$$s_{\mathrm{eff}}(\tau) = \frac{s_{\mathrm{ou}}}{1 - \tau_{\mathrm{ou}}(\langle f'\rangle - \langle fg'/g\rangle)}. \qquad (2.105)$$

Notice that, for globally stable physical systems (i.e., $\langle f'\rangle < 0$) and additive noise (i.e., $g(\phi) = \mathrm{const}$), we obtain $0 < s_{\mathrm{eff}}(\tau_{\mathrm{ou}}) < s_{\mathrm{eff}}(\tau_{\mathrm{ou}} = 0) = s_{\mathrm{ou}}$; therefore, in these cases, noise correlation entails a reduction of the effective noise strength with the consequent sharpening of the pdf with respect to the white-noise case.

We recall that Eq. (2.104) is valid not only for small-noise-correlation scales but also for moderate-to-strong values of τ_{ou}; however, it requires small values of the noise strength s_{ou}. The decoupling approximation is homogeneous for all values of ϕ and no artificial boundaries are introduced.

- **Unified colored-noise approximation (UCNA).** A weak point of the decoupling approximation approach is that it works for small values of s_{ou}. Therefore it is not well suited to investigate noise-induced transitions (see Chapter 3) that usually emerge for moderate-to-strong noise intensities. To overcome this restriction, Jung and Hanggi (1987) proposed another approach known as the unified colored-noise approximation (UCNA), which provides the pdf of ϕ in the form (see also Hanggi and Jung, 1995)

$$p(\phi) = C\left|\frac{1 - \tau_{\mathrm{ou}}g(\phi)[f(\phi)/g(\phi)]'}{g(\phi)}\right|$$
$$\times \exp\left[\int_\phi \frac{f(\phi')[1 - \tau_{\mathrm{ou}}g(\phi')[f(\phi')/g(\phi')]']}{s_{\mathrm{ou}}g^2(\phi')}\mathrm{d}\phi'\right]. \qquad (2.106)$$

This expression is valid for both small- and moderate-to-large-correlation times τ_{ou} with the constraint that

$$\sqrt{\frac{1}{\tau_{\mathrm{ou}}}} + \sqrt{\tau_{\mathrm{ou}}}\left[-f'(\phi) + f(\phi)\frac{g'(\phi)}{g(\phi)}\right] > 0. \qquad (2.107)$$

Notice that, even though the theory developed by these authors (as well as the resulting time-dependent Fokker–Planck equations) is substantially different from those based on the small-autocorrelation-scale approximations, the steady-state distribution of ϕ given by (2.106) is precisely the same as the one obtained by Fox within the small-correlation theory

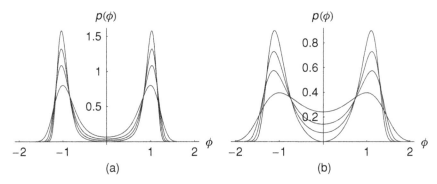

Figure 2.16. Effect of the noise correlation on the Ginzburg–Landau system forced by an O-U process: (a) $s_{ou} = 0.1$, (b) $s_{ou} = 0.5$. In each panel the four curves correspond to $\tau_{ou} = 0$ (i.e., white noise), $\tau_{ou} = 0.3$, $\tau_{ou} = 0.6$, and $\tau_{ou} = 0.99$; the peaks become more pronounced when the noise correlation increases.

[i.e., Eq. (2.102)]. Therefore the effective range of validity of Fox's expression is not limited only to small values of τ_{ou}.

In conclusion, both Eqs. (2.104) and (2.106) [or (2.102)] give good results even for moderate-to-large-correlation scales. However, Eq. (2.104) requires small values of s_{ou}, whereas (2.106) requires that condition (2.107) be satisfied.

Figure 2.16 shows an example of effect of colored noise on the dynamics of ϕ. The model refers to the case of a bistable potential $V(\phi) = \phi^4/4 - \phi^2/2$ driven by an additive O-U process, $\xi_{ou}(t)$, i.e., the so-called Ginzburg–Landau model:

$$\frac{d\phi}{dt} = \phi - \phi^3 + \xi_{ou}(t). \tag{2.108}$$

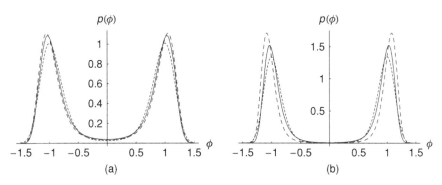

Figure 2.17. Comparison among different approximations of steady-state pdf for a system driven by the O-U process: small-correlation expansion [Eq. (2.103), dashed curve], decoupling approximation [Eq. (2.104), dotted curve], and UCNA [Eq. (2.106), continuous curve]. (a) $\tau_{ou} = 0.3$, (b) $\tau_{ou} = 0.9$ ($s_{ou} = 0.1$ in both panels).

The pdfs are calculated with relation (2.106), which is valid [see condition (2.107)] for

$$1 - \tau_{ou}(1 - 3\phi^2) > 0 \quad \Rightarrow \quad \tau_{ou} < 1. \tag{2.109}$$

Two noise intensities and different correlation scales are shown. The figure clearly shows that in this case the effect of noise correlation cannot be neglected. In fact, remarkable differences exist with respect to the white-noise case (i.e., $\tau_{ou} = 0$), including the emergence of sharper peaks in $p(\phi)$ and the (weak) shift of the modes with increasing values of the correlation scale. In Chapters 3 and 4 we show how the autocorrelation of noise is also able to induce more structural effects on the shape of $p(\phi)$.

To show and compare the effect of the different approximations presented in this section, we calculated $p(\phi)$ for the same model (2.108) by using Eqs. (2.103), (2.104), and (2.106). In this case the three expressions for $p(\phi)$ give very similar results (see Fig. 2.17).

3

Noise-induced phenomena
in zero-dimensional systems

3.1 Introduction

Noise is able to dramatically change the temporal dynamics of zero-dimensional systems. Noise-induced phenomena are fascinating, often counterintuitive, and can play an important constructive role in the behavior of several dynamical systems in the biogeosciences. In fact, noise is not only able to generate disorder, but it may also give rise to new ordered states and induce unexpected dynamical behaviors, which would not be found in the deterministic counterpart of the system.

In this chapter we review some important noise-induced phenomena that were reported in the literature, including noise-induced structural variations of the steady-state probability density function of the state variable (noise-induced transitions), noise-induced phenomena modifying the time correlation and power spectrum of the temporal dynamics (stochastic resonance and stochastic coherence), and noise-induced drift in the oscillations of the state variable.

3.2 Noise-induced transitions

The presence of noise in a stochastic dynamical system is generally associated with disorganized random fluctuations around the stable states of the underlying deterministic system. In these cases the deterministic behavior remains the skeleton of the dynamics, while noise fades out and partly hides the characteristic features of the deterministic system. In these conditions, as the noise intensity increases, the system becomes more disorganized and the features of the underlying deterministic dynamics tend to become more difficult to recognize. However, there are also systems in which suitable noise components can give rise to new dynamical behaviors and new ordered states that do not exist in the deterministic counterpart of the dynamics. Known as *noise-induced transitions*, these behaviors are associated with structural changes in the steady-state probability distribution of the process. These qualitative changes of the steady-state pdf demonstrate how noise can create order and then have a "constructive" role in the dynamics.

The first studies on noise-induced transitions emerged within the context of radio circuits (Kuznetsov et al., 1965), population dynamics (May, 1973), and enzyme systems (Hahn et al., 1974). Theoretical and experimental evidence of this effect of noise was then provided in these and other fields (Horsthemke and Lefever, 1984).

The expression *transitions* is used to recall the phase transitions typical of the equilibrium systems. In systems out of equilibrium, such as those considered in this chapter, the maxima of the steady-state pdf of the state variable correspond to the most probable and preferentially observed states of the system (i.e., the modes of the state variable). These states can be interpreted as the "phases," i.e., as macroscopic stable steady states of the system. The aim of this section is to show how some stochastic dynamics may undergo (noise-induced) nonequilibrium transitions as the variance or autocorrelation scales of noise vary across suitable threshold values. These transitions are evidenced by qualitative changes in the macroscopic behavior of the system and by the consequent emergence of new phases created by noise.

The most important indicators of these transitions are changes in the maxima and minima – i.e., the modes and antimodes – of the probability density function of the state variable.[1] In fact, modes and antimodes provide important information about the shape of the pdf and the preferential states of the system. Another useful tool to detect noise-induced transitions is the *probabilistic potential*, defined in Box 3.1.

The basic idea is that the minima of the probabilistic potential, as defined in Eq. (B3.1-5), represent the noise-induced preferential states of the dynamics of ϕ. By comparing these minima with those of the deterministic potential, we are able to detect the occurrence of noise-induced transitions. In the following subsections we discuss the dependence of modes and antimodes (i.e., minima of probabilistic potential) on the properties of random forcing, considering the four types of noise introduced in the previous chapter. In the fourth chapter we capitalize on these properties to explain the fundamental role of noise in some natural systems.

3.2.1 Noise-induced transitions driven by dichotomous Markov noise

To recognize and investigate noise-induced transitions, we first need to analyze the deterministic counterpart of the dynamics and assess how the states of the system change in the presence of the stochastic forcing. This task can be tricky in the case of DMN because of the two different possible applications of DMN. In Subsection 2.2.2 we stressed that stochastic models forced by dichotomous noise can be formulated

---

[1] Moments could be a valid alternative, which in fact are considered in Chapter 5 for spatiotemporal systems. However, in some cases transitions in the modes are not reflected by qualitative changes of the moments, and vice versa (e.g., Garcia-Ojalvo and Sancho, 1999).

Box 3.1: Probabilistic potential

In Chapter 2 we showed that the steady-state pdf for dynamical system governed by a stochastic differential equation forced by multiplicative white Gaussian noise is, under the Stratonovich interpretation,

$$p(\phi) = \frac{C}{g(\phi)} \exp\left[\frac{1}{s_{gn}} \int_\phi \frac{f(\phi')}{g^2(\phi')} d\phi' \right].$$ (B3.1-1)

This pdf is often written in the form

$$p(\phi) = C \exp\left[-\frac{\mathcal{V}(\phi)}{s_{gn}} \right],$$ (B3.1-2)

where $\mathcal{V}(\phi)$ is defined as the *probabilistic potential* and can be expressed as

$$\mathcal{V}(\phi) = -\int_\phi \frac{f(\phi')}{g^2(\phi')} d\phi' + s_{gn} \log g(\phi).$$ (B3.1-3)

In particular, if $g(\phi) = 1$ (i.e., when the noise is additive), the potential becomes

$$\mathcal{V}(\phi) = -\int_\phi f(\phi') d\phi'$$ (B3.1-4)

and coincides – apart from an irrelevant integration constant – with the deterministic potential $V(\phi)$ of the underlying deterministic process [(i.e., $f(\phi) = dV(\phi)/d\phi$)].

The idea of using the probabilistic potential to detect the transitions is based on the following rationale: Once the pdf $p(\phi)$ is obtained for a generic stochastic equation driven by one of the noises considered in this chapter (dichotomous noise, shot noise, etc.), it is possible to obtain the probabilistic potential $\mathcal{V}(\phi)$ of an equivalent stochastic process drinen by additive white Gaussian noise that would give the same pdf. In fact, by using Eq. (B3.1-2), we obtain

$$\mathcal{V}(\phi) \propto -\log p(\phi).$$ (B3.1-5)

In this way the crests and hollows (or *basins of attraction*) of the probabilistic potential $\mathcal{V}(\phi)$ coincide with the modes and antimodes of $p(\phi)$. Therefore the behavior of the stochastic dynamical system can be immediately visualized as the trajectory of a hypothetical ball that moves on a surface that has the same shape as the probabilistic potential. When the ball is forced by additive white Gaussian noise, it will tend to remain within its basin of attraction and it will move to another basin of attraction (if any) as an effect of strong fluctuations in the random driver. Structural changes in the behavior of the system (i.e., noise-induced transitions) are visualized as changes in the number and the location of the basin of attractions.

following two different approaches, depending on the nature of the physical system under examination. In the functional approach, we start from the analysis of the deterministic dynamics $d\phi/dt = f(\phi)$ and investigate the effect of multiplicative noise on these dynamics:

$$\frac{d\phi}{dt} = f(\phi) + g(\phi)\xi_{dn}(t). \tag{3.1}$$

In this case the steady-state pdf is given by Eq. (2.32); we easily find the deterministic counterpart of the process by setting $\xi_{dn}(t) = 0$ in Eq. (3.1). Thus the deterministic steady states ϕ_{st} are the zeros of $f(\phi)$, i.e., $f(\phi_{st}) = 0$.

In the mechanistic approach, the dynamics switch between the two deterministic processes,

$$\frac{d\phi}{dt} = f_1(\phi) > 0, \quad \frac{d\phi}{dt} = f_2(\phi) < 0, \tag{3.2}$$

depending on whether the value of a stochastic external driver q is greater or smaller than a given threshold θ [see Eqs. (2.12) in Chapter 2]. In this case noise is used as a mechanism to switch between these two processes. The expression of the steady-state pdf is given by Eq. (2.31). Also in this case the underlying deterministic dynamics are obtained by turning the noise variance to zero. If we decrease the variance of the driving force q while maintaining its mean as constant, q_*, in the zero-variance limit q becomes a constant deterministic value, $q = q_*$. The deterministic stationary state is determined by the position of q_* relative to θ: If $q_* > \theta$, the dynamics are expressed by the first of Eqs. (3.2). In this case the (constant) resources q are abundant enough to sustain the growth of ϕ at a rate $f_1(\phi)$. The deterministic steady state $\phi_{st,1}$ is obtained as a solution of $f_1(\phi_{st,1}) = 0$. Conversely, if $q_* < \theta$, the available resources are scarce and ϕ decays at a rate expressed by the second equation in (3.2). In this case we find the deterministic steady state $\phi_{st,2}$ by setting $f_2(\phi_{st,2}) = 0$.

Once the deterministic counterparts of the dynamics have been identified, it is possible to investigate how noise modifies the stable states of the system. To this end, we analyze the modes and antimodes ϕ_m of the pdf of the process $\phi(t)$ forced by dichotomous noise. We can obtain these modes by setting equal to zero the first-order derivative of (2.32) or (2.31), depending on the interpretation adopted for the DMN. In the functional interpretation, using Eq. (2.32), we find that the modes and antimodes are solutions of

$$f(\phi_m) + \tau_c \Delta_1 \Delta_2 g(\phi_m) g'(\phi_m) + \tau_c (\Delta_1 + \Delta_2) f'(\phi_m) g(\phi_m)$$
$$+ \tau_c \left[2 f(\phi_m) f'(\phi_m) - \frac{f^2(\phi_m) g'(\phi_m)}{g(\phi_m)} \right] = 0, \tag{3.3}$$

where $\tau_c = 1/(k_1 + k_2)$, and the superscript $'$ denotes the derivative, for example

$$g'(\phi_m) = \left.\frac{dg(\phi)}{d\phi}\right|_{\phi=\phi_m}, \qquad f'(\phi_m) = \left.\frac{df(\phi)}{d\phi}\right|_{\phi=\phi_m}. \qquad (3.4)$$

The impact of noise properties on the shape of the pdf clearly appears in Eq. (3.3). In fact, the first term is independent of the noise parameters and remains even when the noise term in Eq. (3.1) is turned off. In these conditions the modes and antimodes of $p(\phi)$ coincide with the stable states of the underlying deterministic dynamics in that they are given by the condition $f(\phi_m) = 0$. The second term expresses the effect of the multiplicative nature of the noise [i.e., it is present when $g(\phi) \neq$ const]; the third term results from the asymmetry of the noise (i.e, $\Delta_1 \neq -\Delta_2$); and the fourth term is due to the noise autocorrelation.

If the mechanistic interpretation is adopted, it is convenient to rewrite Eq. (3.3) in terms of the functions $f_1(\phi)$ and $f_2(\phi)$; we obtain

$$\frac{f_1^2(\phi_m)f_2'(\phi_m) - f_2^2(\phi_m)f_1'(\phi_m)}{f_2(\phi_m) - f_1(\phi_m)} - k_1 f_2(\phi_m) - k_2 f_1(\phi_m) = 0. \qquad (3.5)$$

It is clear from Eq. (3.5) that the stable points of the dichotomic dynamics ϕ_m can be very different from their deterministic counterparts $\phi_{st,1}$ and $\phi_{st,2}$, which are the zeros of $f_1(\phi)$ and $f_2(\phi)$.

3.2.1.1 Noise-induced transitions for processes driven by additive DMN

To provide an example of how noise may profoundly affect the dynamical properties of a system through noise-induced transitions, we consider the dynamics described in the previous chapter (Subsection 2.2.3, Example 2.1) where DMN is used in the mechanistic framework, i.e., to switch the dynamics between the two functions

$$f_1(\phi) = 1 - \phi, \qquad f_2(\phi) = -\phi, \qquad (3.6)$$

with deterministic steady states $\phi_{st,1} = 1$ and $\phi_{st,2} = 0$, respectively. Thus the stochastic dynamics resulting from the random switching between these two functions are naturally bounded between 0 and 1. In this example, Eq. (3.5) can be solved to give

$$\phi_m = \frac{1 - k_2}{2 - k_1 - k_2}. \qquad (3.7)$$

Thus the mode or antimode ϕ_m is between the boundaries of the interval $]0, 1[$ if $k_1 < 1$ and $k_2 < 1$ or $k_1 > 1$ and $k_2 > 1$. In the first case ϕ_m is an antimode, whereas in the second case ϕ_m is a mode. It is also of interest to explore the behavior of the pdf close to the boundaries. Using approximation (2.41) we obtain

$$\lim_{\phi \to 0} p(\phi) \sim \phi^{k_2 - 1}, \qquad \lim_{\phi \to 1} p(\phi) \sim (1 - \phi)^{k_1 - 1};$$

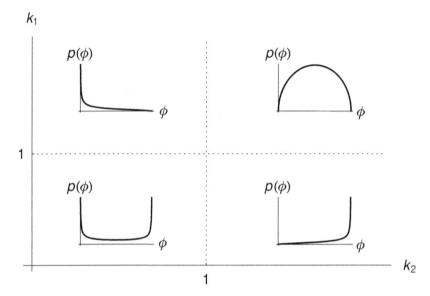

Figure 3.1. Possible shapes of the steady-state pdf as functions of the switching rates k_1 and k_2, with $f_1(\phi) = 1 - \phi$ and $f_2(\phi) = -\phi$.

when $k_1 < 1$ ($k_2 < 1$) the pdf has a vertical asymptote at $\phi = 1$ ($\phi = 0$). We now have the elements to describe the possible shapes of the pdf as a function of the parameters k_1 and k_2 (see Fig. 3.1). The modes and the vertical asymptotes of the pdf are the preferential states of the system. To assess the emergence of noise-induced transitions we need to compare these preferential states with the deterministic stable states. When $k_1 < 1$ and $k_2 > 1$ or $k_1 > 1$ and $k_2 < 1$, noise is unable to create new states, in that the preferential state of the stochastic system coincides with the stable state of the underlying deterministic dynamics. In this case noise creates only disorder in the form of random fluctuations about the stable deterministic state. Conversely, when both the switching rates k_1 and k_2 are greater than one, a new noise-induced state appears at ϕ_m, i.e., a noise-induced transition emerges when the switching rates k_1 and k_2 exceed the threshold $k_1 = k_2 = 1$. Finally, when $k_1 < 1$ and $k_2 < 1$, although the noise does not create any new state, it allows for the coexistence of the two steady states of the underlying deterministic dynamics. Thus noise induces a bistable (i.e., bimodal) behavior that is not observed in the deterministic counterpart of the process, in which only one steady state can exist for a given set of parameters.

It should be noted that in this example the noise is additive in that $g(\phi) = [f_1(\phi) - f_2(\phi)]/(\Delta_2 - \Delta_1) = 1/(\Delta_2 - \Delta_1)$, which is a constant value. Noise-induced transitions emerge even with this simple form of dichotomous noise. These transitions are then due to the autocorrelation of the dichotomous noise. In fact, if we choose a symmetric additive noise, only the correlation term in Eq. (3.3) may cause the emergence of new modes with respect to the deterministic case.

The constructive role of noise autocorrelation is evident also in another example that allows us to discuss some issues related to the use of a functional approach in the stochastic modeling of systems forced by DMN. In this case we start with the study of the deterministic dynamics $d\phi/dt = f(\phi)$, and we use DMN to introduce auto-correlated stochastic fluctuations in these dynamics. Consider the classical Verhulst (or logistic) model (Section 2.2.3, Example 2.4):

$$\frac{d\phi}{dt} = f(\phi) = \phi(\beta - \phi), \quad \phi \in [0, \infty[, \tag{3.9}$$

where $\beta > 0$ is a parameter, called the Malthusian growth parameter or carrying capacity. This nonlinear model was originally proposed to describe the growth of a population and then adopted in many other fields. In dynamical system (3.9), the growth rate of the variable ϕ increases with the value of ϕ itself but is limited by a saturation term that stands for the fact that resources are limited. Despite its apparent simplicity, the model may exhibit a remarkable variety of behaviors with different types of noise. For this reason it is often used to introduce a classical example of noise-induced transitions (e.g., Horsthemke and Lefever, 1984).

For any given positive and constant value of β, deterministic dynamical system (3.9) has two steady states: $\phi_{st,1} = 0$ (unstable) and $\phi_{st,2} = \beta$ (stable). The deterministic dynamics of ϕ can be analytically obtained as a solution of Eq. (3.9):

$$\phi(t) = \frac{\phi(0)e^{\beta t}}{1 + \frac{\phi(0)}{\beta}(e^{\beta t} - 1)}, \tag{3.10}$$

where $\phi(0) > 0$ is the initial condition. The (asymptotic) deterministic steady state is $\phi = \beta$, regardless of the initial condition $\phi(0)$.

We can now assume that an additive dichotomic noise forces the dynamics:

$$\frac{d\phi}{dt} = \phi(\beta - \phi) + \xi_{dn}, \tag{3.11}$$

where, for the sake of simplicity, ξ_{dn} is assumed to be a symmetric noise (i.e., $\Delta_1 = -\Delta_2 = \Delta$ and $k_1 = k_2 = k$). In this case, the pdf of ϕ is obtained by use of Eq. (2.32) and reads

$$p(\phi) = \frac{C}{\phi^2(\beta - \phi)^2 - \Delta^2} \left| \frac{-2\phi + \beta - p}{-2\phi + \beta + p} \right|^{-\frac{k}{p}} \left| \frac{-2\phi + \beta - q}{-2\phi + \beta + q} \right|^{-\frac{k}{q}}, \tag{3.12}$$

where C is the normalization constant, $p = \sqrt{\beta^2 + 4\Delta}$, $q = \sqrt{\beta^2 - 4\Delta}$, and the domain of $p(\phi)$ is

$$\phi \in \left[\frac{\beta + q}{2}, \frac{\beta + p}{2} \right], \tag{3.13}$$

with the constraint that $\Delta < \beta^2/4$ [otherwise the potential corresponding to $f_2(\phi) = \phi(\beta - \phi) - \Delta$ would no longer have a minimum at finite values of ϕ, and the process would diverge].

We can obtain the modes and antimodes of the pdf by applying Eq. (3.3) to model (3.11). In this case only the terms associated with the deterministic dynamics and the noise autocorrelation – i.e., the first and the last addenda on the left-hand side of Eq. (3.3) – are different from zero. Thus it is found that the modes and antimodes of ϕ are solutions of

$$\phi_m(\beta - \phi_m)\left[1 + \frac{(\beta - 2\phi_m)}{k}\right] = 0. \tag{3.14}$$

The deterministic steady state $\phi = \beta$ is always either a minimum or a maximum of $p(\phi)$, depending on the sign of the second-order derivative,

$$\left.\frac{\mathrm{d}^2 p(\phi)}{\mathrm{d}\phi^2}\right|_{\phi=\beta} \propto \frac{\beta(\beta - k)}{k}, \tag{3.15}$$

whereas the other deterministic steady state $\phi = 0$ is outside the domain of $p(\phi)$ [see domain (3.13)]. The value $\phi = (k + \beta)/2$ is another extremum, provided that it is included within the boundaries of domain (3.13); this happens when $q < k < p$; the second-order derivative at this point is

$$\left.\frac{\mathrm{d}^2 p(\phi)}{\mathrm{d}\phi^2}\right|_{\phi=\frac{k+\beta}{2}} \propto \frac{k^2 - \beta^2}{2k}. \tag{3.16}$$

To complete the qualitative study of the shape of the pdf we investigate its behavior near the boundaries as indicated in Chapter 2. We obtain

$$\lim_{\phi\to(\beta+p)/2} p(\phi) \to \infty \ (\to 0) \quad \text{if} \quad k < p \ (k > p),$$

$$\lim_{\phi\to(\beta+q)/2} p(\phi) \to \infty \ (\to 0) \quad \text{if} \quad k < q \ (k > p).$$

We are now able to understand the possible qualitative behaviors of the steady-state pdf of ϕ in the dynamics modeled by Eq. (3.11). As shown in Fig. 3.2, the pdf can assume four different shapes. In particular, we note that (i) the deterministic steady state is a mode of the stochastic dynamics only if $k > \beta$, otherwise it becomes an antimode; and (ii) $p(\phi)$ is unimodal in only one case, whereas in all other cases it is bimodal and exhibits two preferential states, with one of them being at one of the boundaries of the domain. When the dichotomous noise switches frequently (i.e., for high values of k) and the noise amplitude (i.e., Δ) is small, the most visited (i.e., preferential) state coincides with the deterministic steady state (i.e., $\phi_{m,\mathrm{det}} = \beta$), and the stochastic forcing induces random fluctuations around this state. However, when Δ increases, a noise-induced transition occurs, and a second mode appears close to the

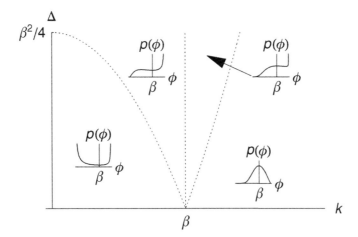

Figure 3.2. Possible shapes of the steady-state pdf for the Verhulst model with additive dichotomous noise as functions of the switching rate k and noise amplitude Δ.

upper boundary of the domain. When the switching rate decreases, the pdf becomes less and less peaked in $\phi = \beta$, until, for $k = \beta$, another noise-induced transition occurs: The shape of the pdf dramatically changes in that it exhibits an antimode corresponding to the deterministic steady state. Finally, for even lower transition rates (i.e., long residence times of the noise in the two states of the DMN), the two modes are at the domain boundaries. All these transitions are due to noise autocorrelation, which is inversely proportional to k and is accounted for by the fourth term in Eq. (3.3). As the autocorrelation increases, the DMN persists longer in each of its two states, and ϕ is more likely to reach a state close to the boundaries of the domain before the DMN switches to the other state. This leads to a bimodal behavior with modes coinciding with the boundaries. Figure 3.3 shows some examples of these nonobvious effects of the additive noise in the Verhulst model.

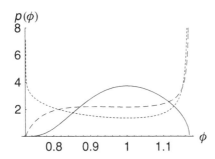

Figure 3.3. Examples of pdf's for the Verhulst model with additive dichotomous noise for $\beta = 1$ and $\Delta = 0.2$. Continuous, dashed, and dotted curves refer to $k = 2, 0.8$, and 0.4, respectively.

3.2.1.2 Noise-induced transitions for processes driven by multiplicative DMN

We now increase the complexity of the system by considering the case of a Verhulst model forced by multiplicative noise (see also Horsthemke and Lefever, 1984). We concentrate on the case in which ξ_{dn} is symmetric and the noise term is a linear function of ϕ, i.e.,

$$\frac{d\phi}{dt} = \phi(\beta - \phi) + \phi\xi_{dn} = \phi\left[(\beta + \xi_{dn}) - \phi\right]. \tag{3.19}$$

In this system the noise randomly modifies the carrying capacity. The corresponding pdf,

$$p(\phi) \propto -\frac{\phi^{\frac{2k\beta}{\Delta^2-\beta^2}}(\Delta + \beta - \phi)^{\frac{k}{\Delta+\beta}}(\phi + \Delta - \beta)^{\frac{k}{\beta-\Delta}}}{\phi[(\phi - \beta)^2 - \Delta^2]}, \tag{3.20}$$

is defined in the domain $[\max(0, \beta - \Delta), \beta + \Delta]$.

In this case the second term in Eq. (3.3) does not vanish. Equation (3.3) becomes

$$\frac{\phi_m[(\phi_m - \beta)(2k - 3\phi_m + \beta) - \Delta^2]}{2k} = 0, \tag{3.21}$$

and the modes and antimodes of ϕ obtained as solutions of (3.21) are

$$\phi_{m,1} = 0, \quad \phi_{m,2,3} = \frac{1}{3}[k \pm \sqrt{3\Delta^2 + (k - \beta)^2} + 2\beta]. \tag{3.22}$$

To sketch the possible shapes of $p(\phi)$ it is again useful to investigate the behavior of the process close to the boundaries of the domain. Using approximation (2.41), we find that, at the upper boundary, $\phi = \beta + \Delta$, the possible behaviors are

$$p(\beta + \Delta) \to \infty, \quad \text{if } \beta + \Delta > k;$$
$$p(\beta + \Delta) = 0, \quad \frac{dp(\phi)}{d\phi}\bigg|_{\phi=\beta+\Delta} \to \infty, \quad \text{if } \frac{k}{2} < \beta + \Delta < k;$$
$$p(\beta + \Delta) = 0, \quad \frac{dp(\phi)}{d\phi}\bigg|_{\phi=\beta+\Delta} = 0, \quad \text{if } \beta + \Delta < \frac{k}{2}.$$

If the lower boundary is $\phi = \beta - \Delta$, the possible behaviors at this boundary are

$$p(\beta - \Delta) \to \infty, \quad \text{if } \beta - \Delta > k;$$
$$p(\beta - \Delta) = 0, \quad \frac{dp(\phi)}{d\phi}\bigg|_{\phi=\beta-\Delta} \to \infty, \quad \text{if } \frac{k}{2} < \beta - \Delta < k;$$
$$p(\beta - \Delta) = 0, \quad \frac{dp(\phi)}{d\phi}\bigg|_{\phi=\beta-\Delta} = 0, \quad \text{if } \beta - \Delta < \frac{k}{2}.$$

In contrast, when the lower boundary is $\phi = 0$, it is not possible to use limit expression (2.41) because $\phi = 0$ is a deterministic steady state. However, studying

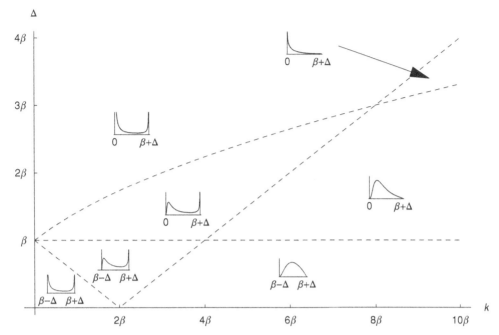

Figure 3.4. Scenario of the steady-state pdf's for the Verhulst model driven by symmetric multiplicative noise. k is the switching rate, Δ is the amplitude of the noise.

directly the expression of $p(\phi)$ given by (3.20) we obtain

$$
\begin{aligned}
&p(0) = \infty, \quad \text{if } \beta(2k + \beta) < \Delta^2; \\
&p(0) = 0, \quad \left.\frac{\mathrm{d}p(\phi)}{\mathrm{d}\phi}\right|_{\phi=0} \to \infty, \quad \text{if } \beta(k + \beta) < \Delta^2 < \beta(2k + \beta); \\
&p(0) = 0, \quad \left.\frac{\mathrm{d}p(\phi)}{\mathrm{d}\phi}\right|_{\phi=0} = 0, \quad \text{if } \Delta^2 < \beta(k + \beta).
\end{aligned}
$$

The different shapes of $p(\phi)$ are shown in Fig. 3.4. We observe a remarkable variety of possible behaviors (Kitahara et al., 1980), depending on the autocorrelation scale $\tau_c = 1/(2k)$ and amplitude Δ^2 of noise. With respect to the additive case (see Fig. 3.2), we can recognize the impact of the multiplicative form of the noise, which induces a greater variety of dynamical behaviors. In particular, in the case of multiplicative noise the deterministic steady state $\phi = \beta$ is never a mode or antimode of the pdf because $\phi = \beta$ is a solution of Eqs. (3.22) only if $\Delta = 0$, i.e., in the absence of noise.

In the case of asymmetric dichotomous noise (i.e., $\Delta_1 \neq \Delta_2$) the third term in Eq. (3.3) also plays a role, further increasing the variety of possible noise-induced transitions.

3.2.1.3 Periodic forcing as a term of comparison

To understand the real role of noise in systems forced by a random dichotomous driver, it is useful to assess whether qualitatively similar phenomena would appear if

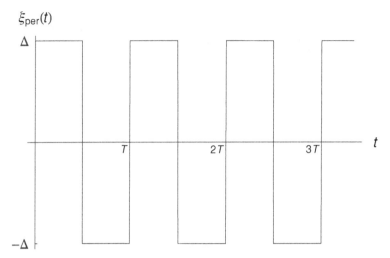

Figure 3.5. Path of the periodic forcing $\xi_{per}(t)$ according to relation (3.23).

the DMN forcing were replaced with deterministic periodic switching. In other words, can a periodic forcing induce similar transitions? The question is an important one (Bena, 2006), even though it is often neglected in the investigation of noise-induced phenomena in physical, chemical, or environmental systems. In fact, if a deterministic periodic component was sufficient to induce the same qualitative transitions, it would mean that only the switching nature of the dichotomous noise actually induces these transitions, rather than the randomness inherent to the noise term. On the other hand, if the periodic forcing is unable to reproduce a similar transition scenario, the randomness of the dichotomous forcing is the real cause of these transitions. In this case the transitions can be properly attributed to noise (i.e., they are noise-induced transitions).

We can address this issue (Horsthemke and Lefever, 1984) by considering a peri-odical, zero-mean, forcing $\xi_{per}(t)$ that switches at regular time intervals of duration $T/2$ between two values $\pm\Delta$ (for the sake of simplicity here we consider a symmetric process), namely,

$$\xi_{per}(t) = \Delta \cdot \left(\Theta\left[\sin\left(\frac{2\pi}{T}t\right)\right] - \Theta\left[\sin\left(2\pi T\left(t - \frac{T}{2}\right)\right)\right] \right), \qquad (3.23)$$

where $\Theta[\cdot]$ is the Heaviside function. Figure 3.5 shows an example of a time series of $\xi_{per}(t)$. For the periodic forcing to be compared with its DMN counterpart, we assume

$$\frac{T}{2} = \tau_1 = \frac{1}{k_1} = \frac{1}{k_2} = \tau_2. \qquad (3.24)$$

It is possible to demonstrate (Doering and Horsthemke, 1985) that, under the influence of this periodic fluctuation, the system

$$\frac{\mathrm{d}\phi}{\mathrm{d}t} = f(\phi) + g(\phi)\,\xi_{\mathrm{per}}(t) \tag{3.25}$$

attains an unique, asymptotically stable steady state, with density function

$$
\begin{aligned}
p(\phi) &= \frac{1}{T}\left[\frac{1}{f(\phi) + g(\phi)\Delta} - \frac{1}{f(\phi) - g(\phi)\Delta}\right] \\
&= \frac{1}{T}\left[\frac{1}{f_1(\phi)} - \frac{1}{f_2(\phi)}\right], \quad \phi \in [\phi_-, \phi_+],
\end{aligned} \tag{3.26}
$$

where the boundaries of the domain ϕ_- and ϕ_+ are the solutions of

$$
\frac{T}{2} = \int_{\phi_-}^{\phi_+} \frac{\mathrm{d}\phi}{f(\phi) + g(\phi)\Delta},
$$

$$
\frac{T}{2} = \int_{\phi_+}^{\phi_-} \frac{\mathrm{d}\phi}{f(\phi) - g(\phi)\Delta}. \tag{3.27}
$$

The rationale underlying Eqs. (3.27) is that the steady state occurs when the process $\phi(t)$ alternately increases and decreases between the same points, ϕ_- and ϕ_+, which become the starting and the ending points, respectively, of the two phases of the dynamics of ϕ forced by ξ_{per}.

We make the following remarks:

- The domain $[\phi_-, \phi_+]$ is always smaller than the support of the corresponding stochastic process. This is because the dynamics persist in each of the two phases, 1 and 2, for only a fixed time $T/2$. Conversely, in the stochastic case, there is always a probability different from zero that the dynamics persist in one of the two phases longer than time $T/2$. Thus the domain explored by $\phi(t)$ is wider than in the deterministic case. As $T \to \infty$ (i.e, $k_{1,2} \to 0$) the domain of $p_{\mathrm{per}}(\phi)$ tends to the domain of the stochastic process. Overall, we observe that the domain $[\phi_-, \phi_+]$ increases with T and depends on Δ.
- There is a remarkable similarity between the mathematical structure of the pdf under stochastic [see Eq. (2.31)] and periodic forcing [Eq. (3.26)]. In fact, when $T \to \infty$, the exponential term in (2.31) tends to one and the two pdf's tend to coincide. In this case, the fluctuations are occurring on a time scale much longer than the typical response time of the system (i.e., longer than the time τ_s needed by the system to approach steady-state conditions while persisting in one of its two phases). When the forcing is periodic, for each value of ξ_{per} the system has time to relax almost completely to the deterministic steady value. The same type of response is observed in the stochastic system with relatively low switching rates (i.e., $k_{1,2} \ll 1/\tau_s$). In fact, in this case there is only a small probability, $k_{1,2}\tau_s$, that the dynamics switch to the other phase before the system approaches the deterministic steady conditions of the current phase. It follows that, when $T \to \infty$, the system spends most of the time near the deterministic steady states regardless of whether the external parameter varies randomly or regularly. Hence the two pdf's tend to coincide.

Conversely, when $T \sim \tau_s$, the stochastic and the deterministic (periodic) dynamics of $\phi(t)$ exhibit more substantial differences in that in the stochastic case a stronger variability exists in the trajectories $\phi(t)$.

We can investigate, in a manner similar to that of the stochastic case, the shape of the pdf and the possible transitions by studying the modes and antimodes of ϕ for the system driven by the periodic forcing. In this case we can set to zero the first-order derivative of Eq. (3.26) and obtain the condition

$$2g(\phi_m)f(\phi_m)f'(\phi_m) - g'(\phi_m)\left[f^2(\phi_m) + \Delta^2 g^2(\phi_m)\right] = 0, \qquad (3.28)$$

where the second addendum exists only when the periodic forcing is multiplicative.

In the additive case, the deterministic steady state [i.e, $f(\phi_{st}) = 0$] is always a mode of the distribution of ϕ. Other modes and antimodes can appear only if the period T or the amplitude Δ of the fluctuation varies in such a way that the interval (ϕ_-, ϕ_+) contains points where $f'(\phi) = 0$. Notice that this is different from the additive stochastic case, in which the modes and antimodes of ϕ are given by solutions [see Eq. (3.3)] of the equation

$$f(\phi)[1 + 2\tau_c f'(\phi)] = 0. \qquad (3.29)$$

In this case the deterministic steady state is always a mode or antimode of ϕ and other modes (or antimodes) can exist if there are values of ϕ that satisfy the condition $f'(\phi) = -1/(2\tau_c)$. Moreover in the stochastic case the domain is always the same (for a fixed value of the amplitude Δ), independent of the other noise characteristics.

To explore the role of randomness and periodicity we consider the same model presented in Subsection 3.2.1.1, but forced by an additive periodic driver,

$$\frac{d\phi}{dt} = \phi(\beta - \phi) + \xi_{per}(t). \qquad (3.30)$$

The steady-state pdf is

$$p(\phi) = \frac{1}{T}\frac{2\Delta}{\Delta^2 - \phi^2(\beta - \phi)^2}, \qquad (3.31)$$

whose boundaries are obtainable (numerically) from Eqs. (3.27). From condition (3.28) we obtain

$$(\beta - \phi_m)\phi_m\left[(\beta - 2\phi_m)\right] = 0. \qquad (3.32)$$

Thus the steady state $\phi = \beta$ of the dynamics in the absence of periodic forcing is a mode (or antimode) of ϕ also in the presence of additive periodic forcing. Another mode (or antimode), $\phi = \beta/2$, may exist if $\phi = \beta/2$ is within the domain (ϕ_-, ϕ_+). Notice that the second-order derivative of the pdf is equal to $2\beta^2$ at $\phi = \beta$, whereas it is equal to $-\beta^2$ at $\phi = \beta/2$. Therefore $\phi = \beta$ is always an antimode, whereas $\phi = \beta/2$ is a mode (if $\phi_- \leq \beta/2 \leq \phi_+$). The two possible shapes of the pdf of ϕ are

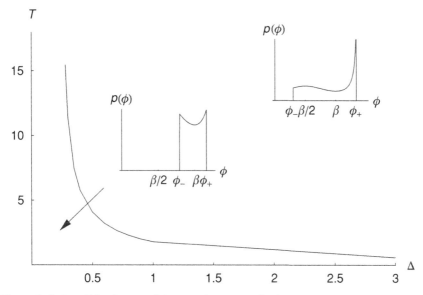

Figure 3.6. Possible shapes of the steady-state pdf when the Verhulst model is periodically forced in an additive way. Δ is the amplitude of the forcing, T is the period.

shown in Fig. 3.6 for the case $\beta = 1$, though different values of this parameter would not change the qualitative features of $p(\phi)$. By comparing Figure 3.2 and 3.6 we can conclude that transitions in the dynamical system forced by DMN are really noise induced and the role of periodicity is a marginal one.

When the periodic forcing is multiplicative, namely,

$$\frac{d\phi}{dt} = \phi(\beta - \phi) + \phi\,\xi_{\text{per}}(t),\tag{3.33}$$

only a U-shaped structure of the pdf is possible. In fact, Eq. (3.28) gives only one solution, $\phi_m = \frac{1}{3}(2\beta - \sqrt{3\Delta^2 + \beta^2})$, which is always a minimum (antimode) of $p(\phi)$. The properties of the periodic forcing, T and Δ, influence only the boundaries of the domain, which are given by the solution of the following set of equations [see (3.27)]:

$$\frac{\phi_-|\phi_+ + \Delta - \beta|}{\phi_+|\phi_- + \Delta - \beta|} = e^{\frac{T(\beta-\Delta)}{2}},\tag{3.34}$$

$$\frac{\phi_+|\phi_- - \Delta - \beta|}{\phi_-|\phi_+ - \Delta - \beta|} = e^{\frac{T(\beta+\Delta)}{2}}.\tag{3.35}$$

Figure 3.7 shows a comparison between two pdf's corresponding to an additive and multiplicative periodic forcing; notice that in the latter case the antimode is larger than the deterministic steady state.

The fact that multiplicative periodic forcing can induce only U-shaped probability distributions of ϕ allows us to conclude that the rich variety of dynamical behaviors

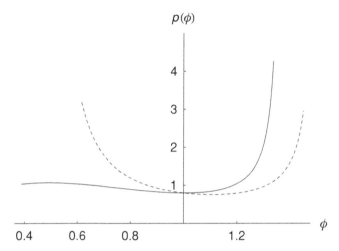

Figure 3.7. Comparison between the pdfs under additive (continuous curve) and multiplicative (dotted curve) periodic forcing of the Verhulst model ($\beta = 1$, $\Delta = 0.5$, $T = 5$).

observed in the case of multiplicative random forcing (see Fig. 3.4) is really noise induced.

3.2.1.4 Noise-induced transitions driven by dichotomous noise with feedback

In Section 2.2, we introduced the case of state-dependent dichotomous noise, i.e., DMN with transition rates k_1 and k_2 dependent on the state ϕ of the system. We also explained how this type of dependency may arise from (and account for) positive or negative feedback between the stochastic driver and the state of the system. In this subsection we show how this feedback (in conjunction with DMN) is able to lead to qualitative changes in the preferential states of a system with respect to those of the underlying deterministic dynamics.

The first step toward the identification of noise-induced transitions is again the recognition of the correct deterministic counterpart of the dynamics. When the functional interpretation is adopted, the deterministic counterpart is the same as the one for the case with no feedback, in that it is obtained from Eq. (3.1), replacing $\xi_{dn}(t)$ with its zero-mean value. In this case the deterministic steady states ϕ_{st}, are the zeros of $f(\phi)$ [i.e., the solutions of $f(\phi_{st}) = 0$].

Conversely, when the mechanistic approach is used, the presence of feedback affects the steady states of the deterministic dynamics. We recall that in the mechanistic usage of the DMN the dynamics switch between the two states, depending on whether a random variable q is above or below a threshold value θ:

$$\frac{d\phi}{dt} = \begin{cases} f_1(\phi) & \text{if} \quad q(t) \geq \theta \\ f_2(\phi) & \text{if} \quad q(t) < \theta \end{cases},$$

(3.36a)
(3.36b)

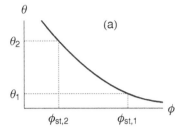
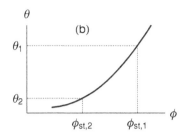

Figure 3.8. Qualitative sketch of the dependency of the threshold θ on the state variable ϕ in the case of (a) positive feedback and (b) negative feedback. The values θ_1 and θ_2 are also shown.

where $f_1(\phi) > 0$ and $f_2(\phi) < 0$ express the state-dependent growth and decay rates, respectively. In the presence of feedback, θ is a function of ϕ. This dependence translates into a state dependence of the switching rates of DMN, $k_1(\phi)$ and $k_2(\phi)$.

As in the case with no feedback (Section 2.1), if we decrease the variance of the driving force q while maintaining constant its mean q_* in the zero-variance limit, q tends to a constant deterministic value, $q = q_*$. We can now analyze two cases of systems with (i) positive feedback (i.e., the threshold θ decreases as ϕ increases), and (ii) negative feedback (i.e., θ increases as ϕ increases).

(i) In the case of positive feedback, the threshold θ is a decreasing function of ϕ [see Fig. 3.8(a)]. In these conditions the deterministic counterpart of the stochastic dynamics depends on the relation between $\theta(\phi)$ and q_*. We first define the maximum and minimum possible values of θ by setting in the relation $\theta(\phi)$ the boundaries of the domain of $p(\phi)$. As noted, these boundaries, $\phi_{\text{st},2}$ (minimum) and $\phi_{\text{st},1}$ (maximum), are the steady states of the two deterministic dynamics, $d\phi/dt = f_{1,2}(\phi)$. Thus two values $\theta_1 = \theta(\phi_{\text{st},1})$ (minimum) and $\theta_2 = \theta(\phi_{\text{st},2})$ (maximum) are obtained [see Fig. 3.8(a)]. If $q_* < \theta_1$ the deterministic dynamics monotonically decrease, converging to the stable state $\phi_{\text{st},2}$. Analogously, if $q_* > \theta_2$ the system persists in state 1 (growth) regardless of the value of ϕ. Thus $\phi(t)$ converges to the deterministic stable state $\phi_{\text{st},1}$. The most interesting situation is with $\theta_1 < q_* < \theta_2$. In this case the deterministic system is bistable: If the threshold value associated with the initial condition ϕ_0 is smaller than q_* [i.e., $\theta(\phi_0) < q_*$], the dynamics of ϕ exhibit a deterministic growth with rate determined by $f_1(\phi)$. As ϕ grows, $\theta(\phi)$ decreases and the system persists in the growth conditions, thereby converging to the steady state, $\phi_{\text{st},1}$. Conversely, if the system is initially in decay (or "stressed") conditions [i.e., $\theta(\phi_0) > q_*$], ϕ decreases with rate $f_2(\phi)$. As ϕ decreases, $\theta(\phi)$ increases. Thus ϕ tends to the steady state $\phi_{\text{st},2}$ in that the system persists in state 2 (decay).

(ii) In the presence of negative feedback, $\theta(\phi)$ increases with ϕ [see Fig. 3.8(b)]. The two limiting values of θ are again defined as $\theta_1 = \theta(\phi_{\text{st},1})$ and $\theta_2 = \theta(\phi_{\text{st},2})$. However, θ_1 is now the maximum and θ_2 the minimum value of θ [see Fig. 3.8(b)].

The deterministic states are then $\phi_{st,1}$ if $q_* > \theta_1$, and $\phi_{st,2}$ if $q_* < \theta_2$. The most interesting dynamics are found when $\theta_2 < q_* < \theta_1$: If the system is initially in the growth state [i.e., $\theta(\phi_0) < q_*$], ϕ increases [hence $\theta(\phi)$ increases] until ϕ reaches the value ϕ_*, with $\theta(\phi_*) = q_*$. In these conditions the system is stable: In fact, if ϕ exceeds ϕ_*, the state variable ϕ decreases with rate $f_2(\phi)$ because $\theta(\phi_*) > q_*$. Conversely, if the system is initially in the stressed (or decay) state [i.e., $\theta(\phi_0) > q_*$], ϕ decreases until it reaches ϕ_* (from above). The stable state of the deterministic system is then ϕ_*.

Once the deterministic counterpart of the dynamics has been identified, it is possible to investigate how the interaction of noise with state dependency modifies the stable states of the system. To this end, we analyze the modes and antimodes ϕ_m of the pdf of the process, $\phi(t)$, forced by state-dependent dichotomous noise. We can obtain these modes by setting equal to zero the first-order derivative of (2.44) or (2.45), depending on the interpretation adopted for the DMN. In the functional interpretation, the modes and antimodes are found from

$$f(\phi_m) + \tau_c \Delta_1 \Delta_2 g(\phi_m) g'(\phi_m) + \tau_c (\Delta_1 + \Delta_2) f'(\phi_m) g(\phi_m) \qquad (3.37)$$

$$+ \tau_c \left[2f(\phi_m)f'(\phi_m) - \frac{f^2(\phi_m)g'(\phi_m)}{g(\phi_m)} \right] + g(\phi_m)\tau_c[\Delta_1 k_2(\phi_m) + \Delta_2 k_1(\phi_m)] = 0.$$

The impact of the state dependency on the shape of the pdf clearly appears from Eq. (3.37). The first four terms in Eq. (3.37) exist also when the dichotomous noise is state independent [see Eq. (3.3)]. When the noise is state dependent, the fifth term in (3.37) appears. This term contributes to the emergence of differences in the stable states (modes) between the stochastic and deterministic dynamics. Notice that when the transition rates are constant (i.e., the noise parameters do not depend on ϕ) the fifth term is zero if the noise is taken, as it is usually done, with a null average value, $\Delta_1 k_2 + \Delta_2 k_1 = 0$. Instead, in the presence of state dependency the mean value of noise is in general a function of ϕ.

If the mechanistic interpretation is adopted, it is convenient to write Eq. (3.37) in terms of the functions $f_1(\phi)$ and $f_2(\phi)$:

$$\frac{f_1^2(\phi_m)f_2'(\phi_m) - f_2^2(\phi_m)f_1'(\phi_m)}{f_2(\phi_m) - f_1(\phi_m)} - k_1(\phi_m)f_2(\phi_m) - k_2(\phi_m)f_1(\phi_m) = 0. \qquad (3.38)$$

It is clear from Eq. (3.38) that the role of the state dependency in modifying the stable states is due to the presence in Eq. (3.38) of the terms $k_1(\phi_m)$ and $k_2(\phi_m)$.

To demonstrate the possible impact of the feedback between noise and the dynamical system, we consider the example described in the previous subsections, in which the alternating processes of growth and decay are expressed by two linear functions,

$$f_1(\phi) = \alpha(1 - \phi), \qquad f_2(\phi) = -\alpha\phi, \qquad (3.39)$$

where $\alpha > 0$ determines the rates of growth and decay. In this system $\phi_{st,1} = 1$ and $\phi_{st,2} = 0$. Similar to the example shown in Subsection 2.2.3.5, here we also assume a linear dependence of θ on ϕ, $\theta(\phi) = \theta_0 + b\phi$, and a logistic distribution $P_Q(q) = \{1 + \exp[-(q - q_*)/\sigma]\}^{-1}$ to represent the variability of the resource q. The steady-state pdf of ϕ is given by Eq. (2.49).

Figure 3.9 summarizes the behavior of $p(\phi)$ as a function of the parameters σ and α for the case of no feedback [$b = 0$, Fig. 3.9(a)], positive feedback [$b < 0$, Fig. 3.9(b)], or negative feedback [$b > 0$, Fig. 3.9(c)]. The case with $b = 0$ [Fig. 3.9(a)] refers to a situation in which $q_* < \theta_0$, which implies that the deterministic stable state is $\phi_{st} = 0$. For small noise intensities, i.e., for small values of σ, the probability distribution is L-shaped, i.e., the most probable state is $\phi = 0$. For increasing σ values, two different kinds of noise transitions occur: If α is small relative to the rate of switching of DMN – i.e., the system responds slowly to the external forcing – a new mode appears at $\phi_m = (k_1 - \alpha)/(1 - \alpha)$; if α is relatively large a bifurcation occurs, i.e., the distribution becomes U-shaped, with two stable states in $\phi = 0$ and $\phi = 1$. In Fig. 3.9(a) the separation among these three regimes is marked by two lines determined through the analysis of the pdf at the boundaries of the domain. It is found that these lines are $\alpha = k_1 = \{1 + \exp[-(\theta_0 - q_*)/\sigma]\}^{-1}$ and $\alpha = k_2 = 1 - k_1$ (Laio et al., 2008).

Figure 3.9(b) shows the case with positive feedback ($b < 0$). For simplicity, we take $b = 2(q_* - \theta_0)$, which implies that the distribution is symmetrical with respect to $\phi = 0.5$ for any value of the parameters [see the expression of pdf (2.49)]. The distribution is U-shaped for low noise intensities, as expected from the bistable behavior of the deterministic counterpart of the dynamics. With increasing values of σ, a first transition occur for $\alpha < 0.5 + b/(8\sigma)$, with two other modes appearing and the distribution assuming an M-shape. If σ is further increased, and $\alpha < k_1(\phi = 1) = k_2(\phi = 0) = \{1 + \exp[(\theta_0 - q_*)/\sigma]\}^{-1}$ (see Laio et al., 2008), a second transition occurs and the system exhibits a new stable state in $\phi_m = 0.5$, and no other stable state exists. The stochastic forcing therefore stabilizes the system around a new statistically stable state. This state is clearly noise induced, in that it does not exist in the deterministic counterpart of the process. The ability of noise to turn a bistable deterministic system into a stochastic process with only one stable state (contained between the two stable deterministic states) is known as *noise-induced stability* (D'Odorico et al., 2005; Borgogno et al., 2007; Ridolfi et al., 2007).

We finally turn to the case with negative feedback [$b > 0$, Fig. 3.9(c)]. We take again $b = 2(q_* - \theta_0)$ to have symmetrical distributions. For low values of σ the distribution has a single mode in $\phi_m = 0.5$, which corresponds to the deterministic stable state. For increasing noise intensity, the distribution becomes first W-shaped, for $\alpha > k_1(\phi = 1) = k_2(\phi = 0) = \{1 + \exp[(\theta_0 - q_*)/\sigma]\}^{-1}$, and then U-shaped, for $\alpha > 0.5 + b/(8\sigma)$. This is a clear example of purely noise-induced bistability, because bistability does not appear in the corresponding deterministic dynamics.

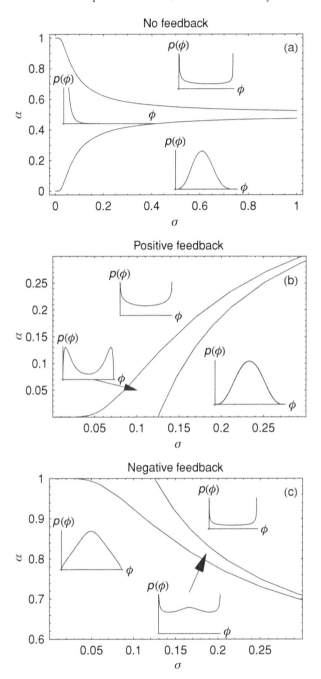

Figure 3.9. Shapes of the probability distributions of ϕ as functions of the parameters σ and α (see text for details): (a) no feedback ($b = 0, \theta_0 = 1.1, q_* = 1$), (b) positive feedback ($b = -0.5, \theta_0 = 1.25, q_* = 1$), and (c) negative feedback ($b = 0.5, \theta_0 = 0.75, q_* = 1$).

3.2.2 Noise-induced transitions for processes driven by shot noise

When a process is driven by WSN, we can obtain the equation for the modes and antimodes of the pdf either by taking the limits (2.56) in Eq. (3.3) or by equating to zero the first-order derivative of Eq. (2.69). We obtain

$$f(\phi_m) - \lambda\alpha^2 g(\phi_m)g'(\phi_m) + \alpha g(\phi_m)f'(\phi_m) = 0. \tag{3.40}$$

A comparison with Eq. (3.3) shows how, in the case of WSN [Eq. (3.40)], the term that is due to the finite autocorrelation of noise vanishes, whereas the terms associated with the multiplicative nature of noise (i.e., the second term) and with its asymmetry (the third term) are still present in Eq. (3.40). In the case of additive noise [i.e., $g(\phi) = \mathrm{const}$], the frequency λ of the jumps does not affect the modes and antimodes of ϕ, except for a drift term that is due to the nonnull average of the shot-noise process. However, additive WSN can affect the number of preferential states (modes) of the system with respect to the deterministic counterpart of the dynamics. In other words, noise-induced transitions may emerge also in systems driven by additive WSN as an effect of the third term in Eq. (3.40).

We consider again as an example the Verhulst model and suppose that it is forced by zero-mean additive WSN ξ'_{sn}:

$$\frac{\mathrm{d}\phi}{\mathrm{d}t} = \phi(\beta - \phi) + \xi_{\mathrm{sn}} = \phi(\beta - \phi) + \alpha\lambda + \xi'_{\mathrm{sn}}, \tag{3.41}$$

where ξ_{sn} is WSN with rate λ, average jump magnitude α, and mean $\langle\xi_{\mathrm{sn}}\rangle = \lambda\alpha$.

In this case the pdf is (see Subsection 2.3.4.2)

$$p(\phi) \propto (\phi - \beta)^{\frac{\lambda}{\beta}-1}\phi^{-\frac{\lambda+\beta}{\beta}}e^{-\frac{\phi}{\alpha}}, \tag{3.42}$$

where domain $\phi \in\ ]\beta, \infty[$. In fact, because the jumps of ξ_{sn} are always positive and $\phi = \beta$ is an attractor of the interjumps' deterministic dynamics, it is sufficient that a jump drive the dynamics to values of ϕ greater than the threshold $\phi = \beta$ for the system to remain confined to values greater than β.

Equation (3.40) gives a mode at

$$\phi_m = \frac{1}{2}\left(\beta - 2\alpha + \sqrt{\beta^2 + 4\alpha^2 + 4\alpha\lambda}\right), \tag{3.43}$$

provided that ϕ_m is within the domain (i.e., $\phi_m > \beta$). This condition is verified if $\lambda \geq \beta$. The behavior of the pdf close to the lower boundary of the domain is

$$\lim_{\phi\to\beta} p(\phi) \to \begin{cases} \infty & \text{if} \quad \lambda < \beta \\ 0 & \text{if} \quad \lambda \geq \beta \end{cases}. \qquad \begin{matrix} (3.44\mathrm{a}) \\ (3.44\mathrm{b}) \end{matrix}$$

Thus the distribution $p(\phi)$ may exhibit two possible shapes, depending on the Poisson rate λ (see Fig. 3.10). Therefore a noise-induced transition appears at $\lambda = \beta$: For $\lambda < \beta$, the main effect of noise is to disturb the deterministic dynamics, whereas,

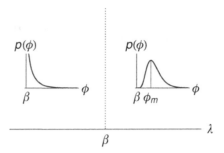

Figure 3.10. Possible steady-state pdf's for the Verhulst model forced by additive WSN.

when $\lambda \geq \beta$, the noise is able to create a new mode, i.e., a new preferential state of the system. We remark that, because the noise is additive, the second term in Eq. (3.40) is zero. Therefore this example demonstrates that transitions may emerge as an effect of the noise asymmetry.

In the case of multiplicative WSN the second term in Eq. (3.40) may or may not contribute to noise-induced transitions. We consider as an example the model

$$\frac{d\phi}{dt} = \phi(\beta - \phi) + \phi\xi_{sn} = \phi[(\beta + \alpha\lambda) - \phi] + \phi\xi'_{sn}. \tag{3.45}$$

The corresponding steady-state pdf is

$$p(\phi) \propto (\phi - \beta)^{\frac{\lambda}{\beta}-1}\phi^{-\frac{\lambda+\beta}{\beta}-\frac{1}{\alpha}}, \tag{3.46}$$

with domain $[\beta, \infty[$.

In this case, Eq. (3.40) gives a mode at

$$\phi_m = \frac{\beta(1 + \alpha) + \alpha\lambda}{1 + 2\alpha}, \tag{3.47}$$

provided that $\lambda \geq \beta$, as in the additive case. It follows that the phase plane is the same as shown in Fig. 3.10. Transitions induced by additive and multiplicative noise are not qualitatively different, in that they involve similar structural changes in the shape of the pdf of ϕ (see the example in Fig. 3.11). Thus, in this example, the key noise property inducing the transition is the asymmetry of the noise rather than its possible multiplicative form.

3.2.3 Noise-induced transitions for processes driven by Gaussian white noise

When the dynamical system is forced by a Gaussian white noise, we can find the modes and antimodes ϕ_m of the pdf by directly deriving Eq. (2.83) or by using the limiting conditions in Subsection 2.4.2 in Eq. (3.3) or (3.40):

$$f(\phi_m) - s_{gn}g(\phi_m)g'(\phi_m) = 0. \tag{3.48}$$

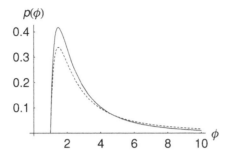

Figure 3.11. Comparison between the steady-state pdf's for the Verhulst model forced by additive (continuous curve) or multiplicative (dotted curve) Poisson noise. The parameters are $\beta = 1$, $\alpha = 10$, and $\lambda = 2$.

The modes and antimodes of the steady-state pdf are then determined only by these two terms. The first term in Eq. (3.48) corresponds to the deterministic dynamics, and the second term accounts for the effect of the multiplicative nature of the external noise. With respect to the more general case of systems driven by dichotomous noise [see Eq. (3.3)], Eq. (3.48) does not exhibit the terms that are due to the autocorrelation and asymmetry of noise. If Eq. (2.80) is interpreted according to Ito's rule, we obtain the equation for the modes and antimodes of the process by deriving Eq. (B2.4-3) with respect to ϕ and equating the result to zero:

$$f(\phi_m) - 2s_{\text{gn}}g(\phi_m)g'(\phi_m) = 0, \tag{3.49}$$

which is the same as Eq. (3.48), except for the fact that the second addendum is multiplied by a factor of 2.

Equation (3.48) clearly shows that when Gaussian noise is additive [i.e., $g(\phi) =$ const], the modes corresponding to the deterministic steady states [i.e., the solutions of $f(\phi) = 0$] exist regardless of the noise intensity. In the case of Gaussian white noise, additive noise has only a disordering effect, in that it induces random fluctuations about the stable states of the deterministic dynamics without being able to qualitatively modify the steady-state behavior of the system. Conversely, when the noise is multiplicative, new dynamical behaviors can appear for suitable forms of $g(\phi)$; for example, if $f(\phi)$ is a polynomial of degree m and $g(\phi)g'(\phi)$ is a polynomial of degree n greater than m, new modes and antimodes may emerge in the presence of multiplicative noise with respect to the deterministic case. Therefore Gaussian white noise can cause noise-induced transitions only if it is multiplicative [i.e., $g'(\phi) \neq 0$]. Similar considerations apply also to the case of transitions in the average $\langle \phi \rangle$ of the process, which can be studied with the methods developed in Box 3.2.

We consider as an example of a system driven by multiplicative Gaussian white noise the case of the Verhulst model:

$$\frac{d\phi}{dt} = \phi(\beta - \phi) + \phi\xi_{\text{gn}}(t) = \phi(\beta + \xi_{\text{gn}} - \phi), \tag{3.50}$$

Box 3.2: Correlation between noise and ϕ

Consider the Langevin equation for a process driven by Gaussian white noise, Eq. (2.80). By taking the expectation of both sides of Eq. (2.80) we obtain

$$\frac{d \langle \phi \rangle}{dt} = \langle f(\phi) \rangle + \langle g(\phi) \xi_{gn} \rangle, \qquad (B3.2\text{-}1)$$

which is sometimes called the *macroscopic equation* of the dynamical system (e.g., van Kampen, 1992). We obtain the stationary states of the deterministic system by setting $d \langle \phi \rangle / dt = 0$ and $\xi_{gn} = 0$ in Eq. (B3.2-1), i.e., as solutions to $\langle f(\phi) \rangle = 0$. When the function $f(\phi)$ is a linear one, a closed-form equation for the average $\langle \phi \rangle$ is obtained; in all other cases, higher-order moments will also influence the solution of the steady-state macroscopic equation (see van Kampen, 1992, pp. 122–127). A Taylor expansion of $f(\phi)$, truncated to the first order, will provide an approximated, closed-form equation for the steady-state average of the process $\langle \phi \rangle$.

We are interested here in considering the role of noise. When the noise is additive, i.e., $g(\phi) = $ const, the term $\langle g(\phi) \xi_{gn} \rangle$ in Eq. (B3.2-1) factorizes and becomes uniformly equal to zero because $\langle \xi_{gn} \rangle = 0$. In formulas, $\langle g(\phi) \xi_{gn} \rangle = \langle g(\phi) \rangle \langle \xi_{gn} \rangle = 0$. As a consequence, additive Gaussian white noise is not able to modify the deterministic stationary states of the system, i.e., to induce transitions in the steady-state average of the process. The same considerations apply when the noise is multiplicative, but the corresponding Langevin equation is interpreted following Ito's convention. We showed in Subsection (2.3.4) that Ito's convention assigns the rule that $g(\phi)$ should be calculated with the value of ϕ just before the jump (van Kampen, 1981). As a consequence, the noise term turns out to be independent of ϕ, and $\langle g(\phi) \xi_{gn} \rangle$ vanishes (see van Kampen, 1992, p. 231).

The situation is different when Statonovich's convention is adopted: In this case the noise turns out to be correlated to ϕ, with (van Kampen, 1992, p. 231)

$$\langle g(\phi) \xi_{gn} \rangle = s_{gn} \left\langle \frac{dg(\phi)}{d\phi} g(\phi) \right\rangle = s_{gn} \langle g'(\phi) g(\phi) \rangle, \qquad (B3.2\text{-}2)$$

which is sometimes called *Novikov's theorem* (e.g., Garcia-Ojalvo and Sancho, 1999). It is clear that in this case the noise may have profound effects on the stationary states of the system, which are now found as solutions of $\langle f(\phi) \rangle + s_{gn} \langle g'(\phi) g(\phi) \rangle = 0$. Noise-induced transitions in the steady-state average of the process are therefore possible in this case. Note that this equation has a form similar to that of Eq. (3.48), except for the sign of the second term. Examples of application of this equation are given in Chapter 5.

This noise-induced effect is not peculiar to Gaussian white noise. In fact, with other types of noise we would find qualitatively similar results (i.e., that multiplicative noise may induce phase transitions in the dynamics). However, with other forms of noise the problem of computing the correlation between noise and ϕ, as in Eq. (B3.2-2), becomes rather intractable. If the noise is white and its cumulants are delta correlated (see van Kampen, 1992, p. 237), the following expression can be obtained after some manipulations of the general expressions provided by Hanggi [1978, Eq. (2.15), p. 409]:

$$\langle g(\phi)\xi \rangle = \sum_{i=1}^{\infty} \frac{\kappa_i}{i!} \langle g_{(i)}(\phi) \rangle, \qquad (B3.2\text{-}3)$$

where κ_i is the cumulant of the ith order of the increments of the integrated process $z(t) = \int_0^t \xi(t)dt$ (see van Kampen, 1992, p. 238), $g_{(1)}(\phi) = g(\phi)$, and $g_{(i+1)}(\phi) = g(\phi)g'_{(i)}(\phi)$ (the same functions were used in Subsection 2.3.4). When the noise is white and Gaussian, the integrated process $z(t)$ is the Wiener process, [see Eq. (2.75)], and $\kappa_i = 0$ for any i, except $\kappa_2 = 2s_{gn}$; Eq. (B3.2-2) is therefore easily recovered from Eq. (B3.2-3).

The corresponding integrated process for shot noise is instead the compound Poisson process in Eq. (2.51), and $\kappa_i = \lambda i! \alpha^i$. We therefore obtain

$$\langle g(\phi)\xi_{sn} \rangle = \lambda \sum_{i=1}^{\infty} \alpha^i \langle g_{(i)}(\phi) \rangle. \qquad (B3.2\text{-}4)$$

If $g(\phi)$ is a linear function of ϕ, $g(\phi) = a + b\phi$ (where a and b are two coefficients), Eq. (B3.2-4) further specifies

$$\langle g(\phi)\xi_{sn} \rangle = \langle (a + b\phi)\xi_{sn} \rangle = \alpha\lambda \frac{a + b\langle \phi \rangle}{1 - b\alpha}, \qquad (B3.2\text{-}5)$$

which is valid for $b\alpha < 1$ (otherwise the process diverges). It is clear from Eqs. (B3.2-4) and (B3.2-5) that the shot noise (when its effect in the Langevin equation is interpreted in the Stratonovich sense) may also play an important role in inducing phase transitions in the dynamics of the system.

To generalize the notation we introduce a function $g_S(\phi)$ of the state variable and write $\langle g(\phi)\xi_m \rangle = s_m\langle g_S(\phi) \rangle$, where s_m is the intensity of the generic multiplicative noise. It is clear that the specific function $g_S(\phi)$ varies depending on the type of noise and its interpretation. $g_S(\phi) = 0$ when $g(\phi) = $ const, or the noise term in the Langevin equation is interpreted in Ito's sense. In the case of Langevin equations with Gaussian white noise, interpreted in the Stratonovich sense, we have $g_S(\phi) = g(\phi)g'(\phi)$.

where $\xi_{gn}(t)$ is Gaussian white noise with intensity s_{gn}. When $\beta > 0$ the domain boundaries are $[0, +\infty[$ and the steady-state pdf is (under Stratonovich's interpretation)

$$p(\phi) = \frac{s_{gn}^{-\beta/s_{gn}}}{\Gamma\left(\frac{\beta}{s_{gn}}\right)} \phi^{\frac{\beta}{s_{gn}}-1} e^{-\frac{\phi}{s_{gn}}}. \qquad (3.51)$$

The pdf in Eq. (3.51) can be recognized to be a gamma distribution, whose modes and antimodes are

$$\phi_{m,1} = 0, \qquad \phi_{m,2} = \beta - s_{gn} \quad (\text{when } \beta > s_{gn}); \qquad (3.52)$$

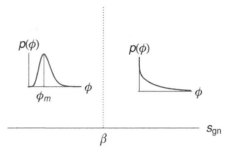

Figure 3.12. Possible shapes of the pdf for the Verhulst model forced by multiplicative Gaussian white noise.

when $\beta > s_{gn}$, $\phi_{m,1}$ is a minimum of $p(\phi)$ and $\phi_{m,2}$ is a maximum, whereas for $\beta < s_{gn}$ the state $\phi_{m,1} = 0$ is a preferential state.

We now have the elements to plot the "phase diagram" (Fig. 3.12) as a function of only one parameter, the noise intensity s_{gn}. When $s_{gn} < \beta$ the pdf is bell shaped with the mode at $\phi = \beta - s_{gn}$ and a minimum at $\phi = 0$, whereas, if the noise intensity exceeds the threshold $s_{gn} = \beta$, the pdf becomes a monotonically decreasing function of ϕ with an asymptote at $\phi = 0$. In the latter case the states close to $\phi = 0$ become the preferential (i.e., most probable) configurations even though $\phi = 0$ is an unstable state of the underlying deterministic dynamics. The abrupt change in the pdf shape for $s_{gn} = \beta$ is a typical example of noise-induced transition. Notice that in this system noise also induces a mode shift when $s_{gn} < \beta$. In fact, for any $s_{gn} > 0$ the mode does not coincide with the deterministic steady state (i.e., $\phi = \beta$). This latter is recovered only in the limit $s_{gn} \to 0$. Figure 3.13 shows the pdf's for two different noise intensities.

It is worth mentioning that the occurrence of the asymptote at $\phi = 0$ is due to the combined action of sufficiently strong noise with the mathematical boundary at $\phi = 0$. In fact, noise drives the system away from the deterministic steady state while the boundary does not allow the system to cross the point $\phi = 0$. This results in a persistence of the system close the boundary itself and then in an increase of mass of probability close to $\phi = 0$. The noisy system behaves differently on the right of the deterministic steady state, where the absence of an upper boundary does not confine the dynamics of ϕ and the only result of an increase in noise strength is an increase in the probability that the system reaches high values of ϕ. The role of the boundary is then crucial to the emergence of phase transitions when Gaussian noise is considered.

3.2.4 Noise-induced transitions for processes driven by Gaussian colored noise

In the previous sections, we showed how the correlation of dichotomous noise plays an important role in the emergence of noise-induced transitions. In this section, we show that correlation is also responsible for noise-induced transition when the system

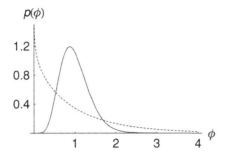

Figure 3.13. Example of pdf's for the Verhulst model ($\beta = 1$) forced by multiplicative Gaussian white noise with intensity $s_{gn} = 0.125$ (continuous curve) and $s_{gn} = 1.125$ (dashed curve). Notice that the maximum of the bell-shaped case is at $\phi_m = 0.875$ and does not coincide with β.

is driven by a colored Gaussian noise. In the following discussion, we consider the UCNA approximation to calculate the steady-state pdf; however, qualitatively similar results are also obtained if the other approximations presented in Section 2.5 are used.

Starting from relation (2.106), the modes ϕ_m are determined as solutions of

$$\left\{1 - \tau_{ou} g(\phi_m)\left[\frac{f(\phi_m)}{g(\phi_m)}\right]'\right\}\left(\left\{1 - \tau_{ou} g(\phi_m)\left[\frac{f(\phi_m)}{g(\phi_m)}\right]'\right\} f(\phi_m) - s_{ou} g'(\phi_m)g(\phi_m)\right)$$

$$+ s_{ou} g^2(\phi_m)\left\{1 - \tau_{ou} g(\phi_m)\left[\frac{f(\phi_m)}{g(\phi_m)}\right]'\right\}' = 0. \tag{3.53}$$

In particular, if $g(\phi) = \text{const}$, we obtain

$$-s_{ou}\tau f''(\phi_m) + [1 - \tau_{ou} f'(\phi_m)]^2 f(\phi_m) = 0, \tag{3.54}$$

showing that even a simple additive colored Gaussian noise is able to induce a transition.

To show an example of transitions driven by an O-U process, consider the Landau model with multiplicative noise:

$$\frac{d\phi}{dt} = \phi - \phi^3 + \phi\xi_{ou}(t). \tag{3.55}$$

The domain of these dynamics is $[0, \infty[$, and the pdf of the state variable ϕ is

$$p(\phi) = C\phi^{\frac{1}{s_{ou}} - 1}(1 + 2\phi^2\tau_{ou})e^{-\frac{\phi^2[1 + (\phi^2 - 2)\tau_{ou}]}{2s_{ou}}}. \tag{3.56}$$

The corresponding deterministic system $d\phi/dt = f(\phi) = \phi - \phi^3$ has two steady states: the unstable state, $\phi = 0$, and the stable state, $\phi = 1$. Therefore the deterministic dynamics always tend to $\phi = 1$ regardless of the initial condition. The presence of noise introduces interesting new dynamical behaviors, as shown in Fig. 3.14, as functions of the two parameters, namely of the intensity s_{ou} and of the autocorrelation scale τ_{ou} of the noise. As long as $s_{ou} < 1$ the pdf is unimodal and tends to zero at

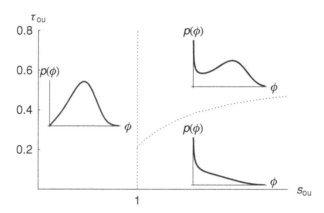

Figure 3.14. Possible shapes of the pdf for the Landau model driven by a multiplicative O-U process as specified in model (3.55).

the boundaries of the domain. In these conditions the only effect of the noise is to shift the position of the mode as in the case discussed in the previous section. As the noise intensity exceeds the threshold $s_{ou} = 1$, an asymptote at $\phi = 0$ appears in the pdf of ϕ and two scenarios are possible: For relatively small autocorrelation scales the pdf remains unimodal, whereas for relatively large autocorrelation scales the pdf becomes bimodal. Figure 3.15 shows some examples of pdf's for increasing values of τ_{ou} according to the UCNA approximation. Notice that condition (2.107) is always verified over the whole domain $]0, \infty[$.

The cases shown in Fig. 3.14 confirm the importance of the deterministic boundaries, as discussed in the case of Gaussian white noise. In addition, here the role of correlation is also visible: When noise correlation is high, the noise is less effective at hiding the features of the underlying deterministic dynamics. Thus in these conditions the pdf is reminiscent of the corresponding deterministic steady state, showing a mode close to ϕ_{st}.

Noise-induced transitions can be also effectively investigated with the stochastic potential defined in Box 3.1. For example, we compare the deterministic, $V(\phi)$ and

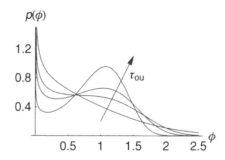

Figure 3.15. Examples of pdf's for the Landau model driven by a multiplicative O-U process as specified in Eq. (3.56), with $s_{ou} = 1.3$ and $\tau_{ou} = 0, 0.3, 0.6$, and 1.5.

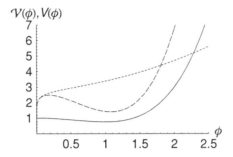

Figure 3.16. Probabilistic potential $\mathcal{V}(\phi)$ (dotted curve with $\tau_{ou} = 0$, dashed curve with $\tau_{ou} = 1.5$) and deterministic potential $V(\phi)$ (continuous curve) for the Landau model driven by a multiplicative O-U process as specified in Eq. (3.56) for $s_{ou} = 1.3$.

probabilistic potential $\mathcal{V}(\phi)$ in Figure 3.16. It is clear that the noise completely modifies the structure of the deterministic potential by inducing a new preferential state at $\phi = 0$. With increasing noise autocorrelation, however, another preferential state close to the deterministic one appears as a minimum of $\mathcal{V}(\phi)$.

3.3 Stochastic resonance

Stochastic resonance is an interesting example of noise-induced phenomena, in which a periodic deterministic forcing and a stochastic driver cooperate to induce more regular transitions in a dynamical system. This noise-induced mechanism was first proposed by Benzi et al. (1982a) to explain the quasi-periodic occurrences of ice ages on Earth during the past 700,000 years. After the seminal paper by Benzi et al. (1982a), this phenomenon has drawn the attention of the science community (e.g., see the reviews by Gammaitoni et al., 1998, and Wellens et al., 2004) and has been invoked to explain several processes observed in neuronal systems, optics, quantum physics, and pattern formation in spatially extended systems (see Section 5.9). In spite of the existence of many environmental periodic and stochastic forcings, the applications to the environmental sciences are relatively rare. Some of these are described in Chapter 4.

To explain the key mechanisms leading to the emergence of stochastic resonance, in this section we start by describing this phenomenon in its simplest form. We will then extend the description to some more complex forms of stochastic resonance in the subsequent subsections.

3.3.1 Basic concepts about stochastic resonance

The typical form of stochastic resonance has three fundamental ingredients: (i) a bistable deterministic dynamical system, (ii) a white random forcing, and (iii) a (weak) deterministic monochromatic (i.e., characterized by a single oscillation frequency) periodic forcing. Let us first consider only the deterministic dynamical system.

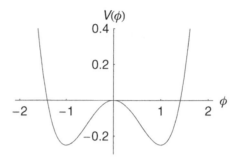

Figure 3.17. Typical shape of a bistable potential. In particular, the figure refers to $V = \phi^4/4 - \phi^2/2$; the minima are at $\phi_{m,1} = -1$ and $\phi_{m,2} = 1$, and the height of the potential barrier, situated at $\phi_b = 0$, is $\Delta V = 1/4$.

Figure 3.17 shows an example of bistable potential, with the two minima, $\phi_{m,1}$ and $\phi_{m,2}$, separated by a potential barrier of height ΔV, situated at $\phi = \phi_b$. In this case the potential is $V(\phi) = \phi^4/4 - \phi^2/2$. In the absence of stochastic and periodic forcings, the dynamics of the variable ϕ are described by

$$\frac{d\phi}{dt} = -\frac{dV}{d\phi} = \phi(1 - \phi^2). \tag{3.57}$$

Depending on the initial condition, the system will tend to either one of the two stable states (see Fig. 3.18) associated with the minima ϕ_m of the potential (i.e., $\phi_{m,1} = -1$ and $\phi_{m,2} = 1$ in this example). The dynamics within each potential well are characterized by the time scale typical of the process close to the minimum; this time scale is proportional to

$$t_{\mathrm{iw}} = \left(\frac{d^2 V}{d\phi^2} \bigg|_{\phi=\phi_m} \right)^{-1} \tag{3.58}$$

and can be assumed as the intrawell relaxation time. In example (3.57) we find $t_{\mathrm{iw}} = 1/2$.

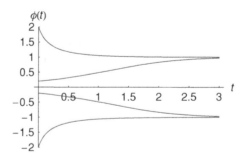

Figure 3.18. Examples of time trajectory for nonforced dynamics regulated by Eq. (3.57). The different curves correspond to different initial conditions.

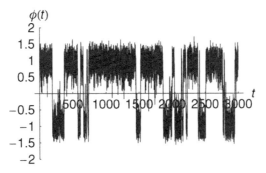

Figure 3.19. Example of temporal path of deterministic dynamics (3.59) driven by white Gaussian noise. The stochastic equation was simulated according to the numerical approach described by Sancho et al. (1982).

We can now introduce the stochastic forcing $\xi(t)$. In this example we use white Gaussian noise with intensity s_{gn}. This noise component is additive, and the dynamics of ϕ are expressed by

$$\frac{d\phi}{dt} = \phi(1 - \phi^2) + \xi_{gn}(t). \tag{3.59}$$

The effect of the stochastic forcing is twofold: It introduces randomness in the temporal dynamics of $\phi(t)$, and, more important, this forcing drives the transitions from a well to another across the potential barrier (which is located at $\phi = \phi_b$). As a result, the dynamics of the variable ϕ exhibit the pattern shown in Fig. 3.19, which shows clear transitions across the barrier. Because these transitions are induced by the random forcing, their occurrence is random, with a mean passage time equal to (see Gardiner, 1983; Wellens et al., 2004)

$$\langle t_c \rangle = \frac{1}{s_{gn}} \int_{\phi_{m,1}}^{\phi_{m,2}} e^{\frac{V(\phi')}{s_{gn}}} \left[\int_{-\infty}^{\phi'} e^{-\frac{V(\phi'')}{s_{gn}}} \, d\phi'' \right] d\phi'. \tag{3.60}$$

In particular in the case shown in Fig. 3.19, we have $\langle t_c \rangle = 100.8$ time units, with $s_{gn} = 0.085$.

If we assume that the transitions occur rarely with respect to the oscillations within each well – i.e., the mean passage time is much longer that the intrawell relaxation time, $\langle t_c \rangle \gg t_{iw}$ – it is possible to use Kramers's formula (Gardiner, 1983),

$$W = \frac{\sqrt{-\frac{d^2 V}{d\phi^2}\Big|_{\phi_b} \frac{d^2 V}{d\phi^2}\Big|_{\phi_m}}}{2\pi} e^{-\frac{\Delta V}{s_{gn}}}, \tag{3.61}$$

to determine an approximation of the mean transition rate, i.e., $W^{-1} \simeq \langle t_c \rangle$. For example, in the case shown in Fig. 3.19 (where $\langle t_c \rangle = 100.8 \gg t_{iw} = 1/2$), application of Eq. (3.61) gives $\langle t_c \rangle = 84.1$ time units.

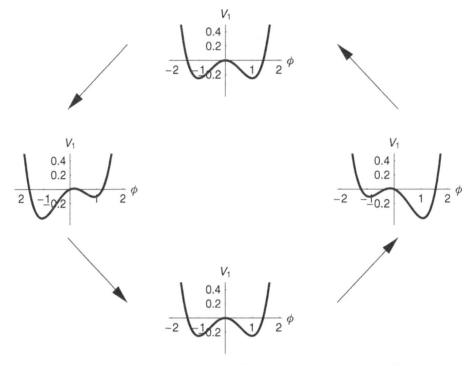

Figure 3.20. Time behavior of the periodic potential $V_1(\phi, t)$ in Eq. (3.62), with $\alpha = 0.15$ and $\omega_p = \pi/100$. The panels refer to $t = 0$, 50, 100, and 150 time units, respectively.

Equation (3.61) shows how the mean transition rate depends on the noise strength, the height of the potential barrier, and the curvature of the potential at the points of maximum and minimum. Moreover, the crossing times are exponentially distributed with mean $\langle t_c \rangle$ (Gardiner, 1983). This property is used in the detection methods presented in the following sections.

If we now add the last ingredient, namely the periodic forcing, the deterministic potential becomes

$$V_1(\phi, t) = V(\phi) + \alpha\phi \sin(\omega_p t), \qquad (3.62)$$

where α and ω_p are the amplitude and frequency of the periodic forcing. In models of stochastic resonance this forcing is typically "weak" in the sense that it is not able to induce transitions between the two wells (i.e., $\alpha < \Delta V$). Therefore the effect of the periodic forcing is to periodically alter the shape of the potential, with the effect of alternately reducing or increasing the height of the potential barrier with respect to each well. Figure 3.20 shows the effect of the periodic forcing on the potential function for the example presented in this section (with $\alpha = 0.15 < \Delta V = 1/4$ and $\omega_p = \pi/100$), and Fig. 3.21 shows an example of the time series generated by the

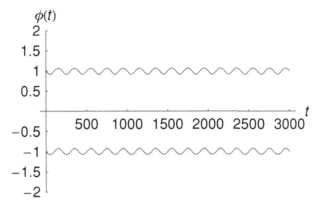

Figure 3.21. Examples of time trajectories corresponding to periodic potential $V_1(\phi, t)$ in Eq. (3.62) with $\alpha = 0.15$ and $\omega_p = \pi/100$. Notice that no transitions occur.

deterministic dynamics expressed

$$\frac{d\phi}{dt} = \phi(1 - \phi^2) - \alpha \sin(\omega_p t). \qquad (3.63)$$

It is observed that in the absence of noise the state of the system remains confined in either one of the two potential wells, depending on the initial condition. The only effect of the deterministic forcing is to induce periodic fluctuations around the steady states, $\phi_{m,1} = 1$ or $\phi_{m,2} = -1$.

As the periodic term in (3.62) is linear in ϕ, it is able to alter the height ΔV but not the curvature of the potential, which is proportional to the second-order derivative. Therefore, in the presence of both the random and the deterministic forcings, i.e., when the dynamics are regulated by

$$\frac{d\phi}{dt} = \phi(1 - \phi^2) - \alpha \sin(\omega_p t) + \xi_{gn}(t), \qquad (3.64)$$

the mean transition rate is

$$\langle t_c \rangle^{-1}(t) \simeq W(t) = \frac{\sqrt{-\left.\frac{d^2 V}{d\phi^2}\right|_{\phi_b} \left.\frac{d^2 V}{d\phi^2}\right|_{\phi_m}}}{2\pi} e^{-\frac{1}{s_{gn}}[\Delta V + \alpha \sin(\omega_p t)]}. \qquad (3.65)$$

We recall that Eq. (3.65) assumes that $\langle t_c \rangle \gg t_{iw}$ (in our example $\omega_p = \pi/100 < t_{iw} = 1/2$).

Equation (3.65) shows how the periodic forcing alternately increases or decreases the probability of transition from a well to the other (see Fig. 3.22). This periodic alternation is a key aspect of stochastic resonance. In fact, when the semiperiod π/ω_p of the periodic modulation of the potential [Eq. (3.62)] is comparable to the mean transition time $\langle t_c \rangle$ of the stochastic process, it is likely that periodic and random forcing cooperate to induce transitions. In other words, the transition probability is

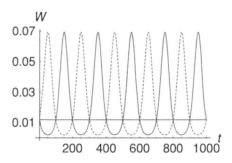

Figure 3.22. Time behavior of the mean transition rates from one well to another, and vice versa (dashed curve) when periodic forcing is present ($\alpha = 0.15$ and $\omega_p = \pi/100$). The horizontal line indicates the constant mean transition rate as given by the Kramers formula in the absence of periodic forcing ($=1/84.1$ in our example).

greater when the transition rate is maximized. Figures 3.23(a)–3.23(c) demonstrate the occurrence of stochastic resonance for suitable noise levels. Figure 3.23(a) refers to the case with $s_{gn} = 0.05$, when $\langle t_c \rangle = 730$. In these conditions the transitions are still random because the noise intensity is either too low or the modulation frequency, ω_p, is too high for the periodic and stochastic forcings to cooperate. In contrast, when $s_{gn} = 0.085$ the mean passage time $\langle t_c \rangle$ is equal to 100 time units, i.e., $\langle t_c \rangle$ is half of the period of the periodic forcing, $2\pi/\omega_p = 200$ time units. This condition is shown in Fig. 3.23(b): It is observed that the occurrence of transitions becomes more regular. This phenomenon of regularization of the fluctuations is known as stochastic resonance. Finally, when the noise intensity further increases ($s_{gn} = 0.15$, $\langle t_c \rangle = 29.5$) the random forcing prevails on the periodic forcing and the transitions become random again.

The key mechanism of stochastic resonance is then summarized as follows: A suitable (weak) periodic forcing can enhance the regularity of transitions between the two states or, vice versa, a random driver is able to activate quite regular transitions in a periodic dynamical system, which in the absence of noise would remain locked in one of its potential wells. This second interpretation shows the counterintuitive aspect of stochastic resonance, i.e., that a suitable noise is able to induce more regular transitions in a dynamical system.

In simple models like the one presented in this section, the occurrence of stochastic resonance can be easy to recognize, and no particular methods for time-series analysis are necessary to detect it, at least qualitatively. However, in most applications, quantitative methods are necessary. In general, two approaches are followed. The first method evaluates the power spectrum of the signal $\phi(t)$ in order to isolate the peak corresponding to the transition frequency, and the second method investigates the probability distribution of the threshold crossing times. These methods are described in Box 3.3.

Box 3.3: Methods for detecting stochastic resonance

The power spectrum $S(\omega)$ is defined as the squared absolute value of the Fourier transform of a signal and indicates the energy contained in each of the frequencies ω present in a given signal (see Appendix A). When all frequencies give the same contribution to the signal, the spectrum is flat. This case corresponds to a white noise. Conversely, peaks in the spectrum are a clear symptom that the corresponding frequencies dominate the signal. This property is particularly suitable for the quantification of stochastic resonance, as it provides a way to extract from the signal the periodicity of the transitions induced by cooperation between noise and deterministic periodic forcing.

The importance of the peak associated with the frequency ω_p of the periodic forcing can be better quantified in terms of the signal-to-noise ratio (SNR),

$$\text{SNR} = \frac{S(\omega_p)}{S_N(\omega_p)}, \tag{B3.3-1}$$

where the numerator indicates the magnitude of the largest peak in the power spectrum and the denominator is the value of S corresponding to the smooth background of the power spectrum at the same frequency. The smooth background can be obtained by a suitable moving average of the spectrum, in which the spike corresponding to ω_p has been preventively eliminated. The rationale underlying the definition of the SNR is that the power spectrum can be written as the sum of a background plus a number of peaks at integer multiples of the driving frequency ω_p. In other words, the SNR quantifies the importance of the peak with respect to the background noise.

We remark that, because some of the frequently used techniques in time-series analysis are based on power-spectrum analysis, a rich body of literature exists on the technical issues emerging in the evaluation of power spectra. These issues include, among others, the effect of the sampling time and the use of suitable weighting windows to reduce biased evaluations (e.g., Papoulis, 1984). This explains why the power spectrum has been the first and most used approach to the study of stochastic resonance, and several theoretical results are available for this method. We refer the interested reader to Gammaitoni et al. (1998) and Wellens et al. (2004) for a review of these methods.

The analysis of the distribution $p(t_c)$ of the crossing times t_c across a suitable threshold is another approach frequently used to quantify the effect of stochastic resonance (Wellens et al., 2004). For example, consider the bistable system discussed in Subsection 3.3.1. In this case, stochastic resonance regularizes the transitions across the potential barrier at $\phi = 0$. Thus it is sensible to choose this value as the threshold for the crossing-time analysis. In fact, when stochastic resonance does not occur the transitions are random and the distribution of crossing times is exponential with a mean equal to $\langle t_c \rangle$ [Eq. (3.60)]; in contrast, as stochastic resonance emerges, the transitions become more regular and the crossing-time distribution departs from the exponential distribution. In particular, when $\langle t_c \rangle \simeq \pi/\omega_p$, a maximum appears at time $t_c = \langle t_c \rangle$, indicating how in these conditions the dynamical system $\phi(t)$ has a remarkably high probability of crossing the potential barrier with a period $\langle t_c \rangle$. Therefore the height of such a mode can be used as a parameter to quantify the occurrence of stochastic resonance. It should be noted that additional (less-pronounced) modes occur also in correspondence to the odd multiples of $\langle t_c \rangle$. In fact, if the system is not able to cross the potential barrier at time $\langle t_c \rangle$, the transition probability is maximized again after one period $2\pi/\omega_p$ and it is therefore more probable that it crosses the potential barrier at time $\langle t_c \rangle + 2\pi/\omega$. In case this transition does not occur, the transition probability has another maximum after two periods, and so on.

Any form of stochastic resonance can be detected through the analysis of the crossing times of the process. However, for other types of stochastic resonance, the crossing-time distribution in the case of random transitions (which was exponential in the previous example) is seldom known analytically and has to be evaluated numerically from realizations of the dynamical process $\phi(t)$ without periodic forcing. We refer the interested reader to the reviews by Gammaitoni et al. (1998), and Wellens et al. (2004) for more details about the use of crossing-time distributions for the detection of other forms of stochastic resonance.

As an example of application of the detection method based on the power spectrum, as described in Box 3.3, we report in Fig. 3.24 the spectrum of the time series plotted in Fig. 3.23(b). We observe a strong peak corresponding to the external deterministic forcing (i.e., for $\omega = \omega_p$), which indicates the occurrence of stochastic resonance.

The SNR ratio defined in Eq. (B3.3-1) is reported in Figure 3.25 as a function of the noise intensity s_{gn} for model (3.64). When the noise intensity is low the peak is weak, namely the transitions between the states are substantially random and no significant periodicity occurs [e.g., see Fig. 3.23(a)]. In contrast, for suitable values of s_{gn} (around $s_{\mathrm{gn}} = 0.085$) the SNR has a peak and transitions become regular, as demonstrated in Fig. 3.23(b); in this case stochastic resonance occurs. As noise intensity grows, the random component tends to disturb the regularity of transitions [e.g., see Fig. 3.23(c)], the cooperation between random and deterministic components weakens, and then the SNR starts to decrease.

3.3.2 Other forms of stochastic resonance

The example presented in Subsection 3.3.1 demonstrated the basic features of processes exhibiting stochastic resonance. However, since its initial formulation by Benzi et al. (1982a), (1982b) and especially after the experimental study by McNamara et al. (1988) on a bistable ring laser, several other forms of stochastic resonance have been proposed. For example, the bistability of deterministic dynamics is not indispensable; in fact, the potential barrier can be replaced with another threshold inherent to the system's dynamics. The key point is that threshold crossing is activated by the random

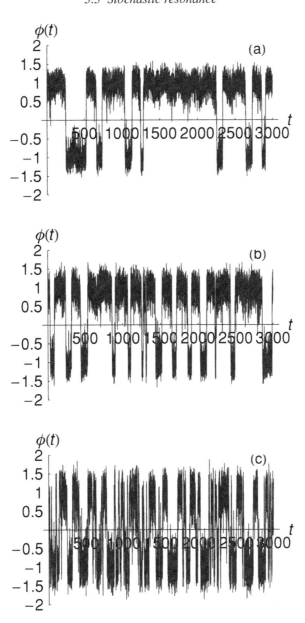

Figure 3.23. Examples of dynamics of ϕ when both stochastic and periodic forcing are present. In all panels the deterministic parameters are the same ($\alpha = 0.15$ and $\omega_p = \pi/100$) whereas the noise intensity increases from (a) ($s_{gn} = 0.05$) to (b) ($s_{gn} = 0.085$) to (c) ($s_{gn} = 0.15$).

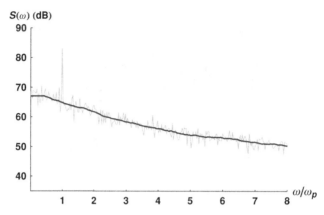

Figure 3.24. Power spectrum of the time series shown in Fig. 3.23(b). The black line indicates the smooth background.

forcing in a way that this noise-induced threshold crossing has a typical time scale $\langle t_c \rangle$. A suitable synchronization with weak periodic forcing can then activate stochastic resonance.

An interesting example of stochastic resonance emerging in systems that are not bistable can be found in excitable systems. Their dynamics are characterized by periods of time at the "ground" (or "rest") state randomly interrupted by quite impulsive noise-induced excitation events that occur when a certain threshold is crossed. Therefore excitable systems have only one stable state (the ground state) and one unstable state (the so-called "excited" or "firing" state) from which the system decays after a large excursion in the phase space. Remarkable examples of excitable systems include chemical reactions, lasers, ion channels, and neuronal systems. In all these fields the concept of stochastic resonance has opened new avenues for the interpretation and the control of the relevant dynamics (Lindner et al., 2004).

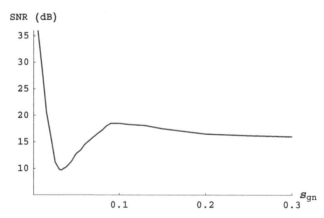

Figure 3.25. Behavior of the SNR as a function of the noise intensity s_{gn} for model (3.64).

Another generalization of stochastic resonance involves the stochastic forcing, which can be non-Gaussian, colored, or both. In fact, noise plays the role of causing the crossing of a threshold of the deterministic dynamics, thereby determining the typical time scale of the crossings. Because neither Gaussianity nor whiteness is fundamental to induce threshold crossing, different and more complex forms of noise can be adopted. However, when white Gaussian noise is not used the analytical methods for the calculation of $\langle t_c \rangle$ become less straightforward (Gammaitoni et al., 1998). For example, interesting recent applications concentrated on dichotomous noise (Barzykin et al., 1998; Berdichevsky and Gitterman, 1999; Zhou et al., 2008) and colored Gaussian noise (Berdichevsky and Gitterman, 1999; Han et al., 2005).

Similarly, even the periodic forcing does not need to be monochromatic. In fact, it can contain different frequencies; stochastic resonance emerges when these frequencies synchronize with the average crossing time associated with suitable noise levels (e.g., Braun et al., 2005a).

In this section we provided a synthesis of only some of the main mechanisms capable of inducing stochastic resonance. However, this is a very active and fast-moving research field: new applications and new types of stochastic resonance are continuously discovered.

3.4 Coherence resonance

Another interesting example of noise-induced coherent response in the time domain is the so-called *coherence resonance* or *stochastic coherence*. This phenomenon can be observed in some excitable systems whose dynamics have a deterministic component, which converges to a stable fixed point close to a Hopf bifurcation.[2] In these conditions the system is close to a transition to a limit cycle. A suitable amount of additive noise is able to induce a quasi-oscillatory behavior, showing a remarkable coherence (i.e., regularity in temporal dynamics) without requiring an external periodic forcing.

A classical example is given by the FitzHugh–Nagumo model, originally proposed for the description of nerve pulses (see Scott, 1975) but used also to model spiral waves in excitable media. The model is

$$\frac{d\phi_1}{dt} = \phi_2 + a + \xi_{gn}(t), \tag{3.66}$$

$$\epsilon \frac{d\phi_2}{dt} = \phi_2 - \frac{\phi_2^3}{3} - \phi_1, \tag{3.67}$$

where a is a parameter, $\xi_{gn}(t)$ is zero-mean Gaussian white noise, and $\epsilon \ll 1$ is a small parameter that allows one to separate the fast time scale (i.e., with only the variable ϕ_1 changing with time) and the slow time scale (with $\phi_1 \approx \phi_2 - \phi_2^3/3$).

---

[2] The Hopf bifurcation is a bifurcation in which a fixed point of a dynamical system loses its stability and a limit cycle branches from the fixed point (Argyris et al., 1994).

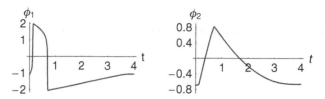

Figure 3.26. Evidence of the excitability of dynamical system (3.66) and (3.67). The parameters are $a = 1.05$ and $\epsilon = 0.01$ (with steady state $\phi_{1,\mathrm{st}} = -1.05$ and $\phi_{2,\mathrm{st}} = -0.66$), and the initial condition corresponds to a 5% perturbation of the steady state.

When noise is absent (i.e., $s_{\mathrm{gn}} = 0$), a Hopf bifurcation occurs at $|a| = 1$: If $|a| > 1$, the stable attractor of the deterministic dynamics is the fixed point $\{\phi_{1,\mathrm{st}}, \phi_{2,\mathrm{st}}\} = \{-a, -a + a^3/3\}$, whereas if $|a| < 1$ a limit cycle appears (Pikovsky and Kurths, 1997).

Let us choose a value of a slightly larger than one. In this condition, the deterministic system tends to the fixed point $\{\phi_{1,\mathrm{st}}, \phi_{2,\mathrm{st}}\}$ but it is excitable, i.e., it exhibits a highly nonlinear response to disturbances. Thus, if the dynamical system is slightly perturbed, the deterministic dynamics exhibit large excursions before returning to the fixed point. For example, Fig. 3.26 shows the behavior of the system when the steady state is slightly perturbed. If we now add the random component, $\xi_{\mathrm{gn}}(t)$ (i.e., $s_{\mathrm{gn}} \neq 0$), interesting behaviors may emerge, as shown in Fig. 3.27: When the noise is relatively weak the spikes are sporadic and occur randomly, whereas if the noise intensity is relatively high the spikes become frequent but their shape (hence the whole signal) is irregular. Conversely, a quite regular (i.e., coherent) oscillatory behavior occurs for intermediate noise intensities. The regularity of the signal for intermediate noise intensity is manifest if we consider the signal in the frequency domain (right-hand panels in Fig. 3.27): In fact, a sharp peak appears in the power spectrum for increasing noise intensities. This behavior can be better understood if we consider two time scales characteristic of the dynamics: the activation time t_a and the excursion time t_e. The activation time is related to the intensity of the disturbance necessary to activate the excursions; therefore t_a decreases when the noise intensity increases. On the other hand the excursion time depends mainly on the deterministic structure of the dynamics and has only a limited dependence on noise. Therefore, when the noise intensity is low, the activation time is longer than the typical duration of an excursion (i.e., $t_a \gg t_e$); in these conditions noise randomly activates the excursions. Conversely, with relatively high noise intensity (i.e., $t_e \gg t_a$) the excursions continuously occur one after another. However, in these conditions the noise is also able to induce remarkable fluctuations in the dynamics of ϕ in the course of each excursion. It is only when the noise has an intermediate intensity that we can have $t_e \gg t_a$ while maintaining quite regular excursions. In this case the signal has a quasi-oscillatory behavior and stochastic coherence occurs. It should be noticed that noise in Eq. (3.66) could be interpreted as

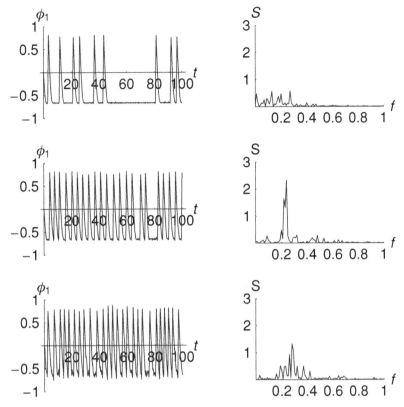

Figure 3.27. The dynamics of the FitzHugh–Nagumo system (3.66) and (3.67) for different noise amplitudes ($a = 1.05$, $\epsilon = 0.01$). $s_{gn} = 0.02$, $s_{gn} = 0.07$, and $s_{gn} = 0.25$, from top to bottom, respectively. The panels in the left column report the time series, the panels in the right column the corresponding power spectra.

a random modulation of the bifurcation parameter a. However, this modulation is not essential for the occurrence of stochastic coherence. In fact, this behavior emerges even if the noise component is moved from Eq. (3.66) to Eq. (3.67).

Apart from the details related to this specific model, the key point of stochastic coherence is that noise can cooperate with an intrinsic time scale of the deterministic dynamics (t_e in the FitzHugh–Nagumo model) to induce regular oscillatory behavior. In this sense, noise is able to unveil a time scale that would be hidden in a deterministic dynamics. In fact, in the absence of noise $\phi(t)$ would just tend to a point attractor (Sagues et al., 2007).

3.5 Noise-induced net transport

One of the most counterintuitive noise-induced phenomena investigated in the past few decades is associated with the ability of unbiased random fluctuations to generate a preferential direction of motion, i.e., a drift in the oscillations of the state variable. This process is known under different names (e.g., *Brownian motor*, *ratchet effect*, *Brownian ratchet*, or *stochastic ratchet*), depending on the context in which the phenomenon is reported or investigated. The first studies on noise-induced transport go back to work by Smoluchowski (1912) and Feynman et al. (1963) in the context of intracellular transport processes (the so-called *molecular motors*; see the review by Howard, 1997), and to some research by physicists in the 1970s (see Reimann, 2002). Some key contributions in the early 1990s (e.g., Ajdai and Prost, 1992; and Magnasco, 1993) gave a new impetus to studies on noise-induced transport, with an increasing number of experimental and theoretical results along with several technological applications (Reimann, 2002; Astunian and Hanggi, 2002). We use the same approach as in the previous two sections and describe the basic aspects of Brownian motors; we refer the reader to Reimann (2002) and Reimann and Hanggi (2002) for a more detailed discussion.

Brownian motors have three basic ingredients: (i) an asymmetric periodic potential, (ii) a noise source, and (iii) a temporal modulation, either of the potential or of the noise intensity. Here we focus on the case in which the noise intensity is modulated, though a similar mechanism is common to all Brownian motors, including those relying on a modulation of the potential.

Let us first consider the deterministic dynamics driven by an asymmetric periodic potential $V(\phi)$. An example of a potential satisfying these two properties (see Reimann, 2002),

$$V(\phi) = V_0 \left[\sin \left(\frac{2\pi \phi}{L} \right) + \frac{1}{4} \sin \left(\frac{4\pi \phi}{L} \right) \right], \tag{3.68}$$

is shown in Fig. 3.28 (L is the coefficient determining the period of the periodic potential). As a consequence, in the deterministic dynamics described by

$$\frac{d\phi}{dt} = -\frac{dV}{d\phi}, \tag{3.69}$$

$\phi(t)$ tends to one of the minima of the potential [i.e., one of the steady states of (3.69)], depending on the initial condition. In particular, in example (3.68) the minima are at

$$\phi_m = Ln + \frac{L}{\pi} \arccos \left(-\frac{\sqrt{1 + \sqrt{3}}}{2} \right), \tag{3.70}$$

where n is an integer number. The corresponding attraction basins are

$$\left[Ln + \frac{L}{\pi} \arccos \left(\frac{\sqrt{1 + \sqrt{3}}}{2} \right), L(n + 1) + \frac{L}{\pi} \arccos \left(\frac{\sqrt{1 + \sqrt{3}}}{2} \right) \right]. \tag{3.71}$$

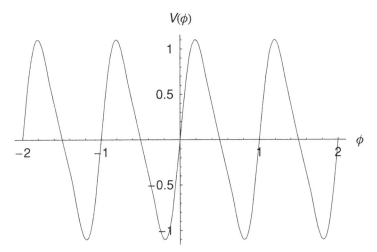

Figure 3.28. Example of periodic asymmetric potential $V(\phi)$. In particular, this example is obtained with Eq. (3.68).

Now, let us introduce an additive, time-independent, and uncorrelated Gaussian noise ξ_{gn}, with zero mean and constant intensity s_{gn}. The corresponding stochastic dynamics become

$$\frac{d\phi}{dt} = -\frac{dV}{d\phi} + \xi_{gn}. \tag{3.72}$$

Even though the potential exhibits an asymmetry, it is possible to demonstrate that the so-called average current in the long-time limit,

$$\langle\dot{\phi}\rangle = \left\langle \lim_{t\to\infty} \frac{\phi(t) - \phi(0)}{t} \right\rangle, \tag{3.73}$$

is zero, i.e., no preferential direction (or net transport) emerges in the random Brownian motion. At first glance, this behavior may seem to be strange because of the asymmetry of the potential. However, it is a consequence of the second law of thermodynamics that establishes zero current for a Brownian motor under the influence of a thermal-equilibrium heat bath, i.e., a random forcing with steady properties (Reimann, 2002).

Completely different dynamics can emerge if we let the noise be time dependent. For example, we can assume that the noise intensity is periodically modulated:

$$\langle\xi_{gn}(t)\xi_{gn}(t + \tau)\rangle = 2s_{gn}T(t)\delta(\tau), \tag{3.74}$$

where $T(t) = T(t + \mathcal{T})$ is a periodic modulation function with period $\mathcal{T}$. A typical case of temporal modulation of $T(t)$ is

$$T(t) = \bar{T}\left\{1 + A\,\text{sign}\left[\sin\left(\frac{2\pi t}{\mathcal{T}}\right)\right]\right\}, \tag{3.75}$$

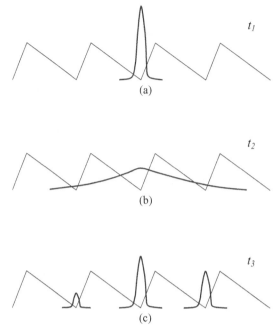

Figure 3.29. Schematic representation of noise-induced transport. At time t_1 the noise is weak and particles are concentrated in the well A. When the noise becomes dominant (time t_2), particles spread across the potential barriers. Finally, when the noise is again reduced (time t_3), the asymmetry of the potential means that particles tend to accumulate more in the well located to the right of well A. Thus a net positive current $\langle \dot\phi \rangle$ occurs.

where $A < 1$ and sign($\cdot$) is the *signum function* [i.e., sign(z) = 1 if $z > 0$; sign(z) = -1 if $z < 0$]. Equation (3.75) expresses a steplike behavior, with noise intensity jumping between $2s_{gn}\bar{T}(1+A)$ and $2s_{gn}\bar{T}(1-A)$ every half-period $T/2$. It is possible to demonstrate (Reimann, 2002) that, in this case, for a large set of parameters,

$$\langle \dot\phi \rangle \neq 0. \tag{3.76}$$

These conditions are associated with the occurrence of net transport.

To understand the physical mechanisms causing net transport, we consider the case in which (i) $\bar{T}(1-A) \ll \Delta V \ll \bar{T}(1+A)$, where ΔV is the height of the potential barrier (defined as in Fig. 3.17), and (ii) the noise intensity has a period T greater than the time scale of the intrawell deterministic dynamics (see Subsection 3.3.1). Under these conditions, the dynamics of ϕ can be schematized as an alternation of the three phases, sketched in Fig. 3.29: When the noise level is low, the deterministic component of the dynamics prevails. Thus the values of ϕ tend to concentrate in correspondence to the minima of the potential [see Fig. 3.29(a)]. Conversely, during the phase with high noise levels, the dynamics are dominated by the random component, whereas the shape of the potential has only a weak effect on the process. In these conditions

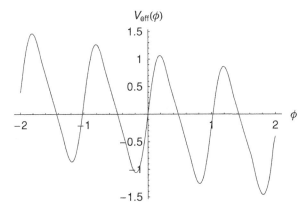

Figure 3.30. Tilting effect on the effective potential that is due to static force F (in this example $F = 0.2$).

the distribution $p(\phi)$ of ϕ tends to spread out similarly to the case of free diffusion [see Fig. 3.29(b)]. When the noise switches again to a low-intensity state, the shape of the potential returns to play an important role. In particular, its asymmetry causes an asymmetric redistribution of $p(\phi)$ in the wells of $V(\phi)$ starting from the conditions (i.e., the positions along the ϕ axis) reached in the previous noise-dominated phase. This redistribution gives rise to net transport. In the case of the example shown in Fig. 3.29, it is immediate to recognize a net transport in the positive (from left to right) direction of the ϕ axis (i.e., $\langle \dot{\phi} \rangle > 0$). However, with more complex combinations of different noise modulations and shapes of the potential, it can be less difficult to predict the direction of net transport (Reimann, 2002).

A frequent and important variation of this simple scheme of Brownian motors is obtained in the presence of an additive, homogeneous, deterministic force F,

$$\frac{d\phi}{dt} = -\frac{dV}{d\phi} + F + \xi_{gn}(t) = -\frac{dV_{eff}}{d\phi} + \xi_{gn}(t), \qquad (3.77)$$

where $V_{eff}(\phi) = V(\phi) - \phi F$ is the associated new potential. In this case it is easy to recognize the effect of F on the drift. For example, Fig. 3.30 shows how the presence of the additive deterministic component F tilts the potential $V(\phi)$, thereby imposing a preferential direction to the random Brownian motion even when the random component is time independent. Therefore, if $F \neq 0$ there is a net transport (i.e., $\langle \dot{\phi} \rangle \neq 0$) even when $T(t) = \text{const}$. In this case the direction of the net transport coincides with the sign of F. However, a less-obvious behavior emerges as an effect of the time modulation in the noise term in dynamics (3.77). In fact, in this case it is possible that noise-induced net transport prevails on the F-induced drift, thereby allowing the dynamics of ϕ to remount (on average) the "staircase" of the potential, as shown in Fig. 3.31. In this example positive values of $\langle \dot{\phi} \rangle$ exist in spite of the negative sign of the force F. If ϕ is interpreted as the position of a particle subjected

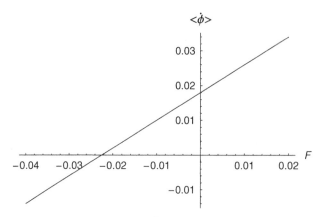

Figure 3.31. Dependence of the drift $\langle \dot{\phi} \rangle$ on static deterministic force F. The parameter values are $k = L = T = 1$, $\bar{T} = 0.5$, $A = 0.8$, $V_0 = 1/2\pi$, and the variables are made dimensionless according to Reimann (2002). Note that $\langle \dot{\phi} \rangle \neq 0$ also when $F = 0$.

to stochastic dynamics (3.77) with a time-dependent random component, this result demonstrates that particles may perform work against the load force F; such work is performed by white noise $\xi(t)$ (Reimann, 2002).

Moreover, in general, the effect of the static force F is shown in Fig. 3.31: The case with $F = 0$ corresponds in fact to dynamics (3.72) with periodic noise (3.74) (see subsequent discussion). We observe that noise modulation, along with an asymmetric potential, is able to induce a positive drift.

Several variations of the "basic" scheme of the Brownian motors presented in this section were proposed. In particular, a number of shapes of the potential and the temporal modulation of noise have turned out to be able to give rise to net transport. In particular, it was demonstrated that noise modulation can also be random instead of deterministic and that the noise term can be substituted with a chaotic forcing (Reimann, 2002). The key mechanism remains the same: Net transport may arise from the cooperation between an asymmetric deterministic potential and a random forcing that alternates (periodically or stochastically) between phases of greater and lower intensity. Thus the dynamics repeatedly switch between phases dominated by noise and phases dominated by the deterministic terms. In this system, noise plays a fundamental role in determining net transport. When the rate of noise modulation is either relatively slow or fast (with respect to the intrawell deterministic dynamics), the net transport tends to zero.

The type of Brownian motors described in this section is usually called *thermal ratchet* because noise and its modulation are generally introduced to model thermal fluctuations. However, we can obtain the same drift effect also by modulating the deterministic potential instead of noise, i.e., when the dynamics are expressed as

$$\frac{d\phi}{dt} = -\frac{dV}{d\phi}[1 + f(t)] + \xi_{gn}, \tag{3.78}$$

where $f(t)$ is a suitable function (either deterministic or stochastic) of time, which determines the interplay between more ordered and more disordered phases required by the Brownian motor mechanism. These types of Brownian motors are called *fluctuating-potential ratchets* (Wellens et al., 2004) and fall into the more general family of the *pulsating ratchets* (Reimann, 2002), which includes also the *traveling-potential ratchets*. In this case the potential has the form $V[\phi - f(t)]$, and an asymmetry of the potential is not necessary for the occurrence of net transport. Drift is always in the same direction as the movement of the potential; however, noise is able to reduce the current $\langle \dot{\phi} \rangle$ with respect to drift velocity of the potential.

Another important family of Brownian motors are the so-called *fluctuating-force ratchets* (Reimann, 2002). In this case a temporal fluctuation $y(t)$ is added to the force F:

$$\frac{d\phi}{dt} = -\frac{dV}{d\phi} + y(t) + F + \xi_{gn}. \tag{3.79}$$

Thus the deterministic dynamics have an effective potential $V_{eff}(\phi, t) = V(\phi) - \phi[y(t) + F]$, namely the basic potential $v(\phi)$ is tilted with intensity $y(t) + F$. This type of motors belongs to the class of time-dependent potentials; however, different from the case of fluctuating ratchets, here the temporal modulation does not affect the amplitude of the potential but rather the tilt angle, which varies either randomly or periodically.

4

Noise-induced phenomena in environmental systems

4.1 Introduction

Environmental systems are typically forced by a number of drivers, such as climate and natural or anthropogenic disturbances, that are not constant in time but fluctuate. With the exception of processes dominated by deterministic oscillations (e.g., daily and seasonal cycles), a significant part of environmental variability is random because of the uncertainty inherent in weather patterns, climate fluctuations and episodic disturbances such as hurricanes, landslides, earthquakes, fires, insect outbreaks, and epidemics (e.g., Ludwig et al., 1978; Benda and Dunne, 1997; D'Odorico et al., 2006b; Gilligan and van den Bosch, 2008). The recurrence of random drivers in biogeophysical processes motivates the study of how a stochastic environment may affect and determine the dynamics of natural systems.

Over the past few decades a number of studies have contributed to the observation, understanding, and modeling of stochastic processes in different areas of the biological (e.g., Bharucha-Reid, 1960; May, 1973; Tuckwell, 1988) and earth sciences (e.g., Krumbein and Graybill, 1965; Yevjevich, 1970; Bras and Rodriguez-Iturbe, 1994; Hipel and McLeod, 1994), leading to the development and application of several modeling frameworks for the study of random environmental fluctuations. This rather rich body of literature can be divided into two major classes, depending on the role played by noise in the dynamics of natural systems. In Chapter 3 we showed how random environmental drivers may either cause stochastic fluctuations of the system around the stable state(s) of the underlying deterministic dynamics (*disorganizing effect* of noise) or induce new dynamical behaviors and new ordered states (May, 1973; Horsthemke and Lefever, 1984) that do not exist in the deterministic counterpart of the process (*organizing effect* of noise). For example, noise could cause the emergence of new stable states, determine new bifurcations, or destabilize the stable states existing in the deterministic system. Known as *noise-induced transitions* (Horsthemke and Lefever, 1984), these noise-induced phenomena suggest that noise may play a more fundamental role than causing random fluctuations in the temporal variability of the state variables. In addition to modifying the number of possible steady states, noise can affect the way the system approaches these states, its sensitivity to disturbances, or the ability to recover after disturbances (i.e., the resilience of the system). Noise can also enhance the regularity of temporal fluctuations and induce a variety of other counterintuitive dynamical behaviors (see Chapter 3). This chapter concentrates on this nontrivial effect of noise in biogeophysical systems, and it investigates conditions leading to the emergence of noise-induced transitions, coherent fluctuations, and other phenomena in the environmental sciences (e.g., May, 1973; Benzi et al., 1982a; Horsthemke and Lefever, 1984; Spagnolo et al., 2004; D'Odorico et al., 2005). In particular, we focus on the temporal dynamics of spatially lumped (i.e., zero-dimensional) univariate systems; the case of noise in spatial dynamics is discussed in Chapter 5. Thus, consistent with the general framework of this book (see Chapters 2 and 3), we consider systems whose dynamics can be expressed as

$$\frac{d\phi}{dt} = f(\phi) + g(\phi)\xi(t), \tag{4.1}$$

where $\xi(t)$ is a noise term. We consider noise-induced behaviors induced by different types of noise, including the case of dichotomous, shot, and Gaussian noise. Some remarks on the properties of bistable systems, reported in Box 4.1, may help in understanding some of the results obtained in this chapter.

4.2 Dichotomous Markov noise in ecosystem dynamics

In Section 2.2 we showed how the effect of environmental variability on ecosystem dynamics may result in the random alternation between stressed and unstressed conditions. The ecosystem's total biomass (or other suitable state variables indicative of ecosystem health or productivity) decreases or increases, depending on whether the system is stressed or unstressed, respectively. The random alternation of these two states is determined by fluctuations of the environmental variables (e.g., available resources, disturbance pressure) about a tolerance threshold marking the transition between favorable and unfavorable conditions for the ecosystem. These dynamics are typically modeled (e.g., D'Odorico et al., 2005; Camporeale and Ridolfi, 2006; Borgogno et al., 2007) as stochastic process $\phi(t)$ driven by DMN.

4.2.1 Noise-induced transitions due to random alternations between stressed and unstressed conditions in ecosystems

We consider the exemplifying case of an ecosystem in which the biomass B (e.g., plant biomass) randomly switches between a growth and a decay state, depending on whether the level $q(t)$ of fluctuating resources (e.g., soil water) is above or below a

Box 4.1: Properties of bistable deterministic systems

Bistable systems are by definition characterized by the existence of two alternative stable states. Ecosystem bistability is often induced by positive feedbacks (Wilson and Agnew, 1992) between the state of the system and the dynamics of its limiting resources (Walker et al., 1981; Walker and Noy-Meir, 1982; Rietkerk and van de Koppel, 1997; D'Odorico et al., 2007a; DeLonge et al., 2008) or disturbance regime (Anderies et al., 2002; D'Odorico et al., 2006b; Ridolfi et al., 2006).

The dynamics of these systems are typically visualized by use of plots of the potential function, which resembles the vertical section of a landscape (Fig. B4.1-1) with two valleys, s_1 and s_2, separated by a "hill ridge," u. In this landscape a ball would roll down toward either one of the two valley bottoms (i.e., stable states in this gravitational-field analog), depending on the initial condition, i.e., on whether the ball (i.e., state of the system) is initially located to the right or the left of the hilltop (unstable state u). The two stable states are also known as *attractors*, and each of the two "valleys" is the *attraction basin* of the corresponding stable state.

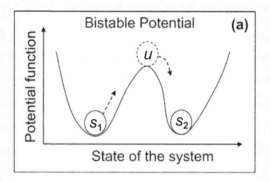

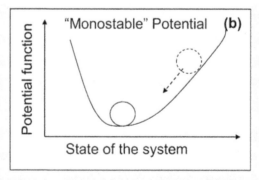

Figure B4.1-1. Potentials associated with (a) bistable and (b) monostable dynamics.

Bistable deterministic systems [Fig. B4.1-1(a)] have dynamical properties that are profoundly different from those of their monostable counterparts [Fig. B4.1-1(b)].

A major difference can be observed in the way these systems respond to disturbances. Assume that a stable configuration of the system is disturbed: In the case of dynamics with only one stable state, once the disturbance ceases, the system returns back to its stable state, regardless of the amplitude of the perturbation. In bistable dynamics the system recovers its initial stable state, [s_1 in Fig. B4.1-1(a)] only if the magnitude of the perturbation is relatively small, i.e., if the perturbation is unable to induce a transition to the basin of attraction of the other stable state (s_2). In fact, if such a transition occurs, the state of the system diverges toward the alternative state s_2. Because of the stability of the state s_2, the dynamics remain locked in this state even once the disturbance is removed. Thus bistable dynamics are prone to highly *irreversible* transitions. Moreover, these dynamics are *highly nonlinear* in the sense that in the proximity of the unstable state u, even small perturbations may lead the system far away from its initial state. These dynamics are clearly characterized by the presence of a *threshold effect* associated with the unstable state, u.

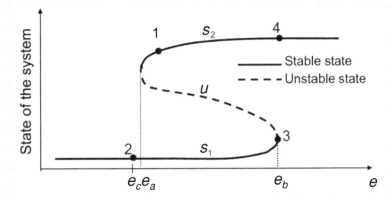

Figure B4.1-2. An example of bifurcation diagram. e is an environmental parameter governing the state of the system.

Ecologists define as *resilience* (Holling, 1973; Gunderson, 2000) of a stable state, say s_1, the ability of the dynamics to recover that state after a disturbance. In Fig. B4.1-1(b) the resilience of state s_1 is measured by the maximum magnitude of disturbances that the system may withstand without escaping from the basin of attraction of s_1. On the basis of the previous discussion, the states of bistable ecosystems have only limited resilience.

In most cases, bistable behavior exists within only a certain interval of the parameter space. Outside this interval one of the stable states disappears. The stable states are often plotted as a function of parameters representative of environmental drivers. Figure B4.1-2 shows a typical example of the dependence of stable (solid curve) and unstable (dashed curve) states as a function of an environmental parameter e. We observe that the number of stable states varies as e crosses the boundaries e_a and e_b of the bistability range. Known as *bifurcation diagrams*, these plots show how nonlinearities and thresholds may contribute to the complexity of dynamical systems. As noted, when the magnitude of a perturbation of an initial stable configuration is gradually increased, at some point a threshold is reached and a further infinitesimal increase in the intensity of the disturbance may lead to great changes in the way the system responds to these disturbances. Moreover, when the parameter e is close to either e_a or e_b, small changes in environmental conditions may lead to big changes (or *phase transitions*) in the state of the system. In the ecology literature these phase transitions are sometimes referred to as *catastrophic shifts* (e.g., Scheffer et al., 2001). We can consider for example the case of the system at state 1 on the branch s_2 of the bifurcation diagram (see Fig. B4.1-2). If e decreases to the value e_c with $e_c < e_a$, branch s_2 disappears. In these conditions the only stable state of the system is point 2 on branch s_1. The transition $1 \rightarrow 2$ is not easily reversible in the sense that, to move the system back to branch s_2, it is not sufficient to increase e above e_a. Indeed, e would need to be increased way above e_a until the e_b value is exceeded (point 3) before branch s_1 is destabilized and the system is able to shift to point 4 on branch s_2. The fact that the dynamics undergo two different pathways depending on whether e decreases or increases within the same interval is an effect known as *hysteresis* (Holling, 1973; Gunderson, 2000).

certain threshold θ. We focus on the case in which both the growth and the decay rates are expressed by linear functions

$$f_1(B) = a_1(1 - B), \qquad f_2(B) = -a_2 B, \qquad (4.2)$$

where a_1 and a_2 are two positive coefficients determining the rates of growth and decay, respectively. With probability P_1 the dynamics are in *state 1* [i.e., $q(t) \geq \theta$] with $dB/dt = f_1(B)$, whereas with probability $1 - P_1$ the dynamics are in *state 2* [i.e., $q(t) \leq \theta$] with $dB/dt = f_2(B)$. Dichotomous noise determines the rate of switching between these two states. When $a_1 = a_2 = 1$ the process is the same as in Example 2.1 in Subsection 2.2.3.1.

In Chapter 2 we showed that the roots of $f_1(B) = 0$ and $f_2(B) = 0$ [i.e., $B = 1$ and $B = 0$, in the case of Eqs. (4.2)] are natural boundaries, in that, if the dynamics start within the interval [0, 1] at time $t = 0$, they remain confined within this interval for any $t > 0$. The pdf of B can be determined with Eq. (2.31) in Chapter 2:

$$p(B) = C \left[a_1(1 - B) + a_2 B \right] (1 - B)^{-1 + \frac{1-P_1}{a_1}} B^{-1 + \frac{P_1}{a_2}}, \qquad (4.3)$$

where C is the normalization constant.

It is interesting to investigate how the properties of the probability distribution of B change in the parameter space, using the methods described in Chapter 3. The distribution $p(B)$ has a singularity at $B = 0$ for $a_2 > P_1$; in these conditions $p(B)$ is L-shaped (Fig. 4.1, case I). Similarly, $p(B)$ is J-shaped (i.e., has a singularity in

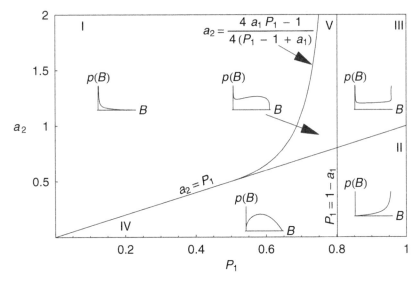

Figure 4.1. Shapes of the probability distributions of biomass B as functions of the parameters P_1 and a_2 of the state-independent dichotomous process (calculated for $a_1 = 0.2$). A variety of dynamics can be obtained, includings L-shaped distributions with preferential state at $B = 0$ (case I), J-shaped distributions with preferential state at $B = 1$ (case II), bistable dynamics with bimodal (U-shaped) distribution (case III), dynamics with only one stable state located between the extremes of the domain of B (case IV), bimodal distributions with a preferential state at $B = 0$ and the other for $B < 1$ (case V).

$B = 1$) for $P_1 > 1 - a_1$ (Fig. 4.1, case II). When both conditions are met, $p(B)$ is U-shaped with two spikes of probability at $B = 0$ and $B = 1$ (Fig. 4.1, case III). When none of these conditions is met, the probability distribution of B has only one mode within the interval $[0, 1]$ and no spikes of probability at $B = 0$ or at $B = 1$ (Fig. 4.1, case IV). It can be shown that when $p(B)$ has a singularity at $B = 0$ (but not at $B = 1$) and $a_2 < (4a_1 P_1 - 1)/[4(P_1 - 1 - a_1)]$, $p(B)$ has both a mode and an antimode in $[0, 1]$, as in Fig. 4.1 (case V). This mode and the spikes of $p(B)$ in cases I, II, and III are preferential (i.e., more probable) states of the system, and we interpret them as statistically stable states of the dynamics.

Because the probability distribution of B exhibits different shapes depending on the values of the parameters a_1, a_2, and P_1 (Fig. 4.1), the preferential states of B vary across the parameter space. For relatively low (high) rates of decay a_2 and high (low) probability P_1 of occurrence of unstressed conditions, the dynamics have a preferential state (i.e., spike of probability) in $B = 1$ ($B = 0$). In intermediate conditions the system may exhibit either one (case IV) or two (case III and V) statistically stable states. This bistability [i.e., bimodality in $p(B)$] emerges as a noise-induced effect (see Chapter 3) and is an example of the ability of noise to induce new states, which do not exist in the underlying deterministic system (Horsthemke and Lefever, 1984).

The deterministic counterpart of these dynamics is a system that is either always unstressed or always stressed, depending on whether the constant level q of available resources is greater or smaller than the minimum value θ required for survival. Thus the dynamics become $dB/dt = f_1(B)$ if in the deterministic system $q > \theta$ (always unstressed conditions, i.e., $P_1 = 1$), or $dB/dt = f_2(B)$, otherwise (always stressed, i.e., $P_1 = 0$). In the specific case of the example in Fig. 4.1 [Eqs. (4.2)] both of these deterministic dynamics have only one stable state, $B_m = 1$ [from the condition $f_1(B_m) = 0$] or $B_m = 0$ [from $f_2(B_m) = 0$], and the deterministic system converges to either of them, depending on whether the (constant) level of resources q is above or below the threshold θ. Thus the deterministic dynamics are not bistable, and it is the random driver that induces bistability [i.e., bimodality in $p(B)$] in the stochastic dynamics of B (Fig. 4.1).

The emergence of bistable dynamics has important implications for the way ecosystems respond to changes in environmental conditions and disturbance regimes (e.g., Holling, 1973; Gunderson, 2000). In fact, the existence of alternative stable states is often associated with the possible occurrence of abrupt and highly irreversible shifts in the state of the system (see Box 4.1), with consequent important limitations to its resilience (e.g., Walker and Salt, 2006). Ecosystem bistability is often induced by positive feedbacks between the state of the system and environmental drivers such as resource dynamics or disturbance regime (see Box 4.1).

This section showed that bistability may also result as an effect of noise (Horsthemke and Lefever, 1984). In this case, the randomness of the switching between stressed and unstressed conditions is able to profoundly affect the dynamical properties of the system by inducing bifurcations that did not exist in the underlying deterministic dynamics: As the noise intensity (i.e., the variance) exceeds a critical threshold, new ordered states are observed to emerge. In this chapter we present other examples showing how the ability of random environmental fluctuations to induce bistability can be found in continental-scale land–atmosphere interactions, landscape-scale vegetation–fire dynamics, hill-slope-scale soil development, population dynamics, and in a few other contexts.

The following subsection shows that the opposite effect may also occur: Noise can turn a deterministic bistable system into a stochastic system with only one stable state (D'Odorico et al., 2005). In either case, noise does not merely induce random fluctuations about the stable states of the system; rather, it creates order by determining the number of stable and unstable states.

4.2.2 Noise-induced stability in dryland ecosystems

In the previous section we analyzed an example of bistable ecosystem dynamics driven by dichotomous Markov noise. In that system bistability emerged as a noise-induced effect, resulting from the interaction of the random driver with the underlying (nonlinear) deterministic dynamics. We now consider a different type of stochastic system: In this case the deterministic counterpart of the process is bistable, and we investigate the effect of noise on these bistable dynamics.

We refer to the case of dryland plant ecosystems, which have been shown to exhibit bistable behavior, with two stable states corresponding to unvegetated- and vegetated-land surface conditions (Walker et al., 1981; Zeng and Neelin, 2000). The existence of these two stable states is usually ascribed to positive feedbacks between vegetation and its most limiting resource: water (e.g., Walker et al., 1981; Rietkerk and van de Koppel, 1997; Zeng et al., 2004; D'Odorico et al., 2007a).

Two different mechanisms are often invoked to explain these feedbacks at different scales. At the regional or subcontinental scales, vegetation may affect the rainfall regime (Zeng et al., 1999) as suggested by global- and regional-climate models (Charney, 1975; Xue and Shukla, 1993; Xue, 1997; Brovkin et al., 1998; Zeng et al., 1999; Wang and Eltahir, 2000). At smaller (patch-to-landscape) scales, soils beneath vegetation canopies are generally moister than adjacent bare-soil plots (e.g., Greene et al., 1994). This feedback has often been attributed to the larger infiltration capacity of vegetated soils (Walker et al., 1981), which are less exposed to rain-splash compaction and exhibit higher hydraulic conductivity resulting from the presence of roots. This mechanism is invoked by a number of competition–facilitation models of pattern formation (von Hardenberg et al., 2001; Rietkerk et al., 2002). Other authors (Zeng and Zeng, 1996; Scholes and Archer, 1997; Zeng et al., 2004; D'Odorico et al., 2007a) explain soil-moisture–vegetation feedbacks as an effect of the lower evapo-transpirational losses from subcanopy soils, compared with bare-soil evaporation. As a result of this feedback in some regions, seedling establishment can occur only in vegetated-soil plots.

Natural and anthropogenic disturbances acting on bistable dynamics may induce abrupt transitions from the stable vegetated state to the alternative unvegetated ("desert") state (e.g., Scheffer et al., 2001). After this transition has taken place, a significant increase in resource availability (i.e., rainfall) is required to destabilize the desert state and reestablish a vegetation cover. This view of drylands as deterministic bistable systems appears to contrast with the existence of a middle ground between desert and completely vegetated landscapes. It has been shown (von Hardenberg et al., 2001; Rietkerk et al., 2002) that spatial heterogeneities and lateral redistribution of resources can explain the emergence of patchy distributions of vegetation and the consequent existence of an intermediate stable condition of vegetation between the two stable states of the zero-dimensional (i.e., nonspatially explicit) system (van de Koppel and Rietkerk, 2004).[1] In this section we show that a similar result can be induced by temporal fluctuations in environmental conditions rather than by the spatial

---

[1] These studies refer to systems in which vegetation bistability is induced by positive feedbacks between vegetation and soil moisture at the patch scale, rather than by regional-scale feedbacks between vegetation and the rainfall regime.

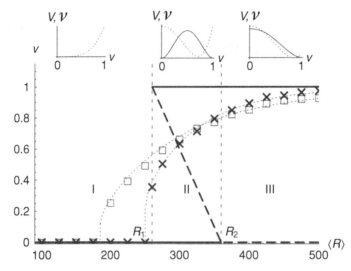

Figure 4.2. Deterministic stable (solid thick lines) and unstable (dashed thick lines) states of Eqs. (4.4) with $R_1 = 260$ mm and $R_2 = 360$ mm. Noise-induced statistically stable states of the stochastic dynamics are shown as a dotted curve (analytically calculated). Numerically calculated values of the modes of v are shown with crosses ($\sigma_R = 0.4\langle R \rangle$) and squares ($\sigma_R = 0.6\langle R \rangle$). The agreement between numerical and analytical solutions suggests that $c = c^+$ is a suitable mathematical approximation for the calculation of the modes of v. The plots in the top inset represent the shapes of the deterministic (V) and stochastic ($\mathcal{V}$) potentials associated with the different regimes of the deterministic (solid line) and stochastic (dotted line) dynamics, respectively. These potentials are calculated for the cases of $\langle R \rangle = 150$, 310, and 410 mm, with $\sigma_R = 0.6\langle R \rangle$. Changes in the shapes of these potentials between the deterministic and stochastic dynamics indicate the emergence of noise-induced stability (after D'Odorico et al., 2005).

heterogeneities: Following D'Odorico et al. (2005), we start from a minimalist model of the bistable vegetation dynamics and show how random interannual climate (rainfall) fluctuations typical of arid climates (e.g., Noy-Meir, 1973; Nicholson, 1980) may lead to the emergence of an intermediate statistically stable configuration between the two stable states of the underlying bistable deterministic dynamics.

We express the dynamics of dryland vegetation as

$$\frac{dv}{dt} = -\frac{dV}{dv} = \begin{cases} -v^3 & \text{(if } R < R_1) & (4.4\text{a}) \\ v(1-v)(v-c) & \text{(if } R \geq R_1) \end{cases}, \quad (4.4\text{b})$$

where v is the normalized vegetation biomass ($0 \leq v \leq 1$; $v = 1$ in the case of completely vegetated region), R is a fluctuating limiting resource (e.g., annual precipitation), and V is the potential function (see inset in Fig. 4.2). As an effect of the vegetation–soil-moisture feedbacks, these dynamics can be either monostable or bistable, depending on the value of annual precipitation R: For small values of R

(e.g., R lower than a threshold, R_1), vegetation establishment is inhibited and only the bare-soil state (i.e., $v = 0$) is stable; for large values of R (e.g., R greater than another threshold R_2), prolonged periods of water stress do not occur, regardless of the amount of existing biomass; thus the state $v = 0$ is unstable, whereas $v = 1$ is stable. In intermediate conditions (i.e., $R_1 \leq R \leq R_2$) the system is bistable: Both bare and completely vegetated soils are stable states of the system. In fact, soil moisture is too low for the establishment of vegetation in bare soil, whereas in completely vegetated plots ($v = 1$) the water available in the soil is sufficient to maintain vegetation cover. To account for this effect of the feedback, we express the coefficient c in Eq. (4.4b) as a function of R, $c = (R_2 - R)/(R_2 - R_1)$ for $R > R_1$ (Fig. 4.2).

The cubic polynomial on the right-hand side of Eq. (4.4b) induces these three distinct regimes. It can be shown that this mathematical representation of vegetation dynamics can be obtained from simple growth–death models of dryland vegetation driven by soil-moisture dynamics (D'Odorico et al., 2005; Borgogno et al., 2007).

To investigate the effect of interannual rainfall fluctuations on vegetation dynamics, we treat R as an uncorrelated random variable, with mean $\langle R \rangle$, standard deviation σ_R, and gamma distribution $p(R)$, though the use of other distributions would not alter the dynamical behavior of the system. As a result of these fluctuations the dynamics alternate between two different regimes similar to the case discussed in the previous section: With probability $P_1 = \int_0^{R_1} p(R)\,dR$, R is lower than R_1, and the dynamics are expressed by Eq. (4.4a). With probability $(1 - P_1)$, R exceeds R_1, and the process is governed by Eq. (4.4b) with c depending on R.

By numerical integration of Eqs. (4.4), we find that random interannual fluctuations of R stabilize the system around an unstable state of the underlying deterministic dynamics (region II in Fig. 4.2). Within a relatively broad range of values of R the probability distribution of v exhibits only one mode (Fig. 4.2) between 0 and 1. This stable state would not exist without the random forcing, as indicated by the comparison between the shapes of the deterministic and stochastic potentials $V(v)$ (inset in Fig. 4.2).

The modes of v (Fig. 4.2, squares and crosses) are the preferential states of the system and can be also determined analytically through a simplified stochastic model. To this end, we take c as a constant and replace it with its average value c^+, conditioned on $R > R_1$. Following the approach described in the previous section, we can represent the temporal dynamics of vegetation by using a stochastic differential equation driven by DMN:

$$\frac{dv}{d\tau} = f(v) + g(v)\xi_{dn} = -v^3 + \frac{v(c^+ - v - c^+ v)\Delta_1}{\Delta_2 - \Delta_1} + \xi_{dn}\frac{v(c^+ v + v - c^+)}{\Delta_2 - \Delta_1},$$

$$(4.5)$$

where ξ_{dn} is a zero-mean dichotomous Markov process (see Chapter 2), assuming values Δ_1 and Δ_2. As noted in Subsection 2.2.2, the two functions $g(v)$ and $f(v)$

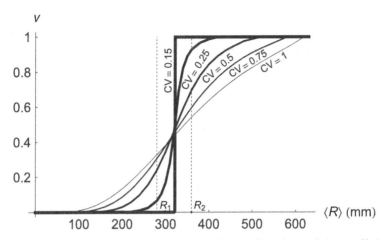

Figure 4.3. Statistically stable states (i.e., modes) as functions of the coefficient of variation (CV) of the random driver R (after Borgogno et al., 2007).

are determined in such a way that $dv/dt = -v^3$ when $\xi_{dn} = \Delta_1$ and $dv/dt = v(1 - v)(v - c^+)$ when $\xi_{dn} = \Delta_2$. The transition probabilities between the two states of ξ_{dn} are $k_1 = (1 - P_1)$ and $k_2 = P_1$. Moreover, the states Δ_1 and Δ_2 need to satisfy the condition $\Delta_1 k_2 + \Delta_2 k_1 = 0$ for ξ_{dn} to be a zero-mean process. The analytical solution of (4.5) provided in Chapter 2 is here used to calculate the modes of v and the stochastic potential $\mathcal{V}(v)$ associated with Eq. (4.5) (Fig. 4.2, dotted lines). This analysis shows that there is a range of values of $\langle R \rangle$ in which the stochastic dynamics have only one preferential state while the stable states $v = 0$ and $v = 1$ of the deterministic dynamics become unstable.

This finding has important ecological implications because climate fluctuations are typically considered as a source of disturbance and are believed to control the transitions between preferential states in bistable dynamics. The emergence of an intermediate noise-induced stable configuration suggests that rainfall fluctuations unlock the system from these preferential states and stabilize the dynamics halfway between bare-soil and full-vegetation cover conditions.

To stress how the emergence of this intermediate stable state can be explained as a noise-induced effect, Borgogno et al. (2007) investigated the dependence of the stable states of the stochastic dynamics on the coefficient of variation (CV), $CV = \sigma_R / \langle R \rangle$, of the interannual rainfall fluctuations. Using a model similar to (4.4), it was found that, as the CV increases, the width of the interval of values of $\langle R \rangle$ in which the system is bistable decreases. For a critical value of CV (CV = 0.15, in Fig. 4.3) the width of the bistability range is zero. For larger values of CV, noise-induced stability emerges.

For larger values of the CV, intermediate noise-induced statistically stable states may even exist within a range of values of $\langle R \rangle$ wider than the bistability range $[R_1, R_2]$, of the underlying deterministic dynamics (Fig. 4.2). This range broadens with increasing noise intensities (i.e., with increasing CV). The existence of noise-induced (partly) vegetated states for values of $\langle R \rangle$ lower than R_1 indicates that rainfall fluctuations prevent the occurrence of desert conditions. Conversely, for $\langle R \rangle > R_2$, rainfall fluctuations act as a disturbance on vegetation in that the mode of v is smaller than the stable deterministic state $v = 1$. The disappearance of bistability for $R_1 < \langle R \rangle < R_2$ reduces the likelihood of catastrophic shifts to the desert state and enhances the resilience of dryland ecosystems.

These results are consistent with the findings obtained by Zeng and Neelin (2000) in the study of the coupled atmosphere–biosphere dynamics of the African continent by use of tropical circulation models coupled with interactive vegetation dynamics. These authors investigated the effect of interannual fluctuations in sea-surface temperatures (SSTs) on the stability of vegetation–ecosystems in Africa. Simulations with no interannual variability in SSTs showed that the coupled vegetation–climate dynamics are bistable in the savanna regions of sub-Saharan and southern Africa [Figs. 4.4(a)–4.4(c)]. In these regions the system converges either to a vegetated or unvegetated steady state, depending on the initial conditions. Figure 4.4(c) shows the areas where, in the absence of SST fluctuations, the vegetation–climate dynamics are bistable.

However, when the system is forced by interannual fluctuations in SSTs the spatial extent of these bistable areas significantly decreases [Fig. 4.4(d)] and the variability of the SSTs stabilizes the coupled system in an intermediate state. In fact, Fig. 4.5 clearly shows that, although in the absence of SST fluctuations the system exhibits the alternative stable states of desert and forest, interannual fluctuations in SSTs allow the system to converge to a state with intermediate vegetation cover. In this case the dynamics are not bistable but converge to this intermediate state regardless of the initial conditions. Moreover, interannual rainfall variability resulting from oscillations in the SSTs enhance vegetation cover in areas that would be desert in the absence of climate fluctuation. A similar effect has also been found in the dynamics of other ecosystems (Rennermalm et al., 2005).

4.2.3 Noise-induced biodiversity

In the previous subsection we showed how random climate fluctuations may convert bistable deterministic dynamics into a system with only one stable state. This effect of noise-induced stability enhances ecosystem resilience, i.e., the ability to recover after a disturbance (see Box 4.1). Here we focus on another possible mechanism of noise-enhanced resilience, which is based on the increase in biodiversity, as explained in Box 4.2.

Recent studies (Hughes et al., 2007) investigated the two-way interaction between species diversity and disturbances. Biodiversity has been related to the way ecosystems respond to environmental fluctuations (Yachi and Loreau, 1999). At the same time, the effect of environmental variability on biodiversity has been

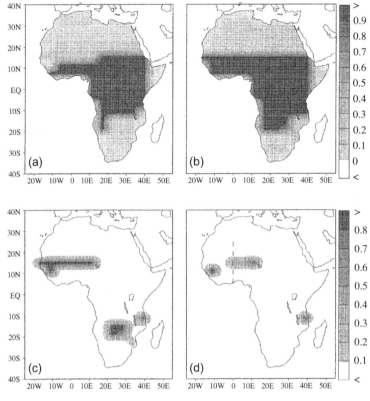

Figure 4.4. Equilibrium vegetation distribution from a coupled atmosphere–vegetation model. Dark shaded areas correspond to forest cover. Light gray areas correspond to sparse vegetation and desert conditions. (a) Simulation starting from desert (i.e., unvegetated) conditions in the whole African continent; (b) simulation starting with forest vegetation across the whole continent; (c) map of the difference between (b) and (a); (d) difference between equilibrium vegetation (obtained with fluctuating SST and initial desert conditions) and the equilibrium state in (a) (after Zeng and Neelin, 2000).

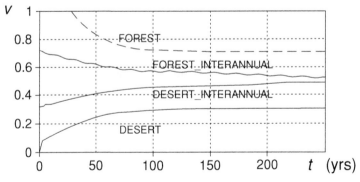

Figure 4.5. Average vegetation cover in the Sahel region as a function of time. Calculated with model simulations using different initial conditions (after Zeng and Neelin, 2000).

Box 4.2: Biodiversity and ecosystem resilience

Recent experimental (e.g., Steneck et al., 2002; Bellwood et al., 2004; Folke et al., 2004) and theoretical studies (e.g., Chapin et al., 1997; Walker and Salt, 2006) determined a number of ecosystem attributes able to enhance resilience. Among them, biodiversity was found to have the potential to increase resilience by increasing species redundancy (Lawton and Brown, 1993), i.e., the coexistence of species that have a similar effect on the ecosystem function (Chapin et al., 1997; Folke et al., 2004) but different sensitivities to disturbances. By definition, a redundant system could function with a lower number of species. However, systems that minimize their diversity tend to be less resilient (Chapin et al., 1997; Walker and Salt, 2006). In fact, in the absence of disturbances they have the same functioning as their redundant counterparts. However, in a redundant system, if environmental stressors cause the extinction of some species, other species with similar functional traits but different sensitivity to disturbance would be able to provide the same service to the ecosystem. This effect, or *insurance hypothesis* (McNaughton, 1977; Naeem and Li, 1997), indicates that redundancy provides the system with a "buffer" that favors postdisturbance recovery.

   This effect of resilience enhancement is due in particular to the *response diversity* (Elmqvist et al., 2003), which exists when the organisms of the same functional group have different response traits, i.e., different sensitivities to disturbance (Bengtsson et al., 2003; Folke et al., 2004). Although the functional diversity ensures the existence of groups of organisms that perform different functions in the ecosystem, the response diversity ensures that the system is redundant and therefore resilient.

   Some authors have shown that a more diverse system is more resilient, i.e., is able to respond better to environmental fluctuations (McNaughton, 1977; Yachi and Loreau, 1999). In Subsection 4.2.3 we show that random environmental fluctuations may, in turn, augment the resilience of the system by enhancing diversity.

---

investigated both experimentally (see Mackey and Currie, 2001) and theoretically (e.g., May, 1973; Chesson, 1994). For example, May (1973) studied the effect of environmental fluctuations on niche overlap and species packing and showed how environmental variability allows for the existence of higher degrees of niche overlap, thereby favoring biodiversity. Tree–grass co-dominance in savannas has been explained as the result of nonequilibrium dynamics induced by interannual climate variability or random fire occurrences (Higgins et al., 2000; D'Odorico et al., 2006b), whereas in riparian ecosystems the coexistence of species with different water tolerances has been related to random water-table fluctuations (Ridolfi et al., 2007). All these studies suggest that random environmental variability may favor the coexistence of species with different functional traits and demonstrate how biodiversity can be enhanced by environmental variance (e.g., Chesson, 1994). This notion of noise-enhanced biodiversity is supported by some experimental observations:

Benedetti-Cecchi et al. (2006) and Bertocci et al. (2007) recently found an increase in the diversity of rock-shore populations of algae and invertebrates exposed to an increase in the variance of environmental fluctuations (with constant mean). Using a framework based on the theory of stochastic differential equations with dichotomous noise, this subsection puts these experimental results into the broader context of how the state of an ecosystem may benefit from moderate environmental variability through the enhancement of biodiversity.

Following D'Odorico et al. (2008), we consider the case of a system in which all species are controlled by the same environmental variable R (e.g., water, energy, light, or nutrients). We assume that R is a positive random variable with gamma distribution $p(R)$ (though the use of other distributions would not alter the results presented in this subsection), mean $\langle R \rangle$, and standard deviation σ_R. The species coexisting within the ecosystem are from a pool determined by the ecosystem's biogeographical conditions and evolutionary history. Each species is unstressed (i.e., able to live, grow, and reproduce) when R remains within a certain *fitness interval* I_δ, or *niche*, whereas its biomass decays when the environmental conditions keep R outside this interval. All species have in general fitness intervals I_δ with different amplitudes (or *fitness ranges*) δ and different midpoint values. However, for the sake of simplicity we first assume that all fitness intervals have the same amplitude δ and that no mutual interaction (e.g., competition–facilitation) exists among species. We then consider more general conditions. In the absence of interactions we can model the temporal dynamics of each species independently of the others and assume that fluctuations in R determine the switching between growth (unstressed conditions) and decay (stressed conditions) in species biomass, depending on whether R falls within or outside of the fitness interval I_δ of that species. We use a linear decay and a logistic growth for the stressed and unstressed conditions, respectively:

$$\frac{dB}{dt} = \begin{cases} a_1 B(\beta - B) & \text{if} \quad R \in I_\delta & (4.6a) \\ -a_2 B & \text{otherwise} & (4.6b) \end{cases},$$

where B is the species' biomass, β is the carrying capacity (i.e., the maximum sustainable value of B), and the coefficients a_1 and a_2 determine the decay and growth rates, respectively.

Random fluctuations in R determine the switching between these two dynamics, depending on whether R is contained within or falls outside the fitness window. Following the approach presented in the previous subsection the stochastic dynamics resulting from the random switching between Eqs. (4.6a) and (4.6b) are modeled as a dichotomous Markov process. With probability P_1 the environmental variable R is contained within the fitness interval, where $P_1 = \int_{R_0}^{R_0+\delta} p(R)\, dR$, and R_0 is the lower limit of the fitness interval I_δ. In these conditions, the species is not stressed and its growth is expressed by (4.6a). Conversely, with probability $1 - P_1$ the species is

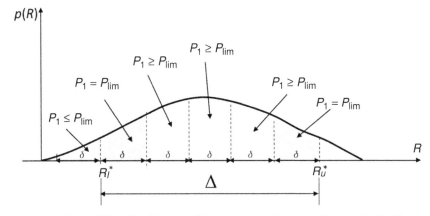

Figure 4.6. Probability distribution of the resource R. For a given δ the biodiversity potential Δ is defined as the interval (R_l, R_u) in which all species with fitness range δ remain unstressed for a sufficient fraction of time to avoid extinction (after D'Odorico et al., 2008).

stressed, and its dynamics are modeled by (4.6b). The solution of the stochastic differential equation associated with these dynamics provides the probability distribution $p(B)$. It can be shown (Camporeale and Ridolfi, 2006) that when

$$P_1 \leq P_{\lim} = \frac{a}{a + \beta} \tag{4.7}$$

(with $a = a_2/a_1$), the species goes extinct because B is zero with probability tending to one. This condition is independent of the values of Δ_1 and Δ_2 but depends on the parameters β and a of the dynamics. Relatively low values of P_1 correspond to conditions in which the environmental variable remains too often outside the fitness interval to allow for the survival of that species. Thus a species can survive only when $P_1 \geq P_{\lim}$. As shown in Fig. 4.6 for a given distribution of resources $p(R)$ and fitness range δ, there are two limit positions [if $p(R)$ is unimodal] along the R axis in which the condition $P_1 = P_{\lim}$ can be met. These two limit positions, R_l^*, and R_u^*, are determined by the conditions

$$\int_{R_l^*}^{R_l^* + \delta} p(R)\, \mathrm{d}R = P_{\lim}, \quad \int_{R_u^* - \delta}^{R_u^*} p(R)\, \mathrm{d}R = P_{\lim}. \tag{4.8}$$

Species with fitness range δ can survive when their fitness interval is contained within the interval in that they remain unstressed for a sufficient fraction of time that condition (4.7) is never met. Thus for a given distribution $p(R)$ of the environmental variable (Fig. 4.6) we can determine the interval (R_l^*, R_u^*) on the R axis in which

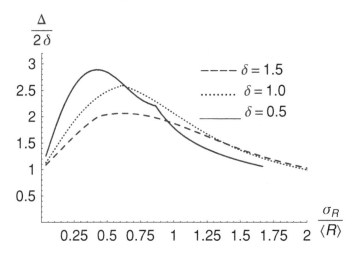

Figure 4.7. Biodiversity potential as a function of the CV, $CV = \sigma_R/\langle R \rangle$, of environmental noise, calculated with $\beta = 1$, R gamma distributed with mean, $\langle R \rangle = 3$, and $a = 0.11$, i.e., $P_{\text{lim}} = a/(a + \beta) = 0.1$ (after D'Odorico et al., 2008).

species with fitness range δ remain unstressed for a sufficient fraction of time to avoid extinction. If Δ is the interval width (i.e., $\Delta = R_u^* - R_l^*$), Δ/δ is an indicator of the biodiversity that could be sustained in the ecosystem, i.e., of the *potential for biodiversity*. Larger values of Δ/δ are associated with a broader range of species that would be able to have access to environmental conditions favorable to growth and survival. The exact number of species, of course, depends on a number of other factors, including the value of $\langle R \rangle$, the presence of other limiting factors, and the competitive and facilitative interactions among species. However, with all the other factors being the same, species biodiversity is expected to increase with Δ/δ. Thus Δ/δ is not exactly the number of species coexisting in the system (as it would be in the case of nonoverlapping niches) but a measure of the ability of the system to maintain biodiversity. The biodiversity potential Δ depends in general on the width of the fitness window δ, the probability distribution of R, and the parameters a_1, a_2, and β of vegetation dynamics. Notice how when the variance of R is zero the process becomes deterministic with $R = \langle R \rangle$. In this case the amplitude Δ tends to 2δ.

We can first investigate the effect of noise (i.e., environmental variability) on biodiversity in the absence of interspecific interactions. We keep constant the values of the parameters a, β, and $\langle R \rangle$, and study the dependence of biodiversity potential on the standard deviation σ_R of R. The results are shown in Fig. 4.7 for different values of the fitness range δ. It is found that environmental noise has two major effects: (i) with moderate levels of noise intensity (i.e., of $CV = \sigma_R/\langle R \rangle$) environmental fluctuations enhance the system's potential for biodiversity with respect to the deterministic case $\Delta = 2\delta$ (i.e., $\Delta > 2\delta$); (ii) with relatively large noise intensities, environmental fluctuations limit the ability of the system to support diverse communities of individuals (i.e., $\Delta < 2\delta$). In the first case, by favoring biodiversity, noise plays a "constructive" role in the dynamics of the system, whereas in the second case it has a "destructive" effect, in that it acts as a source of disturbance resulting in the noise-induced extinction of some species (e.g., May, 1973). This finding is consistent with the idea underlying the *intermediate-disturbance hypothesis* (Connell, 1978; Huston, 1979), i.e., that moderate disturbances can be beneficial to an ecosystem. According to this hypothesis, disturbances of moderate strength would not allow the attainment of equilibrium conditions by preventing competitive elimination of the less-adapted species that leads to an equilibrium with less diverse species composition than in nonequilibrium conditions. However, Fig. 4.7 shows that environmental variability may cause diversity enhancement even in the absence of competition. Figure 4.7 shows also that generalist species (i.e., species with high δ) are better adapted than specialists (i.e., species with low δ) to withstand and benefit from environmental fluctuations. The effect of these fluctuations depends also on the interplay between the scales of growth or decay and the intensity of the stochastic drivers (Loreau, 1992).

As noted, the preceding framework assumes that all species have the same δ and do not interact with one another. However, the results presented in Fig. 4.7 are not affected by these assumptions. In fact, the variability of δ among species would not modify the nature of the dependence of Δ/δ on the standard deviation σ_R of R. Moreover, interspecies competition and facilitation–cooperation affect vegetation dynamics by either decreasing (competition) or increasing (facilitation) the species carrying capacity β, which, in turn, affects the potential for biodiversity. To investigate the effect of competition (facilitation) D'Odorico et al. (2008) expressed β as a decreasing (increasing) function of the number of species (hence of the Δ/δ ratio) and found that the positive effect of environmental fluctuations on biodiversity is stronger in the presence of facilitative mechanisms, whereas it is weakened by competition, consistent with the notion that biodiversity is favored by cooperation and reduced by competition (e.g., Mulder et al., 2001), though it could be argued that competition may enhance biodiversity by inducing niche restriction.

The increase of the potential for biodiversity found with low levels of environmental variance is consistent with the notion presented throughout this chapter that noise may have a constructive effect and induce order and stability in environmental systems (Chesson, 1982, 2000; Loreau, 1992; D'Odorico et al., 2005, 2006b), whereas the occurrence of species extinction with relatively strong fluctuations (i.e., high variance) corresponds to the more intuitive destructive role of noise (e.g., May, 1973).

Space-based ecological theories (e.g., Holt, 1984; Tilman, 1994) have shown that the spatial variability of environmental conditions typically found in heterogeneous landscapes offers better opportunities to maintain higher levels of functionally similar species. The results in Fig. 4.7 demonstrate that a similar effect may arise from the temporal variability of resource availability.

4.2.4 State-dependent dichotomous Markov noise in environmental systems

We consider a process in which feedback exists between random environmental drivers and the state of the system. In a number of systems this feedback is mediated either by the dynamics of resource availability or by the disturbance regime. For example, dryland vegetation is typically limited by soil moisture; thus random rainfall inputs affect vegetation through the dynamics of soil moisture. Vegetation growth is sustainable only when soil moisture exceeds a certain threshold; otherwise a mortality-induced decrease in (live) plant biomass occurs. As a result, rainfall inputs determine the switching between stressed and unstressed conditions. However, these dynamics are often more complex because of the existence of a positive feedback between soil moisture and vegetation: In arid and semiarid environments moister near-surface soils are found beneath vegetation canopies rather than in the surrounding bare-soil areas (Greene et al., 1994; Breman and Kessler, 1995; D'Odorico et al., 2007a). Thus growth requires less rainwater on well-vegetated soils than on soils with only a thin, sparse canopy cover. This fact translates into a state dependency in the threshold θ of the random driver q (precipitation), as indicated in Chapter 3.

Another example is represented by the dynamics of woody vegetation in semiarid, fire-prone environments. In this case the encroachment of woody plants has been found to be limited by fires (e.g., Anderies et al., 2002; van Wilgen et al., 2003), which, in turn, depend both on ignition and on the presence of grass fuel (see also Section 3.2). In the study of the dynamics of woody vegetation in savannas, fire ignition is the random external forcing. Ignition does not act directly on woody plants, in that its effect is mediated by grass-fuel availability, which, in turn, is inversely related to woody-plant biomass B. Relatively high values of B correspond to a system dominated by woody vegetation, in which herbaceous vegetation (i.e., fuel load) is present only in limited amounts. In this system, with the potential for random ignition (e.g., lightning) being the same, a woodland savanna is less prone to fires than an open savanna with relatively low tree density. Thus positive feedback exists between vegetation and fire pressure. The dynamics of woody-plant biomass can be modeled by Eqs. (4.2) with state-dependent threshold $\theta = \theta(B)$. Thus, in this case, state dependency exists in the probabilities P_1 of being in state 1 and in the transition rates of the DMN, $k_1(B) = 1 - P_1(B)$ and $k_2(B) = P_1(B)$ (see Chapter 2). This state dependency implies that the dichotomous noise acts on the dynamics as a multiplicative noise, in that its parameters are state dependent. As noted in Chapter 3, this multiplicative noise cannot be factorized, i.e., it cannot be expressed as the product of a state-independent dichotomous noise with a suitable function of B (Laio et al., 2008).

We now consider the particular case in which the functions $f_1(B)$ and $f_2(B)$ are linear as in Eqs. (4.2). Moreover, we use a linear dependence of k_1 and k_2 on B: $k_2(B) = P_0 + bB$, with b being positive in the case of positive feedback

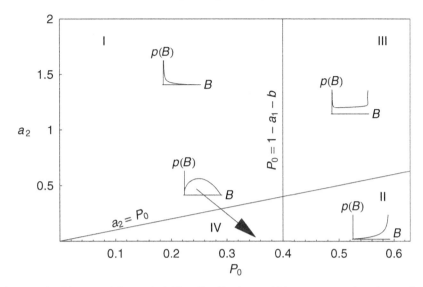

Figure 4.8. Shapes of the probability distributions of biomass B as functions of the parameters P_0 and a_2 of the state-dependent dichotomous process (calculated for $a_1 = 0.2$ and $b = 0.4$).

between B and $k_{1,2}$. Figure 4.8 shows the shape of $p(B)$ for different values of the parameters.

In Subsection 4.2.1 we showed how random fluctuations in the resource q may generate an interesting variety of behaviors, including the emergence of bistable dynamics, which did not exist in the deterministic counterpart of the system. These bistable dynamics are clearly noise induced (Horsthemke and Lefever, 1984; Bena, 2006). As noted, bistability may also emerge as an effect of positive feedbacks (see Box 4.1) between environmental drivers and the state of the system (Wilson and Agnew, 1992). Comparing Figs. 4.1 and 4.8, we observe that case V in Fig. 4.1 disappears in the presence of the state dependency induced by the feedback, whereas the size of the area III in the parameter space increases with the "strength" b of the feedback. We interpret this effect as an enhancement of bistability induced by the positive feedback between resources and state of the system. Thus noise induces a bistable behavior in the dynamics even in the absence of the feedback, and the feedback enhances the bistability of the system by broadening the interval of the parameter space in which the probability distribution of B is bimodal. A similar effect is found in Subsection 4.3.2 in the case of state-dependent Poisson processes.

4.3 Environmental systems forced by white shot noise

Some environmental drivers are not always continuously active but occur as random, episodic events of relatively short duration, which can be modeled as instantaneous random pulses. Typical examples include insect outbreaks, epidemics, landslides (Benda and Dunne, 1997; D'Odorico and Fagherazzi, 2003), fires (Daly and Porporato, 2006), and rainfall (Todorovic and Woolhiser, 1975). Accounting for the intermittent character of these occurrences may be crucial to the correct understanding of the dynamics of environmental systems at time scales smaller than (or comparable to) the recurrence time of the random episodic events. In this case, the random forcing needs to be modeled as a temporal sequence of intermittent events occurring at random discrete times t_i (with $i = 1, 2, \ldots$). When the recurrence times $\tau_i = t_{i+1} - t_i$ are exponentially distributed random variables with mean λ, this sequence is known as *white Poisson noise* of rate λ (see Chapter 2). In several applications each event has a random intensity h with given distribution $p(h)$ and mean α; in this case the random sequence is known as a *marked Poisson process*, and, when h is an exponentially distributed random variable, it is called *white shot noise*. In the case of Poisson noise, Eq. (4.1) can be rewritten in the form

$$\frac{d\phi}{dt} = f(\phi) + g(\phi)\xi_{sn}, \qquad (4.9)$$

where $\xi_{sn}(t)$ is the white-shot-noise term.

4.3.1 Harvest process: Fire-induced tree–grass coexistence in savannas

Harvest processes are a class of simple univariate dynamics (e.g., Kot, 2001) in which a population A undergoing logistic growth is harvested proportionally to its abundance:

$$\frac{dA}{dt} = aA(A_c - A) - kA, \qquad (4.10)$$

where A is, for example, the population's density, A_c is the carrying capacity of A (i.e., the maximum sustainable value of A allowed by the available resources), a is the reproduction rate of the logistic growth, and k is the harvest rate. If we assume that A is a positive-valued variable and $A_c > k/a$, these deterministic dynamics have two equilibria, namely $A = 0$ (unstable state) and $A = A_c - k/a$ (stable state).

In a number of natural processes the harvest rate has a random and episodic character. For instance, fires are clear examples of episodic random disturbances that could be modeled as Poisson noise to account for their intermittent and random character. In this subsection we investigate the properties of a *Poisson harvest process*, i.e., of a stochastic harvest process in which the harvest rate k is modeled as white shot noise and the parameters λ and α of the white shot noise ξ_{sn} represent the average frequency and magnitude of the harvest events, respectively.

This model was used (D'Odorico et al., 2006b) to investigate the effect of stochastic fire dynamics on tree–grass coexistence in mesic and subhumid savannas.[2] We assume that – in the absence of disturbances – these savannas tend to be dominated by woody plants (e.g., van Langevelde et al., 2003). In these systems competition can at most slow down the growth of tree biomass but it is unable to induce a stable state of tree–grass coexistence. Thus the temporal dynamics of tree biomass can be modeled without explicitly accounting for interspecies competition. The relatively slow demographic growth of woody vegetation enables grasses to use space and resources left available by trees after fire occurrences. However, in periods between fire events, trees are able to reclaim space and resources from grasses because no niche separation is assumed to exist and trees are able to outcompete grasses. We characterize the state of the system through the state variable A, representing the total woody biomass. A ranges between 0 and a maximum value A_c, the ecosystem *carrying capacity*, which depends on the existing resources (i.e., water, nutrients, and light). We assume A_c to be constant in time and normalize A with respect to A_c, i.e., $A' = A/A_c$. To simplify the notation we drop the superscript from the normalized biomass and refer to A as the woody biomass normalized between 0 and 1. As noted, the growth of A in periods between fire occurrences is modeled with a logistic equation and the harvest rate is proportional to the existing biomass. These dynamics are expressed by (4.9) with $f(A) = aA(1 - A)$ and harvest rate proportional to $g(A) = -A$. The steady-state probability distribution of A can be determined as described in Chapter 2 [see pdf (2.34)], leading to

$$p(A) = C A^{\frac{1}{\alpha} - \frac{\lambda}{a} - 1}(1 - A)^{\frac{\lambda}{a} - 1}, \tag{4.11}$$

which is a beta distribution with normalization constant (e.g., Johnson et al., 1994)

$$C = \frac{\Gamma\left(\frac{1}{\alpha}\right)}{\Gamma\left(\frac{1}{\alpha} - \frac{\lambda}{a}\right) \Gamma\left(\frac{\lambda}{a}\right)}, \tag{4.12}$$

where $\Gamma(\cdot)$ is the gamma function.

A variety of dynamical behaviors emerge for different combinations of the parameters. Figure 4.9 shows the dependence of the shape of the distribution of A – which is here studied through its modes A_m (see Chapter 3) – on the parameters of fire and vegetation dynamics. In this figure, curves (a), (b), and (c) divide the parameter space into five zones characterized by different dynamical behaviors. For relatively low values of λ/a, the mode of A is $A_m = 1$; thus the preferential state of the system

---

[2] The dynamics of savanna ecosystems were investigated in the recent past with a variety of approaches (e.g., Sankaran et al., 2004), depending on whether (i) vegetation dynamics are studied in lumped (e.g., Walker et al., 1981) or spatially extended systems (e.g., Jeltsch et al., 1996; Rodriguez-Iturbe et al., 1999a; van Wijk and Rodriguez-Iturbe, 2002) and (ii) tree–grass coexistence results from competition (Walter, 1971), demographic dynamics (e.g., Higgins et al., 2000), mixed competition–demographic processes (e.g., Anderies et al., 2002; van Langevelde et al., 2003), or landscape heterogeneities (e.g., Kim and Eltahir, 2004). In this subsection we use a simplistic stochastic model to demonstrate the role of noise in the demographic dynamics resulting from fire–vegetation interactions in spatially lumped systems.

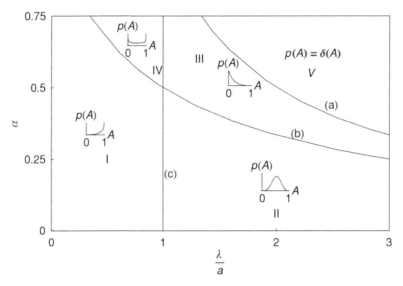

Figure 4.9. Dependence of the shape of the probability distribution $p(A)$ of woody biomass A on parameters of the fire–vegetation dynamics. The solid curves divide the parameter space into five zones corresponding to woodland (I), savanna (II), grassland (III), and a bistable domain (IV) with two likely preferential states of grassland and woodland. Zone (V) represents combinations of parameters corresponding to the complete and permanent disappearance of woody vegetation. The properties of the beta distribution (e.g., Johnson et al., 1994) can be used to determine the expressions of the solid curves: (a) $\lambda/a = 1/\alpha$, (b) $\lambda/a = -1 + 1/\alpha$, (c) $\lambda/a = 1$.

corresponds to a woodland (zone I in Fig. 4.9). For relatively large values of λ/a, the mode of A is $A_m = 0$, suggesting that the system tends to be dominated by grasses (zone III in Fig. 4.9). Thus for most values of α there are two critical threshold values of λ/a marking the transitions from savanna to woodland (i.e., $\lambda/a = 1$) or to grassland (i.e., $\lambda/a = -1 + 1/\alpha$). Conditions favorable to tree–grass coexistence emerge within a broad range of values of λ (zone II in Fig. 4.9). In these conditions, a statistically stable state of vegetation exists in correspondence to partial tree cover {i.e., for values of A between 0 and 1, with $A_m = [a(1 - \alpha) - \lambda\alpha]/[a(1 - 2\alpha) + b\alpha]$}. This finding supports theories of fire-induced savanna dynamics. Notice how the dynamical behavior of the system depends on the ratio λ/a, suggesting that the same behavior can be attained with different combinations of fire frequency and tree or shrub encroachment rates (i.e. of the parameters λ and a).

Figure 4.9 shows also that, within a small region in the parameter space (zone IV), A exhibits U-shaped probability distributions. As noted in Chapter 3, because the modes of $p(A)$ are preferential states of the system we interpret bimodality as a sign of bistability, i.e., in these conditions both woodland and grassland have high probabilities of occurrence.

We notice that this bistable behavior does not emerge in the deterministic counterpart of the system. In fact, as noted, the deterministic harvest process [Eq. (4.10)] has only one stable state. We can obtain the equilibrium points A^* of the underlying deterministic dynamics from (4.9), eliminating the random fluctuations by replacing the noise term with its average value, $\lambda\alpha$. Thus the equilibria A^* are given by the roots of

$$aA^*(1 - A^*) - \lambda\alpha A^* = 0. \qquad (4.13)$$

The solution of (4.13) leads to $A^* = 0$ (unstable state) and $A^* = 1 - \lambda\alpha/a$ (stable state). Thus, for a given value of α, the stable state of the system increases from 0 to 1 as λ/a decreases from $1/\alpha$ to 0, consistent with the behavior of the modes of the stochastic dynamics through zones III, II, and I of Fig. 4.9. For $\lambda/a > 1/\alpha$ the fire-induced disturbances are too frequent and intense to allow the process to leave the state $A = 0$, which is stable both in the deterministic and in the stochastic (zone V in Fig. 4.9; see caption) dynamics. Thus the behaviors found in zones I, II, III, and V exist also in the deterministic system and these transitions are not noise induced. However, the bimodal behavior in zone IV is completely noise induced. In fact, because $f(A)$ is a quadratic function, the deterministic system is not bistable. For the bounds to induce bistability, an unstable deterministic state should exist between 0 and 1, as in the case of the bounded Takacs process (Porporato and D'Odorico, 2004). In the present case, such a state does not exist. Thus we conclude that multiplicative noise is able to create new states that did not exist in the underlying deterministic dynamics.

Within a certain portion of the parameter space, the dynamics become bimodal, i.e., they exhibit two preferential states, whereas the deterministic system [i.e., Eq. (4.10)] has only one stable state. The emergence of bimodality indicates that, in this part of the parameter space (zone IV), conditions favorable to tree–grass coexistence are unstable, whereas woodland and grassland are alternative statistically stable states of the system (D'Odorico et al., 2006b). The emergence of this noise-induced bistability causes a discontinuous response of the system to changes in parameter values, as shown by the existence of threshold effects and hysteresis in the bifurcation diagram in Fig. 4.10. Subsection 4.3.2 shows how feedbacks between fires and vegetation may enhance this noise-induced behavior.

4.3.2 Poisson harvest process with state-dependent harvest rate

In Subsection 4.3.1 we investigated the properties of a stochastic harvest model in relation to the case of fire–vegetation dynamics in mesic savannas. Here we show how a state-dependent rate $\lambda'(A)$ of the Poisson process may induce other interesting noise-induced behaviors in the dynamics. The dependence of the probability of fire occurrence λ' on A has been well documented (e.g., van Wilgen et al., 2003) and is due to the fact that in savannas the fuel load is contributed by grass biomass.

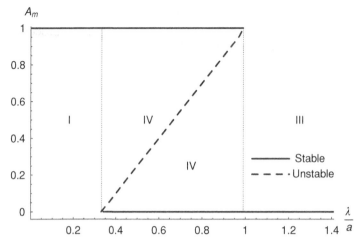

Figure 4.10. Bifurcation diagram showing the stable and unstable states (thick solid and dashed lines) of the dynamics as functions of λ/a and with $\alpha = 0.75$. It is observed that with increasing values of λ/a the preferential state of the system shifts from $A_m = 1$ (zone I of Fig. 4.9) to $A_m = 0$ (zone III). This transition occurs through a bistability domain (zone IV, between the two vertical dotted lines), which exists only for $\alpha > 0.5$, as shown in Fig. 4.9. The dynamics exhibit a hysteresis in that the transitions from $A_m = 1$ to $A_m = 0$ and from $A_m = 0$ to $A_m = 1$ occur for different values of λ/a capable of destabilizing the states $A_m = 1$ and $A_m = 0$, respectively.

Landscapes with denser tree or shrub cover (i.e., higher A) exhibit lower grass biomass and a consequently lower probability of fire occurrence (e.g., van Wilgen et al., 2003; Goldammer and de Ronde, 2004; D'Odorico et al., 2006b). Thus the rate λ' of fire occurrence is a decreasing function of A, which we assume to be linear, $\lambda'(A) = \lambda + bA$ (with $b \leq 0$, $\lambda > -b$). When the rate of (deterministic) growth is expressed by a logistic function, $f(A) = aA(1 - A)$, and the effect of fire occurrences is again taken to be proportional to $g(A) = -A$; using Eq. (2.73) we obtain that the steady-state pdf of A is again a beta distribution (D'Odorico et al., 2006b),

$$p(A) = CA^{\frac{1}{\alpha} - \frac{\lambda}{a} - 1}(1 - A)^{\frac{\lambda}{a} + \frac{b}{a} - 1}, \tag{4.14}$$

with normalization constant

$$C = \frac{\Gamma\left(\frac{1}{\alpha} + \frac{b}{a}\right)}{\Gamma\left(\frac{1}{\alpha} - \frac{\lambda}{a}\right)\Gamma\left(\frac{\lambda}{a} + \frac{b}{a}\right)}. \tag{4.15}$$

Figure 4.11 shows that the feedback (i.e., $b < 0$) is able to increase the size of the portion of the parameter space in which noise-induced bistable dynamics are observed (zone IV). In fact, a comparison with Fig. 4.9 (case with no feedback) shows that curve (c) shifts to the right, thereby allowing for the emergence of bistability, even with lower values of α. By enhancing this bistable behavior, the positive

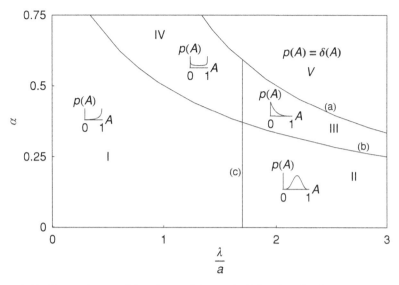

Figure 4.11. Dependence of the shape of the probability distribution $p(A)$ of woody biomass A on parameters of the fire–vegetation dynamics in the presence of positive feedback between fire frequency and grass biomass ($b \neq 0$ with $b < 0$). Calculated with the same parameters as in Fig. 4.9 but with $b/a = -0.7$. With respect to the case $b = 0$ (i.e., Fig. 4.9) line (c) shifts to the right, enabling a wider bistable domain, IV.

fire–vegetation feedback destabilizes conditions favorable to tree–grass coexistence (i.e., the existence of a mode of A between 0 and 1), whereas grassland and woodland become alternative stable states of the system, as shown in Fig. 4.12. Notice that the state dependence $\lambda'(A)$ only enhances this bistable behavior, and its emergence is induced by the multiplicative noise (see previous section).[3]

4.3.3 Stochastic soil-mass balance

In this section we provide an example of systems forced by multiplicative Poisson noise, with a bound imposed on the distribution $p(h)$ of the "jumps" h of the shot noise [see Eq. (4.9)]. It is shown that the state dependency induced by this bound is capable of leading to noise-induced transitions. This example is based on the dynamics of soil production and erosion in landslide-prone landscapes. These dynamics were traditionally investigated (Kirkby, 1971; Ahnert, 1988; Dietrich et al., 1995; Roering et al., 1999) through a soil-mass-balance equation in which the variability in time of

---

[3] In fact, the deterministic counterpart of the process with state-dependent rate of fire occurrence has only one unstable state ($A^* = 0$) and one stable state [$A^* = (1 - \lambda\alpha/a)/(1 + \alpha b/a)$] that can be calculated with (4.13), replacing λ with $\lambda' = \lambda + bA$.

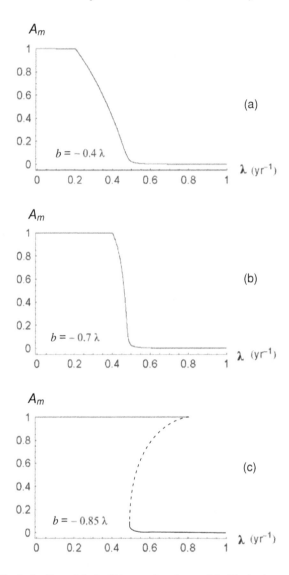

Figure 4.12. Statistically stable (solid curve) and unstable (dashed curve) states of the stochastic dynamics of woody biomass A calculated with $\alpha = 0.20$, $a = 0.12 \text{ yr}^{-1}$, and with different strengths λ/b of the fire–vegetation feedback (after D'Odorico et al., 2006b).

soil thickness H is expressed as the difference between the rates of soil production and erosion (see also Fig. 4.13). Soil production is due to bedrock weathering and can be expressed (Carson and Kirkby, 1972; Heimsath et al., 1997) as a deterministic function $f(H)$ of the soil depth H. Landslide and debris-flow dynamics can be treated as a stochastic (Poisson) process (Wu and Swanston, 1980; D'Odorico, 2000; D'Odorico and Fagherazzi, 2003) to account for the intermittent and episodic random character

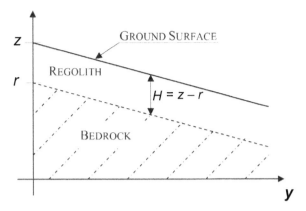

Figure 4.13. Schematic representation of a soil-mantled slope.

of mass-wasting events. We first consider the case in which regolith thickness H does not affect the depth of soil removed by a landslide other than through the bound $h \leq H$. Thus state dependency in the random driver arises only from this bound, i.e., we assume that each landslide removes an exponentially distributed random amount h of soil or the whole soil column H, whichever is less. The overall dynamics of soil development can be expressed by (4.9) with $g(H) = -1$ and a truncated distribution $p(h)$ of h, as in Fig. 4.14. If $H_{\max}$ is the maximum thickness of the regolith above which the rates of soil production are negligible, we normalize H with respect to

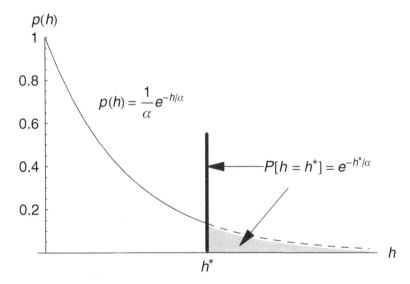

Figure 4.14. Probability distribution $p(h)$ of h truncated at $h^*(H)$. The distribution is exponential for $0 < h < h^*(H)$ and has a probability mass (or "atom") P^* at $h = h^*(H)$ equal to the probability that h exceeds $h^*(H)$ in the nontruncated exponential distribution (shaded area in this figure). $p(h)$ is clearly state dependent.

$H_{\max}$ and investigate the dynamics of the normalized variable, $H' = H/H_{\max}$. To simplify the notation, in what follows we drop the superscript and refer to H as the normalized soil depth. The soil-production function $f(H)$ decreases with the soil thickness because the soil mantle protects the bedrock from weathering agents (e.g., Ahnert, 1988; Dietrich et al., 1995; Heimsath et al., 1997). Carson and Kirkby (1972) speculate that $f(H)$ has a maximum and that for relatively shallow soils the rate of soil production decreases with decreasing values of H because the mechanical weathering by roots is weaker in shallow soils, where only sparse vegetation is able to grow. Thus we model $f(H)$ as

$$f(H) = \frac{\rho_r}{\rho_s} w_b (H + b)(1 - H), \qquad (4.16)$$

where ρ_r and ρ_s are the densities of the parent rock and of the soil, respectively, w_b is the rate of bedrock weathering, and b is a parameter determining the rate of soil production when the bedrock is at the surface. Notice that, if $b = 0$, the dynamics would tend to $H = 0$ and remain locked in this state because on the denuded slope the rate of soil production would be zero. In this case the steady-state probability distribution of H is $p(H) = \delta(H)$. We consider the case $b > 0$ and note that the dynamics have a boundary at $H = 1$. We can use the theory presented in Chapter 2 to determine the steady-state probability distribution of H:

$$p(H) = C e^{\frac{H}{\alpha}} (1 - H)^{\frac{\lambda}{a(1+b)} - 1} (H + b)^{-\frac{\lambda}{a+(1+b)} - 1}, \qquad (4.17)$$

where $a = w_b \rho_r / \rho_s$ and C is the normalization constant. Figure 4.15 shows how the shape of $p(H)$ dramatically changes with different values of the noise parameters λ and α. Depending on these parameters, the system may have only one preferential state contained between 0 and 1 [Fig. 4.15(a), dashed curve], or at the boundaries $H = 0$ [Fig. 4.15(a), dotted curve] and $H = 1$ [Fig. 4.15(b), dotted curve] of the [0,1] interval. Thus relatively high values of λ and α are associated with high erosion rates – i.e., with weathering-limited systems – whereas with relatively low values of these parameters, the system develops thicker soil deposits (transport-limited dynamics). Bimodality may emerge in intermediate conditions [U-shaped distributions in Fig. 4.15(a) and Fig. 4.15(b), solid curve], indicating that the system has two preferential states (either no soil mantle or relatively deep soil deposits), whereas intermediate conditions have a low probability of occurrence. These systems have a high likelihood to be in either transport-limited or weathering-limited conditions as suggested by D'Odorico (2000), who investigated the same dynamics of soil development with a more complex form of the function $f(H)$ obtaining qualitatively similar results. The transition times between the modes of the bistable soil dynamics was investigated by D'Odorico et al. (2001) in terms of mean-first-passage times.

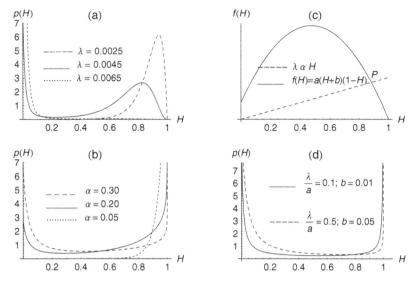

Figure 4.15. (a), (b) Probability distribution of soil thickness H calculated with (4.17) for different values of λ and α: (a) $\alpha = 0.05$, $a = 0.0012$, and $b = 0.05$; (b) $\lambda = 0.001$, $a = 0.0012$, and $b = 0.05$; (c) stable state of the deterministic process with state-dependent erosion rate $\lambda \alpha H$: The stable state is obtained as intersection (point P in the figure) between the soil-production function $f(H)$ and the deterministic erosion rate. (d) Probability distribution of soil thickness calculated with (4.18).

We can investigate the deterministic counterpart of these dynamics by replacing the stochastic rates of erosion with their average value $\lambda \alpha$. Thus these deterministic dynamics have only one stable state between 0 and 1, which is determined as the solution of the equation $f(H) - \lambda \alpha = 0$. In the case of stochastic dynamics, a second preferential state emerges at $H = 0$. This state is induced by the bound of the multiplicative Poisson noise ($h \leq H$), which translates into the existence of a probability "atom" at $h = H$ in the distribution $p(h)$ of h (see Fig. 4.14). Because this bound corresponds to a state dependency in the effect of the random forcing on the process $H(t)$, the dynamics of H are driven by multiplicative noise. Moreover, although the stable state of the deterministic dynamics is always smaller than 1, Fig. 4.15(b) shows examples of bistable stochastic dynamics with a stable state at $H = 1$. Thus the stable states of the stochastic and deterministic dynamics are not the same.

We now consider a similar process in which landslides are modeled as random (Poisson) occurrences, with each landslide removing the whole soil column. In fact, observational evidence indicates that the failure surface usually coincides with the regolith–bedrock interface (e.g., O'Loughlin and Pearce, 1976; Sidle and Swanston, 1982; Trustrum and De Rose, 1988). In this case the distribution of the Poissonian jumps h is $p(h) = \delta(h - H)$. Alternatively, the process could be expressed by (4.9) with $g(H) = -H$ and constant values of $h = 1$. The steady-state pdf of H (D'Odorico and Fagherazzi, 2003) is

$$p(H) = C(H + b)^{-\frac{\lambda}{a}-1}(1 - H)^{\frac{\lambda}{a}-1}, \qquad (4.18)$$

where $C = (\lambda/a)(1 + b)b^{\lambda/a}$. We can obtain the deterministic counterpart of this process by replacing the stochastic rate of erosion with its mean value λH. Thus the underlying deterministic process has only one stable state and no unstable state within the [0, 1] interval [Fig. 4.15(c), dashed curve with $\alpha = 1$], and the stochastic dynamics have a U-shaped distribution [Fig. 4.15(d)], suggesting that there is a statistically unstable state between 0 and 1; 0 and 1 are the stable states of the stochastic process. This result is noise induced and resembles the case of noise-induced bistability discussed in the previous two sections. The Poisson noise is able to convert a deterministic system with only one stable state into a stochastic one with two preferential states and an intermediate unstable state. This effect is due to the term $\lambda \alpha g'(H)$ in the equation, i.e., to the multiplicative character of noise (Eq. 2.69).

4.3.4 Stochastic soil-moisture dynamics

Stochastic differential equations driven by white shot noise have been used in recent years to model the soil-water balance (Rodriguez-Iturbe et al., 1999b; Laio et al., 2001), providing a theoretical framework to investigate the effect of climate, soils, and vegetation on soil-moisture dynamics in the root zone. The probability distributions of the soil-water content determined with this model have been used in a number of ecohydrological applications, including the study of water-stress conditions in vegetation, the modeling of land–atmosphere interactions, and the analysis of the hydrologic controls on photosynthesis and nutrient cycling (see Rodriguez-Iturbe and Porporato, 2005). In this subsection we use this model to investigate the effect of soil-moisture–precipitation feedbacks on the dynamics of soil moisture. Following Laio et al. (2001), we consider the water balance of a surface soil layer of depth Z_r (Fig. 4.16). The only input to the soil-water balance is due to precipitation, which is modeled as a stochastic process. We use a Poisson process of rainfall occurrences at rate λ (average storm frequency), with each storm having an exponentially distributed random depth h with mean storm depth α. The output is due to evapotranspiration and leakage, which are modeled as deterministic functions of soil moisture (Fig. 4.17). We express soil-moisture dynamics through a stochastic differential equation expressing the soil-water balance at a point (Laio et al., 2001) as

$$nZ_r \frac{ds(t)}{dt} = \varphi[s(t), t] - \chi[s(t)], \qquad (4.19)$$

where s is soil moisture, t is time, n is the soil porosity, Z_r is the active soil depth (i.e., the root zone, which is active in the exchange of water with the overlying atmosphere),

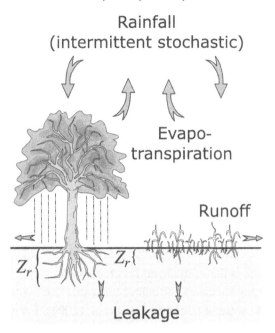

Figure 4.16. Soil-water balance at a point expressed in terms of depth-average soil moisture s in the root zone Z_r.

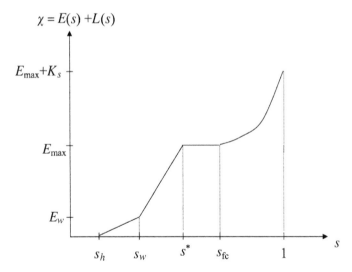

Figure 4.17. Schematic representation of the dependence of evapotranspiration and leakage losses on soil moisture s. Leakage occurs only when soil moisture exceeds field capacity s_{fc}. Transpiration is not limited by s for $s > s^*$. For $s < s^*$ plants are in a state of water stress and transpiration is a decreasing function of s. Transpiration is zero when s is smaller than the wilting point s_w. For s ranging between the hygroscopic point s_h and the wilting point s_w, soil evaporation is the only contributor to soil-moisture losses.

$\varphi[s(t), t]$ is the rate of rainfall infiltration, and $\chi[s(t)]$ represents water losses that are due to evapotranspiration $E[s(t)]$ and leakage $L[s(t)]$. The infiltration rate φ is a state-dependent Poisson process because only a fraction of the rainfall amount infiltrates the ground when the soil is close to saturation. Thus the probability distribution of the depth h' of infiltrating rainfall normalized with respect to nZ_r (i.e., $h' = h/nZ_r$) is (Laio et al., 2001)

$$b(h'; s) = \gamma e^{-\gamma h'} + \delta(h' - 1 + s) \int_{1-s}^{\infty} \gamma e^{-\gamma u} \mathrm{d}u, \qquad (4.20)$$

where $\gamma = nZ_r/\alpha$ and $\delta(\cdot)$ is the Dirac delta function. Equation (4.19) can be rewritten as

$$\frac{\mathrm{d}s}{\mathrm{d}t} = \frac{\varphi(s(t), t)}{nZ_r} - \rho(s), \qquad (4.21)$$

where $\rho(s) = \chi(s)/nZ_r$ is the normalized loss function.

Solutions of (4.21) for the case of λ independent of s (i.e., with no soil-moisture–atmosphere feedbacks) were obtained by Laio et al. (2001). Feedbacks between soil moisture and precipitation can be accounted for (D'Odorico and Porporato, 2004) through a state dependency on storm frequency (see Fig. 4.18). We can determine the pdf of soil moisture by solving (4.21) as shown in Chapter 2:

$$p(s) = \frac{C}{\rho(s)} \exp\left[-\gamma s + \int_s \frac{\lambda(u)}{\rho(u)} \mathrm{d}u\right]. \qquad (4.22)$$

In the soil-water balance, the function $\chi[s(t)]$ is the sum of losses that are due to evapotraspiration and leakage, both of which act as soil-drying processes. Daily evapotraspiration is expressed as (Laio et al., 2001)

$$E(s) = \begin{cases} 0 & 0 < s(t) \leq s_h \\ E_w \frac{s(t) - s_h}{s_w - s_h} & s_h < s(t) \leq s_w \\ E_w + (E_{\max} - E_w) \frac{s(t) - s_w}{s^* - s_w} & s_w < s(t) \leq s^* \\ E_{\max} & s^* < s(t) \leq 1, \end{cases} \qquad (4.23)$$

where s^* is the soil-moisture value below which plant transpiration is reduced by stomatal closure, s_w is the soil-water content at the wilting point, s_h is soil moisture at the hygroscopic point, E_w is the soil-evaporation rate, and $E_{\max}$ is the maximum evapotranspiration reached at s^*. s^* and s_w depend on both soil and vegetation characteristics. Thus, for values of soil moisture exceeding s^*, evapotranspiration is not limited by the soil-water content and occurs at a maximum rate $E_{\max}$. As s decreases below s^*, plants undergo a state of water stress (Porporato et al., 2001) and linearly reduce the rate of evapotranspiration, which becomes zero when s is at the wilting point. For $s < s_w$, no transpiration occurs and all losses are due only to soil evaporation, which becomes zero when s reaches the hygroscopic point s_h. Leakage losses

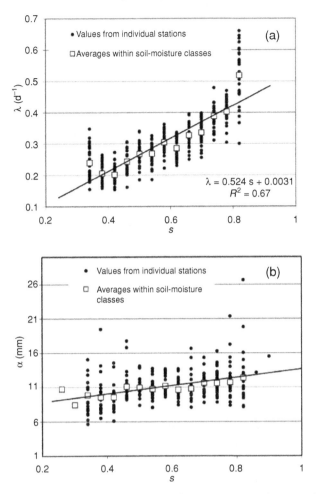

Figure 4.18. Dependence between average soil moisture (top 50 cm) and (a) average storm frequency λ and (b) depth α at 27 rainfall stations uniformly distributed across Illinois (three stations for each climate division). The solid dots indicate λ and α values at individual stations, and the squares represent statewide averages (i.e., $\langle\lambda\rangle$ and $\langle\alpha\rangle$) within soil-moisture classes. The solid lines represent a linear fit (after D'Odorico and Porporato, 2004).

$L(s)$ are null when soil moisture is below field capacity s_{fc}, whereas, when $s > s_{\mathrm{fc}}$, leakage is modeled (Laio et al., 2001) as

$$L(s) = K(s) = \frac{K_s}{e^{\beta(1-s_{\mathrm{fc}})}}\{e^{\beta[s(t)-s_{\mathrm{fc}}]}\}, \quad [s_{\mathrm{fc}} < s(t) \le 1], \qquad (4.24)$$

where β is a parameter of the soil-water retention curves (Laio et al., 2001), and K_s is the saturated hydraulic conductivity. The parameters β, K_s, and s_{fc} depend on only the soil properties. We may obtain total losses by combining Eqs. (4.23) and (4.24), $\rho(s) = [E(s) + L(s)]/nZ_r$ (Fig. 4.17).

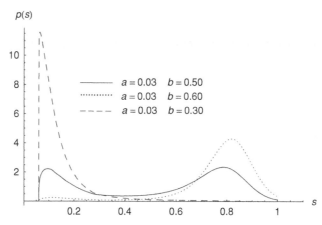

Figure 4.19. Probability distribution of soil moisture calculated with different parameters of the feedback function $\lambda = a + bs$. The bimodal behavior disappears when the sensitivity of storm frequency to soil moisture (i.e., the parameter b) is either too weak (dashed curve) or too strong (dotted curve); in intermediate conditions the probability distribution has a well-defined bimodality. The soil parameters are typical of loamy sand ($s_h = 0.06$, $s_w = 0.11$, $s^* = 0.28$, $s_{fc} = 0.31$, $\beta = 12.76$, $n = 0.44$, $K_s = 2.0$ m d^{-1}). The other parameters are $E_{max} = 2.5$ m d^{-1}, $E_w = 0.05$ m d^{-1}, $\alpha = 11.0$ mm/storm, $Z_r = 50$ cm (after D'Odorico and Porporato, 2004).

In the absence of feedback between soil moisture and precipitation, the probability distribution of soil moisture is typically unimodal. Positive feedbacks between soil moisture and precipitation (e.g., Brubaker and Entekhabi, 1995; Findell and Eltahir, 1997, 2003; Alfieri et al., 2008) can occur during the growing season at midlow latitudes. These feedbacks have been found to affect more the number of storms in the growing season (i.e., the parameter λ) than their magnitude (i.e., the mean storm depth α), as shown in Fig. 4.18 for the case of Illinois. Thus the occurrence of soil-moisture–precipitation feedbacks translates into a state dependency in the parameter λ, which, in turn, has been shown to lead to the emergence of bimodality in $p(s)$ under certain combinations of soil, plant, and climate conditions (Rodriguez-Iturbe et al., 1991; D'Odorico and Porporato, 2004). Thus soil-moisture–precipitation feedbacks could cause the emergence of two preferential states in summer soil-moisture dynamics. Higher (lower) soil-water contents would be associated with higher (lower) probabilities of rainfall occurrence, which would keep the system locked – for a relatively long fraction of the warm season – in a wet (dry) state. The effect of noise (i.e., randomness of precipitation) on these dynamics is to cause transitions between these two preferential states, whereas intermediate conditions would correspond to unstable states with the lowest probability of occurrence (D'Odorico and Porporato, 2004).

Figure 4.19 shows some examples of probability distributions of soil moisture calculated for a loamy sand soil by use of different parameters for the feedback function

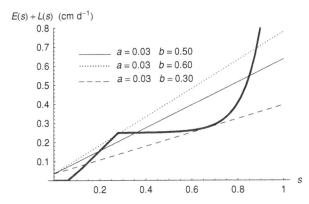

Figure 4.20. Stable and unstable states of the deterministic process shown as intersections of the loss curve (thick curve), $E(s) + L(s)$, with the average input rate $\lambda(s)\alpha^* = (a + bs)\alpha^*$ for different values of b. The parameters are the same as in Fig. 4.19.

$\lambda(s) = a + bs$. The solid-curve plot shows very well-defined bimodal behavior. The emergence of these two preferential states is important from a climatologic perspective because it suggests that, because of soil-moisture–precipitation feedbacks, the system is more likely to be in either "dry" or "wet" conditions, whereas long-term average climatic conditions have the lowest probability of occurrence. Thus in some regions the common use of long-term average parameters to characterize the hydroclimatic regime would not provide meaningful information on soil-moisture dynamics.

In Chapter 3 we have also stressed that, because of the dependency of λ on s, the WSN in Eq. (4.9) is multiplicative. Because multiplicative random drivers are known for their ability to cause noise-induced transitions (see Chapter 3) we could wonder whether the bistability (i.e., bimodality) found in $p(s)$ (Fig. 4.19) emerges as a noise-induced effect or as a result of the soil-moisture–precipitation feedbacks. To address this point, we study the deterministic counterpart of Eq. (4.19), which we obtain by replacing the noise term with its mean value $\lambda(s)\alpha^*$:

$$\frac{ds}{dt} = \frac{\alpha^*}{nZ_r}\lambda(s) - \rho(s), \qquad (4.25)$$

where α^* is the mean of $h = nZ_r h'$ [from Eq. (4.20)], which in general might differ from α because the condition $s \leq 1$ imposes a bound to the amount of water that can infiltrate the ground. The equilibria of the deterministic dynamics are given by the condition $\rho(s) = \alpha^*/nZ_r\lambda(s)$, i.e., by intersections of the $E(s) + L(s)$ curve (Fig. 4.20) with the line $\alpha^*\lambda(s) = \alpha^*(a + bs)$ (Fig. 4.20, thin straight line). In the absence of feedbacks ($b = 0$) this line is horizontal and only one intersection (stable state) exists. With suitable slopes of the $\lambda(s)$ relation, multiple intersections may exist, as shown in Fig. 4.20 (dotted line): In this case the deterministic dynamics already exhibit three equilibria corresponding to two stable states separated by an unstable state. Thus we conclude that the bimodal behavior of $p(s)$ does not emerge as a noise-induced effect but as a result of the positive feedback between soil moisture and precipitation.

These results differ from those reported in Section 4.6, in which we show the emergence of bimodality as a noise-induced effect associated with interannual climate fluctuations, without invoking feedbacks between soil moisture and precipitation (D'Odorico et al., 2000).

4.4 Environmental systems forced by Gaussian white noise

This section presents some examples of noise-induced phenomena in systems forced by Gaussian white noise:

$$\frac{d\phi}{dt} = f(\phi) + g(\phi)\xi_{gn}, \tag{4.26}$$

where ξ_{gn} is a zero-mean white Gaussian noise with intensity s_{gn}. Because the seminal work by Horsthemke and Lefever (1984) concentrated mainly on the case of Gaussian noise, here we discuss only a few examples that are relevant to the biogeosciences and point the reader to that treatise for a more detailed discussion of these models.

4.4.1 Harvest process driven by Gaussian white noise

One of the simplest models of population dynamics is expressed by the logistic equation, or *Verhulst process* (e.g., Murray, 2002):

$$\frac{dA}{dt} = aA(A_c - A), \tag{4.27}$$

where the parameter a (also known as the *reproduction rate*) determines the growth rate and the carrying capacity A_c represents the maximum sustainable value of the state variable A. A_c typically depends on the available resources (energy, water, nutrients, or light) and environmental conditions.

To account for the effect of disturbances on the dynamics of A, a more general model, known as a *harvest process*, is often used, which also accounts for the additional control on population growth exerted by disturbances, emigration, biomass harvesting, or predators. As noted in Subsection 4.3.2, harvest models typically assume that the harvest rate is proportional to A (e.g., Kot, 2001):

$$\frac{dA}{dt} = aA(A_c - A) - kA. \tag{4.28}$$

Equation (4.28) has only one stable state, $A = A_c - k/a$. In this subsection we consider the effect of noise on these dynamics. We consider the case of a harvest model in which parameter a is interpreted as a white Gaussian random variable with mean

a_0 and intensity s_{gn}; i.e., $a = a_0 + \xi_{gn}$. Thus the stochastic dynamics can be expressed as (4.26) with

$$f(A) = a_0 A(A_c - A) - kA, \qquad g(A) = A(A_c - A). \qquad (4.29)$$

This model is well suited to investigate the possible emergence of noise-induced transitions as an effect of random environmental fluctuations in the growth rate of the dynamics. In what follows we normalize A with respect to A_c and use the state variable $A' = A/A_c$. Moreover, to simplify the notation we drop the prime superscript.

The steady-state pdf of A can be calculated with Eq. (2.83) from Chapter 2:

$$p(A) = Ce^{-\frac{1}{s_{gn}}\frac{k}{1-A}} A^{\frac{1}{s_{gn}}(a_0-k)-1}(1 - A)^{-\frac{1}{s_{gn}}(a_0-k)-1}, \qquad (4.30)$$

where C is the normalization constant. It can be shown that when $k = 0$ the normalization constant C is not finite and $p(A) = \delta(A - 1)$. In all the other cases ($k \neq 0$), as the noise intensity (i.e., s_{gn}) increases, the dynamics undergo a noise-induced transition from a system with only one stable state (for the deterministic system or for low values of s_{gn}) to bistable dynamics. Figure 4.21(a) shows the modes of A calculated with Eq. (3.48): As s_{gn} exceeds a critical value $s_c = (a_0 - k)$, a noise-induced transition occurs, in that multiple equilibria emerge and the probability distribution of A becomes bimodal [Fig. 4.21(b)]. Thus the multiplicative noise is able to induce a new stable state $A = 0$ that did not exist in the underlying deterministic system.

4.4.2 Stochastic genetic model

Logistic harvest model (4.29) can be generalized to account for possible inputs of A occurring at constant rate β. Known as a *genetic model* (Horsthemke and Lefever, 1984), this process finds a number of applications to population biology and dynamic ecology. For example, the genetic model could describe the dynamics of a population affected by logistic growth, state-dependent emigration or harvest, and state-independent immigration (β). If, for the sake of simplicity, A is normalized with respect to A_c in a way that it ranges between 0 and 1, these dynamics are expressed by

$$\frac{dA}{dt} = aA(1 - A) + \beta - kA. \qquad (4.31)$$

Dividing both sides of (4.31) by k, normalizing the model's parameters $a' = a/k$, $\beta' = \beta/k$, rescaling the time variable ($t' = kt$), and dropping the prime superscripts, we obtain

$$\frac{dA}{dt} = aA(1 - A) + \beta - A. \qquad (4.32)$$

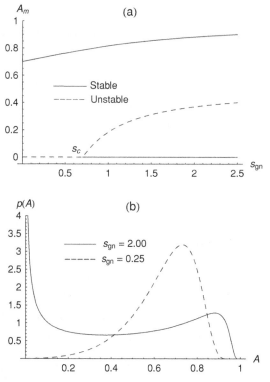

Figure 4.21. (a) Stable and unstable states of the stochastic harvest process with random reproduction rate a with Gaussian distribution, mean a_0, and intensity s_{gn} ($a_0 = 1$, $k = 0.3$, and $A_c = 1$). (b) The pdf of state variable A of the stochastic harvest process with multiplicative noise [same parameters as in (a)].

This deterministic model has at most two equilibrium points (i.e., a stable and an unstable state), which are roots of the second-order polynomial on the right-hand side of (4.32). With parameter a treated as a random variable (mean α_0 and intensity s_{gn}), Eq. (4.31) becomes

$$\frac{dA}{dt} = a_0 A(1 - A) + \beta - A + A(1 - A)\xi_{gn}. \tag{4.33}$$

The properties of this stochastic genetic model were investigated in detail by Horsthemke and Lefever (1984) and Lefever (1990). As in the previous case the right-hand side of Eq. (3.48) is a polynomial, which increases its order by one as s_{gn} increases from zero to $s_{gn} > 0$. Thus noise might be capable of increasing the number of stable states of the system. Figure 4.22(a) shows the equilibrium states of (4.33) as a function of noise intensity (i.e., of s_{gn}). As s_{gn} exceeds a critical value s_c, the system undergoes a noise-induced transition: The stable state of the underlying deterministic dynamics

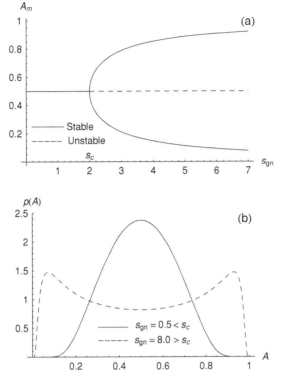

Figure 4.22. Stochastic genetic model: (a) stable and unstable states of the system as a function of the noise intensity (s_{gn}), (b) pdf of ϕ with $a_0 = 0$, $\beta = 0.5$ (in this case $s_c = 2$).

becomes unstable and two new stable states emerge. The steady-state probability distribution of A can be determined from Eq. (2.83):

$$p(A) = Ce^{-\frac{1}{s_{gn}}\left(\frac{1-\beta}{1-A}+\frac{\beta}{A}\right)} A^{\frac{1}{s_{gn}}(a_0+2\beta-1)-1}(1-A)^{-\frac{1}{s_{gn}}(a_0+2\beta-1)-1}, \qquad (4.34)$$

where C is the normalization constant. Figure 4.22(b) shows some examples of probability distributions of A, which become bimodal when s_{gn} exceeds s_c, indicating the occurrence of noise-induced bistability.

We note that in a number of systems the harvest rate is a nonlinear function of A. For example, the consumption by predators is likely to reach saturation for high values of A as the predator needs are met (Ludwig et al., 1978). To account for this saturation effect the harvest rate can be expressed as $kA^b/(1 + A^b)$, as suggested for the case of insect outbreak dynamics (Ludwig et al., 1978). Interesting noise-induced transitions may also emerge in these systems when in Eq. (4.28) the coefficient k (instead of a) is treated as a (Gaussian) random variable to account for the effect of rapidly varying (random) environmental conditions. These dynamics were investigated in detail by Horsthemke and Lefever (1984) in the context of predator–prey systems. We refer the interested reader to this book and to Ludwig et al. (1978) for more details on this process.

4.4.3 Noise-induced extinction

The effect of random fluctuations in the carrying capacity (Levins, 1969; May, 1973) may induce interesting phenomena in the logistic process. May (1973) treated the carrying capacity in Eq. (4.27) as a normally distributed random variable, $A_c = A_0 + \xi_{gn}$, with mean $A_0 > 0$ and intensity s_{gn}, and determined the steady-state probability distribution of A. When the noise intensity (i.e., s_{gn}) is relatively small, the random environment induces stochastic fluctuations of A about its long-term mean $\langle A \rangle$. Moreover, $\langle A \rangle$ is smaller than A_0 and decreases with increasing values of s_{gn}, suggesting that noise has the effect of reducing the mean value of A with respect to the carrying capacity A_0 of the underlying deterministic dynamics. As s_{gn} exceeds a critical-threshold value, the random fluctuations become so intense that no steady-state pdf of A exists, and the system tends to extinction. The occurrence of extinction as the noise intensity exceeds a critical value is clearly a noise-induced phenomenon.

When $A_c = A_0 + \xi_{gn}$, Eq. (4.27) can be rewritten as (4.26) with rescaled time $t' = a\,t$, and

$$f(A) = A(A_0 - A), \quad g(A) = A. \tag{4.35}$$

Notice that, because for $A = 0$ $g(A) = 0$, the dynamics have an intrinsic boundary at $A = 0$; it can be shown (Horsthemke and Lefever, 1984) that this boundary is also a natural boundary (i.e., it cannot be accessed by A) if $s_{gn} < A_0$.

Conditions conducive to noise-induced extinction were determined by May (1973), who used Ito's interpretation of the stochastic equation (see Chapter 2), leading to the steady-state probability distribution of A:

$$p(A) = C^{-1} \left(\frac{A}{s_{gn}} \right)^{\frac{A_0}{s_{gn}} - 2} e^{-\frac{A}{s_{gn}}}, \tag{4.36}$$

with normalization constant

$$C = s_{gn} \Gamma \left(\frac{A_0}{s_{gn}} - 1 \right) \tag{4.37}$$

and where $\Gamma(\cdot)$ is the gamma function. Equation (4.37) is obtained from the integration of $p(A)$ in $[0, +\infty)$. Notice that C is finite only if $A_0/s_{gn} > 1$. When s_{gn} exceeds A_0 there is no finite value of C and no equilibrium solution exists. In this case, $A = 0$ is no longer a natural boundary, in that the whole probability mass is concentrated at this intrinsic boundary; the pdf of A becomes $p(A) = \delta(A)$, where $\delta(\cdot)$ is the Dirac delta function (Horsthemke and Lefever, 1984). Thus when the intensity s_{gn} of the

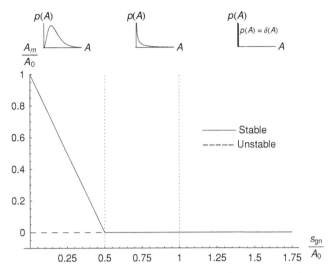

Figure 4.23. Transition points in the stochastic *Verhulst model* (Ito's interpretation), i.e., of a logistic process with random normally distributed carrying capacity.

environmental fluctuations exceeds A_0, A tends to extinction with a probability of one. Another transition point can be found with respect to the shape of $p(A)$. In fact, the mode (i.e., preferential "stable" state) A_m of A [which we obtain by setting equal to zero the first-order derivative of the right-hand-side of (4.36)] is

$$A_m = A_0 - 2s_{gn} \quad \text{for} \quad s_{gn} < A_0/2, \tag{4.38}$$

and $A = 0$ is an unstable state of the system. When s_{gn} exceeds $A_0/2$, the stable state is $A = 0$. Figure 4.23 shows the stable and unstable states of the system as a function of the intensity of the environmental fluctuations: It can be observed that the qualitative properties of the dynamics change as an effect of intensification of random fluctuations with two transition points (Horsthemke and Lefever, 1984). As s_{gn} exceeds $A_0/2$, $A = 0$ becomes the preferential state of the dynamics, whereas for $s_{gn} > A_0$ the steady state of the system $A = 0$ (extinction) occurs with a probability that tends to one.

To be consistent with the other analytical results presented in this chapter, we look at the results obtained when Eqs. (4.26)–(4.35) are integrated by using the Stratonovich interpretation (see Chapter 2). In this case, the pdf of A differs from the one obtained by May (1973):

$$p(A) = C^{-1} \left(\frac{A}{s_{gn}} \right)^{\frac{A_0}{s_{gn}} - 1} e^{-\frac{A}{s_{gn}}} \quad \text{with} \quad C = s_{gn} \Gamma \left(\frac{A_0}{s_{gn}} \right). \tag{4.39}$$

The normalization constant C is finite for $A_0 > 0$ (independent of s_{gn}). The modes of (4.39) can be easily obtained with Eq. (3.48):

$$A_m = A_0 - s_{gn} \quad \text{for} \quad s_{gn} < A_0/2, \tag{4.40}$$

whereas if s_{gn} exceeds $A_0/2$, the mode is zero. Thus, in the Stratonovich interpretation, this second transition in the shape of the distribution still exists, is controlled by the noise intensity, and is induced by the random fluctuations (i.e., it is a noise-induced transition). However, for $A_0 > 0$ (i.e., in the typical ecological applications of this logistic-Verhulst process to population dynamics) no noise-induced extinction (i.e., the noise-induced attainment of $A = 0$ with probability $\rightarrow 1$) occurs. Thus in this case the interpretation rule seems to play a key role in the noise-induced behaviors of these dynamics (e.g., Turelli, 1978; Braumann, 1983).

Similar results can be obtained with a (stochastic) Malthusian growth model

$$\frac{dA}{dt} = (r + \xi_{gn})A, \tag{4.41}$$

where r is the average growth rate. Using Ito's calculus, we can observe that, in the presence of relatively strong fluctuations in the growth rate (i.e., $s_{gn} > r$), extinction occurs with probability tending to one even with positive values of the mean growth rate r, i.e., when in model (4.41) the expected value of A grows indefinitely with time (Lewontin and Cohen, 1969; Braumann, 1983). However, these dynamical properties do not appear when Stratonovich's calculus is used. In this case the probability of extinction tends to one when $r < 0$, i.e., with negative values of the average growth rate. This condition is independent of s_{gn}, i.e., of the intensity of random environmental fluctuations. Thus, with Stratonovich's interpretation of (4.41), extinction does not occur as a noise-induced effect.

Overall, this discussion stresses how the choice of either Ito's or Stratonovich's interpretation may lead to dramatically different results. It is therefore crucial to choose the correct interpretation rule while solving stochastic differential equations. As noted in Chapter 2, Stratonovich's calculus prevents the emergence of biases. However, these biases appear only in the case of systems driven by multiplicative white noise. Therefore, in the case of systems forced by multiplicative white noise, Stratonovich's interpretation should preferably be used (van Kampen, 1981). In fact, in Chapter 2 we showed that Stratonovich's rule is consistent with the interpretation of white Gaussian noise and WSN as limit cases of DMN. Alternatively, Braumann (1983, 2007), showed that the two methods would lead to the same results when the growth-rate parameter r is adequately reinterpreted.

The model of noise-induced extinction presented in the previous section treats the carrying capacity as a random variable to account for stochastic fluctuations inherent in the variability of the limiting resources. In particular, in that case the carrying capacity was modeled as white Gaussian noise. Thus it was implicitly accepted that, although the mean is larger than zero, the random fluctuations may lead to negative values of the carrying capacity. These negative fluctuations could be responsible for the effects of noise-induced extinction previously discussed. These negative values have limited physical meaning when the state variable is defined as a positive quantity (e.g., population density, vegetation cover, or biomass). To assess whether noise-induced extinction can emerge even when the carrying capacity A_c is modeled as a positive random variable, we express A_c as the exponential of a white Gaussian noise. Analytical solutions are difficult to obtain for the stochastic logistic process; however, an analytical expression for the steady-state probability distribution of the state variable can be determined in the case of the *Gompertz process* (e.g., Goel et al., 1971):

$$\frac{dA}{dt} = -aA \log \frac{A}{A_c}. \tag{4.42}$$

This process resembles to some extent Verhulst (logistic) process (4.27) in that they both model the growth of a population with carrying capacity A_c.[4]

To solve stochastic equation (4.42) we consider the auxiliary variable $B = \log(A)$ and define the parameter $B_c = \log(A_c)$. Equation (4.42) can be rewritten as

$$\frac{dB}{dt} = a(B_c - B). \tag{4.45}$$

Thus we can solve Eq. (4.42) by finding a solution for Eq. (4.45) using the analytical solution for Eq. (4.26), driven by Gaussian white noise. Using Stratonovich's interpretation, we obtain the probability distribution of B [see Eq. (2.83)]. The distribution of A is then determined as a derived distribution of $p(B)$. It is found that the mode of A decreases with increasing noise intensities. However, this mode always remains positive and different from zero. Thus the stochastic Gompertz process does not exhibit the noise-induced transitions that were reported for the case of the logistic process.

---

[4] In both processes the growth rate is proportional to A and to a function $G(A, A_c)$, which accounts for a decrease in the growth rate (*saturation effect*) as the population approaches its carrying capacity:

$$\frac{dA}{dt} = aAG(A, A_c), \tag{4.43}$$

with

$$G(A, A_c) = \frac{\left(1 - \frac{A}{A_c}\right)^{1-q}}{1 - q}. \tag{4.44}$$

The parameter q is zero in the case of the logistic process, whereas it tends to 1 in the Gompertz process (e.g., Goel et al., 1971).

4.4.4 *Noise-induced bistability in climate dynamics:*
Effect of land–atmosphere interactions

The interaction of multiplicative Gaussian noise with the nonlinear dynamics of the soil-water balance was investigated by Rodriguez-Iturbe et al. (1991) through a minimalist model of soil-moisture dynamics at the regional-to-semicontinental scale:

$$nZ_r \frac{\mathrm{d}s}{\mathrm{d}t} = P\Phi(s) - E(s), \tag{4.46}$$

where n is the soil porosity, Z_r is the depth of the root zone (i.e., of the surface layer of soil that is active in water exchanges with the atmosphere), s is the relative soil moisture ($0 < s \leq 1$), P is the rainfall rate, $\Phi(s)$ is an infiltration function expressing the fraction of P that infiltrates the ground, and $E(s)$ is the evapotranspiration rate. E depends on the soil-water content and decreases from its maximum value $E_{\max}$ to zero as s tends to zero. Known as potential evapotranspiration, $E_{\max}$ represents the maximum rate of evapotranspiration when soil moisture is not limiting and depends on atmospheric conditions and solar irradiance. The infiltration function $\Phi(s)$ accounts for the increase in infiltration capacity with increasing soil storage capacity. Thus $\Phi(s)$ is a decreasing function of soil moisture. Rodriguez-Iturbe et al. (1991) modeled $E(s)$ and $\Phi(s)$ as

$$E(s) = E_{\max} s^c, \qquad \Phi(s) = 1 - \epsilon s^r, \tag{4.47}$$

where c depends on the land cover, whereas ϵ and r are nonnegative constant parameters. Interesting dynamics emerge from Eq. (4.46) when the rainfall rate is expressed as a function of soil moisture, i.e., $P = P(s)$, to account for the state dependence of precipitation induced by land–atmosphere interactions. This dependence would be due to the fact that in a given region a fraction P_m of P is contributed by locally recycled moisture, i.e., by moisture from local evapotranspiration, and the remaining precipitation P_a is from atmospheric moisture advected from outside of the region (e.g., Salati et al., 1979; Eltahir and Bras, 1996; Trenberth, 1998). Thus the precipitation P_m from local recycling is a (linear) function of the evapotranspiration rate $P_m = kE$, where k is a proportionality constant that depends on the region's size, the mean wind velocity, and precipitable moisture. Because the evapotranspiration rate depends on soil moisture [Eqs. (4.47)], precipitation recycling induces a state dependency in the overall rainfall regime:

$$P(s) = P_a + P_m = P_a(1 + \alpha s^c), \tag{4.48}$$

where $\alpha = kE_{\max}$ expresses the strength of the feedback. The rainfall rate is affected by climate fluctuations. Rodriguez-Iturbe et al. (1991) noted that, because advected precipitation is controlled by evaporation from the oceans, it has a relatively small CV, whereas precipitation recycling undergoes stronger environmental fluctuations.

Thus the coupling parameter α is treated as a Gaussian random variable with mean α_0 and intensity s_{gn}:

$$\alpha = \alpha_0 + \xi_{gn}. \tag{4.49}$$

Inserting Eqs. (4.47)–(4.49) into Eq. (4.46), we obtain the analog of Eq. (4.26) for the soil-moisture variable s:

$$\frac{ds}{dt} = f(s) + g(s)\xi_{gn}, \tag{4.50}$$

where

$$f(s) = a(1 + \alpha_0 s^c)(1 - \epsilon s^r) - bs^c, \quad g(s) = as^c(1 - \epsilon s^r), \tag{4.51}$$

and a and b are the rates of advective precipitation and potential evapotranspiration normalized with respect to the effective soil thickness nZ_r, i.e., $a = P_a/(nZ_r)$ and $b = E_{max}/(nZ_r)$. Notice that the dynamics are intrinsically bounded between 0 and 1.

The solution of stochastic differential Eq. (4.50) leads to the steady-state probability distribution $p(s)$ of soil moisture. Rodriguez-Iturbe et al. (1991) calculated $p(s)$ by using Ito's interpretation. To be consistent with the other results presented in this chapter, here we use Stratonovich's interpretation and note that in this case the results do not qualitatively depend on the interpretation as the system undergoes the same noise-induced transitions regardless of the interpretation rule for the multiplicative-noise term. Thus $p(s)$ can be calculated with Eq. (2.83). An analytical expression of $p(s)$ can be determined, though it is not reported here as it is fairly long and cumbersome. Rather, we show the results for the particular case in which $r = 2$ and $c = 1$:

$$p(s) = C s^{\frac{1}{as_{gn}}(a-\frac{b}{a})-1}(1-s)^{-\frac{1}{2as_{gn}}(1+a-\frac{b}{a})-1}$$
$$\times (1+s)^{\frac{1}{2as_{gn}}(1-\alpha_0+\frac{b}{a})-1} e^{-\frac{1}{2as_{gn}}\left[\frac{2}{s}+\frac{b}{a(1-s^2)}\right]}, \tag{4.52}$$

where C is the normalization constant. Figure 4.24 shows $p(s)$ calculated with the same parameters as in Rodriguez-Iturbe et al. (1991). The modes of the distribution are obtained with Eq. (3.48). When $r = 2$ and $c = 1$, Eq. (3.48) has five roots. In the numerical example of Fig. 4.24 only one root is real for $s_{gn} = 0$, indicating that the deterministic dynamics have only one stable state. As s_{gn} increases above a critical value (i.e., $s_{gn} > 0.475$ in this example), Eq. (3.48) has three real roots corresponding to one minimum and two maxima of $p(s)$. In these conditions the distribution of s becomes bimodal. For even larger values of s_{gn} (i.e., $s_{gn} > 0.82$), Eq. (3.48) has two more real roots; however, because both of them are negative (hence are outside the domain $[0, 1]$ of s) their emergence is not associated with any other noise-induced transition. Similar results can be obtained with Ito's interpretation of the multiplicative noise, though in that case the transition occurs for a smaller value of s_{gn}.

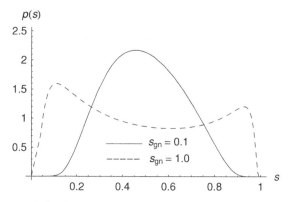

Figure 4.24. The pdf of soil moisture [Eq. (4.52)] calculated with $c = 1, r = 2, \epsilon = 1$, $nZ_r = 0.10$ m, $E_p = 1.6$ m/yr, $P_a = 0.8$ m/yr, and $\alpha = 0.637$ (same parameters as in Rodriguez-Iturbe et al., 1991).

Thus environmental fluctuations acting on soil-moisture dynamics as multiplicative noise are capable of inducing new preferential states that did not exist in the deterministic counterpart of the process (see also Demaree and Nicolis, 1990). The emergence of bimodality in the dynamics of soil moisture indicates that the system has two preferred states corresponding to relatively dry and relatively wet states, whereas intermediate conditions have low probability of occurrence. Notice how, unlike the case presented in Subsection 4.3.4, the emergence of bimodality in Eq. (4.52) is completely induced by environmental fluctuations (Rodriguez-Iturbe et al., 1991). The positive feedback between precipitation and soil moisture tends to sustain and enhance anomalously dry and wet conditions, leading to the persistence of these anomalies for relatively long times, whereas noise induces shifts from a state to the other. This minimalist model has been used to explain the persistence of climatic anomalies in low-latitude and midlatitude continental regions (e.g., Nicholson, 2000).

4.5 Environmental systems forced by colored Gaussian noise

In the previous chapters we stressed how the emergence of noise-induced transitions depends in general on both the noise intensity (i.e., the variance) and on its autocorrelation. Exact expressions for the modes of stochastic processes driven by autocorrelated (or *colored*) noise exist in the case of DMN, whereas in the case of autocorrelated Gaussian noise (or the Ornstein–Uhlenbeck process) only approximate relations are available (see Chapter 2). In this section we show how the autocorrelation of the random driver may affect the properties of environmental dynamics. To this end, we consider the case of noise-induced bistability in coupled soil-moisture dynamics described in the previous section and investigate how the results shown in Fig. 4.24

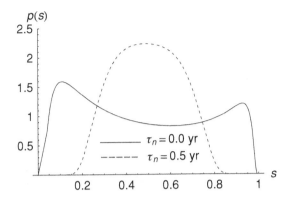

Figure 4.25. Effect of autocorrelation on the emergence of noise-induced transitions in soil-moisture dynamics in systems forced by colored Gaussian noise (same parameters as in Fig. 4.24, with $s_{gn} = 1$).

would change in the case of colored noise, i.e., if $\xi_{gn}(t)$ in Eq. (4.50) is autocorrelated as a result of *memory* in the synoptic forcings determining convective precipitation.

As noted in Chapter 2, to study the combined effect of intensity (s_{gn}) and auto-correlation scale (τ_n) on noise-induced transitions, we need to use expressions of the probability distribution of the state variable that are valid also for relatively high values of s_{gn} and τ_n. Thus we can use here an approximate expression based on the the united colored-noise approximation (see Chapter 2). This expression provides a suitable representation of the probability distribution of s in that condition (2.107) in Chapter 2 is verified. The results of this analysis show (Fig. 4.25) that even relatively small-correlation scales can have dramatic effects on the shape of the probability distribution of s. In this specific case, an increase in τ_n may turn a stochastic bistable system into a system with only one statistically stable state, suggesting that the memory of large-scale synoptic forcings may prevent the emergence of multiple preferential states in the regional climate.

The role of colored noise in multivariate environmental dynamics was recently investigated in the context of trophic interactions in population dynamics (Petchey et al., 1997; Heino, 1998; Heino et al., 2000; Laakso et al., 2001; Greenman and Benton, 2003; Xu and Li, 2003; Vasseur, 2007). Most of these studies concentrated on a particular case of autocorrelated noise, which is characterized by power spectra with power-law behavior, $1/f^\beta$. The exponent β of the spectrum is zero in the case of white noise, and values of β greater than zero correspond to autocorrelated (or colored) $1/f^\beta$ noise. Numerical simulations have shown that populations that are either more sensitive to environmental fluctuations or more prone to extreme events are less exposed to extinction risk in a fluctuating environment if the environmental noise is autocorrelated (Schwager et al., 2006). Moreover, it was found (e.g., Vasseur and Fox, 2007) that in consumer–producer populations resource variability decreases, while consumer variability increases as the spectral exponent β increases from zero (white noise) to positive values (colored noise).

4.6 Environmental systems forced by other types of noise

Noise-induced transitions may emerge also in processes forced by other types of random drivers. These transitions are typically investigated through numerical simulations rather than through exact solutions of stochastic differential equations, either because of the lack of a theory for the solution of these equations or because of the inability of the dynamics to be expressed through stochastic differential equations. Here we present the case of noise-induced transitions emerging in the distribution of average seasonal soil moisture as an effect of interannual rainfall fluctuations (D'Odorico et al., 2000). We consider the daily dynamics of soil moisture s, expressed by (4.19). To simplify this analysis we assume that no feedback exists between soil moisture and precipitation (i.e., λ independent of soil moisture). As noted, the analytical solution of (4.19) provides the probability distribution of s [Eq. (4.22)] and of its moments (not shown). We use the average value of soil-moisture during the growing season, $\langle s \rangle$, as an indicator of the seasonal soil-water content. $\langle s \rangle$ depends on the parameters representative of vegetation, soils, and rainfall regime (Rodriguez-Iturbe et al., 1999b). Most of these parameters either remain constant with time or undergo small variability, except for those characterizing the rainfall regime, i.e., the parameters λ and α expressing the average rainstorm frequency and depth. Interannual climate fluctuations determine year-to-year changes in these parameters, particularly in arid and semiarid regions, which are known for the higher variability and unpredictability of precipitation (D'Odorico et al., 2000; D'Odorico and Porporato, 2006).

To evaluate the impact of these interannual rainfall fluctuations on the year-to-year variability of $\langle s \rangle$, we treat the seasonal values of λ and α as random variables that vary from year to year. The effect of fluctuations in λ and α on the distribution of $\langle s \rangle$ is an interesting case of superstatistics (Beck, 2004) that can be solved with a Monte Carlo procedure, i.e., sampling values of λ and α from their respective distributions (assuming that these two variables are independent) and calculating the value of $\langle s \rangle$ associated with those values. A population of values of $\langle s \rangle$ is obtained when this procedure is repeated several times. The distribution of these values of $\langle s \rangle$ is the derived distribution of $\langle s \rangle$ as a function of the distributions of λ and α. Figure 4.26 shows an example of the results of this analysis: It is observed that for small variability (i.e., CV) of the parameters λ and α, the distribution is unimodal. However, as variability increases a bimodal behavior emerges. Because bimodality is observed when the noise intensity increases above a critical value, this is an example of noise-induced transition.

The results in Fig. 4.26 show that the system tends to select two preferential states and to switch between them as a result of interannual fluctuations in rainfall

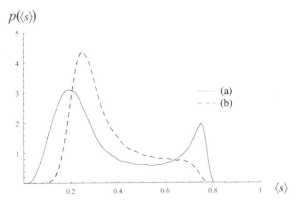

Figure 4.26. The pdf of average soil moisture during the growing season. The parameters for soil and vegetation are as follows: $n = 0.43$, $Z_r = 1.40$ m, $K_s = 9.5$ 10^{-6} m/s, $s_{\text{fc}} = 0.8$, $s^* = 0.36$, $E_{\max} = 3.2$ mm/day. The rainfall is characterized by $\langle \alpha \rangle = 12.4$ mm/storm and $\langle \lambda \rangle = 0.21$ d^{-1}, with CVs as follows: (a) CV$[\alpha] = 0.45$, CV$[\lambda] = 0.23$; (b) CV$[\alpha] = 0.22$, CV$[\lambda] = 0.11$ (after D'Odorico et al., 2000).

parameters. The two preferential states correspond to "dry" and "wet" seasonal soil-moisture conditions. The implications of the emergence of bimodal behavior is of foremost importance for ecohydrologic dynamics because it implies that the system is more likely to be found in two states that are far from the long-term average, whereas the long-term average conditions occur with low probability (D'Odorico et al., 2000). Moreover, the bimodal behavior enhances the likelihood of occurrence of dry conditions and the effect of disturbance of climate fluctuations on terrestrial ecosystems (Ridolfi et al., 2000).

4.7 Noise-induced phenomena in multivariate systems

The study of the effect of noise on the dynamics of multivariate environmental systems is limited by the lack of a general analytical framework for the solution of sets of nonlinear stochastic differential equations. In most cases, these systems are investigated through numerical simulations, though some approximate methods were also developed (e.g., May, 1973; Horsthemke and Lefever, 1984). In this section we present the case of multispecies population dynamics, investigated through either a linearization of the stochastic equations (May, 1973) or with methods based on the mean-field theory[5] (Mankin et al., 2004). Other multivariate systems are presented in this chapter in the context of populations with random interspecies interactions (Subsection 4.7.3) or with coherence-resonance behavior (Subsection 4.8.4).

---

[5] A comprehensive description of the mean-field approach is given in Chapter 5 (in particular, see Box 5.4), in which this technique is used to solve systems of stochastic ordinary differential equations obtained by spatial discretization of stochastic partial differential equations.

4.7.1 Stochastic dynamics of two competing species

In this subsection we consider the case discussed by May (1973) of two species whose populations, $A_1(t)$ and $A_2(t)$, compete for the same pool of resources according to the dynamics

$$\frac{dA_1}{dt} = A_1(A_{c_1} - A_1 - \alpha A_2),$$

$$\frac{dA_2}{dt} = A_2(A_{c_2} - A_2 - \alpha A_1), \qquad (4.53)$$

where A_{c_1} and A_{c_2} are the carrying capacities of A_1 and A_2, respectively. The steady state $[A_{1,st} = (A_{c_1} - \alpha A_{c_2})/(1 + \alpha), A_{2,st} = (A_{c_2} - \alpha A_{c_1})/(1 + \alpha)]$ is stable when the highest eigenvalue of the community matrix (see Box 4.3) has a negative real part. If $A_{c_1} = A_{c_2} = A_c$, the steady state $A_{1,st} = A_{2,st} = A_{st}$ is stable when $\Lambda = A_{st}(1 - \alpha) > 0$, i.e., when $\alpha < 1$, which is the Gause–Lotka–Volterra criterion (May, 1973).

We can now consider the case in which the parameters A_{c_1} and A_{c_2} are two white-Gaussian-noise processes with mean A_0 and intensity s_{gn}. The randomization of these parameters turns (4.53) into a set of nonlinear stochastic differential equations, which does not lend itself to exact solutions. However, we can obtain an approximate solution by expressing the state variables in (4.53) in terms of normalized deviation from the steady state, $\nu_1 = (A_1 - A_{st})/A_{st}$ and $\nu_2 = (A_2 - A_{st})/A_{st}$ and linearizing these equations (i.e., neglecting all the quadratic terms). In this case, if $\alpha < 1$, the steady-state probability distributions of ν_1 and ν_2 are both Gaussian, with zero mean and variance (May, 1973):

$$\sigma_{\nu_{1,2}}^2 = \frac{2s_{gn}}{A_0(1 - \alpha)}. \qquad (4.54)$$

Thus the effect of environmental variability (i.e., $s_{gn} \neq 0$) on the fluctuations of the state variables is enhanced as $\alpha \to 1$. It is intuitive to expect that extinction occurs when s_{gn} exceeds a threshold value, similar to the case of the univariate system discussed in Section 4.3. Similar to the univariate case, random environmental fluctuations in the carrying capacity are able to cause noise-induced extinctions. Thus, using Eq. (4.54), May (1973) suggested that extinction does not occur if $s_{gn} \ll A_0(1 - \alpha)/2$. As α increases from 0 to 1 (i.e., the strength of the competitive interactions between species increases), weaker environmental fluctuations are needed to cause noise-induced extinctions. A similar analysis can be applied to predator–prey systems with two or more species (May, 1973). More recently, the role of noise in predator–prey systems was investigated in detail with respect to the emergence of stochastic resonance and spatial patterns (Spagnolo et al., 2004). Some of these studies are reviewed in this chapter and in Chapter 6.

Box 4.3: Stability of multivariate deterministic systems

In this box we review the conditions determining the linear stability of a deterministic n-dimensional system. We indicate with $\boldsymbol{\phi} = (\phi_1, \phi_2, \dots, \phi_n)$ the state variables and express their dynamics with a set of first-order differential equations:

$$\frac{d\phi_i}{dt} = f_i(\phi_1, \phi_2, \dots, \phi_n), \qquad (i = 1, 2, \dots, n). \tag{B4.3-1}$$

To evaluate the stability of the equilibrium state $\boldsymbol{\phi}_{\text{st}} = (\phi_{1,\text{st}}, \phi_{2,\text{st}}, \dots, \phi_{n,\text{st}})$ with respect to a small perturbation of amplitude $\boldsymbol{\epsilon} = (\epsilon_1, \epsilon_2, \dots, \epsilon_n)$, we investigate the properties of the dynamics when its state $\boldsymbol{\phi}$ has a small displacement $\boldsymbol{\epsilon}$ from the equilibrium point:

$$\boldsymbol{\phi} = \boldsymbol{\phi}_{\text{st}} + \boldsymbol{\epsilon}. \tag{B4.3-2}$$

For (B4.3-2) to be a state of the system it has be a solution of (B4.3-1). Inserting (B4.3-2) into (B4.3-1), applying a Taylor expansion (truncated to the first order) of $\boldsymbol{\phi}$ about $\boldsymbol{\phi}_{\text{st}}$, and using equilibrium condition $f_i(\boldsymbol{\phi}_{\text{st}}) = 0$ $(i = 1, 2, \dots, n)$, we obtain

$$\frac{d\epsilon_i}{dt} = \sum_{j=1}^{n} a_{i,j} \, \epsilon_j, \tag{B4.3-3}$$

where $a_{i,j}$ are the coefficients of the Jacobian matrix $\mathbf{A}$ [or *community matrix* in the ecology literature (May, 1973)]:

$$a_{i,j} = \left(\frac{\partial f_i}{\partial \phi_j} \right)_{\boldsymbol{\phi} = \boldsymbol{\phi}_{\text{st}}}. \tag{B4.3-4}$$

The solution of the set of differential equations (B4.3-3) is in the form

$$\epsilon_i(t) = \sum_{j=1}^{n} C_{i,j} e^{\lambda_j t}, \tag{B4.3-5}$$

where coefficients $C_{i,j}$ depend on the initial conditions and λ_j expresses how the perturbation varies with time. The equilibrium point $\boldsymbol{\phi} = \boldsymbol{\phi}_{\text{st}}$ is stable if the perturbation decays with time, i.e., if all the values of λ_j $(j = 1, 2, \dots, n)$ are negative.

We find the values of λ_j by inserting (B4.3-5) into (B4.3-3):

$$\lambda \epsilon_i(t) = \sum_{j=1}^{n} a_{i,j} \epsilon_j(t). \tag{B4.3-6}$$

This system of equations can be rewritten in the form

$$(\mathbf{A} - \lambda \mathbf{I}) \boldsymbol{\epsilon}(t) = 0, \tag{B4.3-7}$$

where $\mathbf{I}$ is the identity matrix. This homogeneous, linear set of equations has nontrivial solutions only if the determinant of its matrix is zero:

$$|\mathbf{A} - \lambda \mathbf{I}| = 0. \tag{B4.3-8}$$

Thus λ_j $(j = 1, 2, \dots, n)$ are eigenvalues of the matrix $\mathbf{A}$. The stability of the equilibrium point $\boldsymbol{\phi} = \boldsymbol{\phi}_{\text{st}}$ requires that the real parts of these eigenvalues be negative.

4.7.2 *Phase transitions in multivariate systems driven by dichotomous noise*

We now consider the case of transitions that are not associated with the emergence of new ordered states (modes) in the probability distribution of the state variables, but with the noise-induced coexistence of two alternative stable phases represented by two different steady-state probability distributions. In this case the dynamics select either one of these phases, depending on the initial condition.

We follow the framework by Mankin et al. (2004), who investigated the dynamics of an N-species symbiotic ecological system with a generalized Verhulst self-regulation mechanism (i.e., density-dependent population growth) and symbiotic interspecies interactions. The abundance A_i ($A_i > 0$) of species i varies with time as

$$\frac{dA_i}{dt} = A_i \left\{ \alpha_i \left[1 - \left(\frac{A_i}{K_i} \right)^\beta \right] + \sum_{j \neq i} J_{i,j} A_j \right\}, \qquad (4.55)$$

where $\beta \geq 0$. In Eq. (4.55) α_i is the growth-rate parameter and K_i is the carrying capacity for species i. $(J_{i,j})$ is the interaction matrix, with coefficients $J_{i,j}$ expressing the effect of species j on the dynamics of species i. Following Mankin et al. (2004), we concentrate on the case of symbiotic interactions, i.e., with $J_{i,j} > 0$ both for $i > j$ and $j > i$. Moreover, to simplify the notation and the mathematical analysis of Eq. (4.55) we refer to the case of a system with species-independent parameters. Thus we take $\alpha_i = \alpha$ and $J_{i,j} = J/N$.

Environmental variability (e.g., climate fluctuations) determines random changes in the availability of the limiting resources, which translates into fluctuations in the carrying capacity. To investigate the effect of environmental fluctuations on the system's dynamics, we treat K_i (i.e., the carrying capacities) as an autocorrelated random variable:

$$K_i = K[1 + a_0 \xi_{dn,i}(t)], \qquad (4.56)$$

where $a_0 < 1$ and $\xi_{dn,i}(t)$ is a zero-mean DMN, with $\xi = \Delta_{1,2} = \pm 1$ with autocorrelation scale τ_c (see Chapter 2). Following Mankin et al. (2004), we can rewrite the term $K_i^{-\beta}$ in Eq. (4.55) as

$$K_i^{-\beta} = K^{-\beta} \gamma [1 + a \xi_{dn,i}(t)], \qquad (4.57)$$

with

$$\gamma = \frac{1}{2(1 - a_0^2)} \left[(1 + a_0)^\beta + (1 - a_0)^\beta \right], \qquad (4.58)$$

$$a = \frac{(1 + a_0)^\beta - (1 - a_0)^\beta}{(1 + a_0)^\beta + (1 - a_0)^\beta}. \qquad (4.59)$$

In what follows we use the dimensionless variables

$$A_i' = \frac{A_i}{K}, \quad t' = \alpha t, \quad J' = \frac{J\,K}{\alpha}, \quad \tau_c' = \alpha \tau_c, \tag{4.60}$$

and drop the prime superscript to simplify the notation. To study the effect of noise and interspecies interactions on the dynamics of this system, some analytical expressions can be obtained for the distribution of the state variables A_i by use of the mean-field assumption $(1/N)\sum_{j \neq i} A_j \approx \langle A \rangle$. Inserting Eq. (4.57) into Eq. (4.55), we can express the dynamics of A_i as a function of the mean-field conditions $\langle A \rangle$ and of the other parameters of the system:

$$\frac{\mathrm{d}A_i}{\mathrm{d}t} = A_i \left\{ 1 + J\langle A \rangle - \gamma A_i^\beta [1 + a\xi_{\mathrm{dn},i}(t)] \right\}. \tag{4.61}$$

The solution of Eq. (4.61) can be determined with the methods presented in Chapter 2:

$$p(A, \langle A \rangle) = \frac{\beta A^{-(1+\beta)}(1 + J\langle A \rangle)}{\gamma a B \left[\frac{1}{2}, \frac{1}{2\beta\tau_c(1+J\langle A \rangle)} \right]} \tag{4.62}$$

$$\times \left| 1 - \frac{(1 + J\langle A \rangle)^2}{\gamma^2 a^2} \left[\frac{\gamma}{(1 + J\langle A \rangle)} - \frac{1}{A^\beta} \right]^2 \right|^{\frac{1}{2\beta\tau_c(1+J\langle A \rangle)} - 1},$$

where $B[\cdot]$ is the beta function. Using the conditions presented in Chapter 2, we find that the extremes A_1 and A_2 of the domain of A are

$$A_{1,2} = \left[\frac{1 + J\langle A \rangle}{\gamma(1 \mp a)} \right]^{\frac{1}{\beta}}. \tag{4.63}$$

The steady-state probability distribution of A expressed by (4.62) depends on $\langle A \rangle$, which in turn depends on the distribution of A. Using the Weiss mean-field approach, Mankin et al. (2004) imposed the self-consistency condition

$$\langle A \rangle = \int_{A_1}^{A_2} A\, p(A, \langle A \rangle)\, \mathrm{d}A. \tag{4.64}$$

We can then calculate the probability distribution of A by inserting into Eq. (4.62) values of $\langle A \rangle$ obtained as solutions of Eq. (4.64). Multiple solutions of (4.64) correspond to dynamical systems, which exhibit two alternative stable phases, i.e., two statistical regimes characterized by distinct probability distributions.

To evaluate the effect of noise on these dynamics, we first consider the deterministic counterpart of the system [i.e., Eq. (4.61) with $\xi_{\mathrm{dn},i}(t) = 0$]. In the case $\beta = 1$ the dynamics are stable (see Box 4.3) for $J < \gamma$, with the system converging to the state $\langle A \rangle \to 1/(\gamma - J)$. However, as J exceeds γ, the dynamics become unstable and $\langle A \rangle$ diverges to ∞. Thus for $\beta = 1$ the deterministic system exhibits transitions from stable to unstable conditions induced by the coupling (expressed by the parameter J)

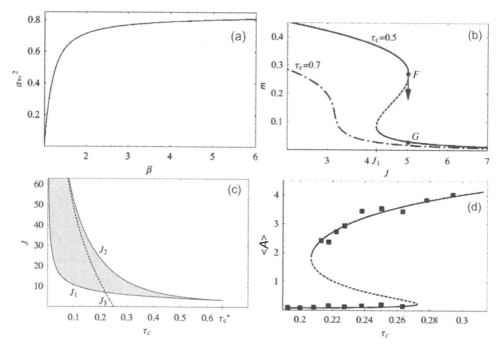

Figure 4.27. (a) Critical noise intensity a_{0c}^2 as a function of the parameter β, (b) order parameter m as a function of the coupling strength J and the correlation scale τ_c (with $\beta = 2$ and $a_0 = 0.99$), (c) region of the parameter space (J, τ_c) in which the dynamics are bistable, i.e., multiple solutions of the self-consistency equation exist (shaded area). The dashed curve, $J = J_3$, corresponds to a change in the shape of the probability distribution of A. The solid curves $J = J_1$ and $J = J_2$ correspond to changes in the number of possible probability distributions of A (with $\beta = 2$ and $a_0 = 0.98$). (d) Mean value of A as a function of the correlation scale τ_c (with $\beta = 2$ and $a_0 = 0.939$) (after Mankin et al., 2004).

among state variables. When $\beta > 1$, the deterministic system is always stable (with only one stable state), and no transition to instability occurs. However, transitions to a bistable regime may emerge as an effect of environmental fluctuations, which may induce bistability and abrupt first-order-like transitions of $\langle A \rangle$ from a state to another. The bistable dynamics we are referring to in this section are associated with existence of multiple solutions of the self-consistency equation, which correspond to the existence of more than one steady-state probability distribution of the state variable A. The dynamics will evolve toward either one of these stable statistical regimes, depending on the initial conditions.

The existence of bistability depends on the noise parameters, namely on the amplitude a_0 [see Eq. (4.56)] and on the autocorrelation scale τ_c, as well as on the coupling parameter J and the parameter β of the generalized Verhulst process. Mankin et al. (2004) showed that bistability may emerge only when the noise intensity exceeds a critical value a_{0c}. The dependence of a_{0c} on β is shown in Fig. 4.27(a). For values of

a_0 smaller than this critical value, the system has only one stable state. As the noise intensity (i.e., the standard deviation expressed by the amplitude a_0 of the fluctuations) increases above a_{0c}, multiple stable states may exist. Thus this is clearly an example of a noise-induced phase transition (see Chapter 2). The dependence of these phase transitions on the coupling parameter was investigated by Mankin et al. (2004) through the *order parameter*, $m = 1/[2\beta\tau_c(1 + J\langle A\rangle)]$. Figure 4.27(b) shows a plot of m (obtained as solution of the self-consistency equation) as a function of J. In the case $\tau_c = 0.5$ the dynamics exhibit a range (J_1, J_2) of values of J in which the system is bistable, i.e., the self-consistency equation shows three possible equilibrium states, with two of them being stable and the other unstable. The dynamics undergo discontinuous transitions (i.e., abrupt changes in $\langle A\rangle$) as J varies across the two bifurcation points J_1 and J_2. For example, if the state of the system is represented by point F in Fig. 4.27(b), a slight increase in J induces a catastrophic transition to the stable state G on the other branch of the bifurcation diagram.

The combinations of values of τ_c and J corresponding to bistable conditions are shown in the shaded area of Fig. 4.27(c) (Mankin et al., 2004). The dashed curve $J_3(\tau_c)$ separates two parts of the parameter space in which the pdf $p(A)$ of A has a qualitatively different shape: It is observed that $p(A)$ has a bell-shaped distribution for $J < J_3$ and a U-shaped distribution as J exceeds J_3. This change in the shape of $p(A)$ (and hence in the preferential states of the system) with changing values of τ_c is an example of noise-induced transitions induced by changes in the correlation scale of the noise term (see Chapter 2). However, it is important to stress that these transitions are different from those associated with the appearance or disappearance of multiple solutions of the self-consistency equation. In the latter case noise does not simply induce a transition in the shape of the distribution (which for example may change from unimodal to bimodal) but affects the number of steady-state probability distributions of A. As a result, the mean value of A may abruptly and discontinuously pass to the other phase, i.e, the other branch of the bifurcation diagram.

Figure 4.27(c) shows that the bistability range may exist only when the scale of autocorrelation is smaller than a critical value τ_c^*. An example of bifurcation diagram of $\langle A\rangle$ as a function of τ_c can be obtained with the self-consistency equation [Fig. 4.27(d)]. It is observed that there is a clear hysteresis (see Box 4.1) with respect to changes in τ_c: The mean value of A can abruptly shift to zero when τ_c decreases below a critical value [≈ 0.21 in the case of Fig. 4.27(d)]. Once this shift occurs, it is not sufficient to increase τ_c above that same value for $\langle A\rangle$ to recover its initial state. Rather, the correlation scale needs to increase above ≈ 0.27 [in the case of Fig. 4.27(d)] for the state $\langle A\rangle = 0$ to be destabilized and the dynamics to shift to the upper branch of the bifurcation diagram. Similar hystereses can be observed with respect to the coupling parameter J [Fig. 4.27(b)] and to the noise amplitude a_0 (not shown). Thus changes in the amplitude and correlation scales may induce interesting phase transitions associated with the emergence and disappearance of bistable behavior and with abrupt shifts to an alternative state of $\langle A \rangle$. Because the deterministic counterpart of this system is monostable, the discontinuous transitions found in this model of symbiotic interactions are purely noise induced. These phase transitions are often defined as *reentrant* in that bistability appears as the noise amplitude (or the correlation scale) exceeds a critical value, but it also disappears as another (larger) critical value of a_0 is exceeded.

Similar results were obtained (Mankin et al., 2002) when the random driver was a trichotomous Markov noise (i.e., three-state Markov noise). The same authors also provided some examples of bistable systems with unidirectional transitions (Sauga and Mankin, 2005), i.e., with transitions that may occur in only one direction. Other studies investigated the effect of noise on multivariate ecological systems in which random forcing does not induce stochastic fluctuations in the carrying capacity as in Mankin et al. (2002, 2004) and Sauga and Mankin (2005) but in the interaction coefficients $J_{i,j}$. The following subsection shows the relation between random interspecies interactions and the stability and complexity of ecological systems with a large number of species (Gardner and Ashby, 1970; May, 1972).

4.7.3 Stability of multivariate ecological systems with random interspecies interactions

We consider the case of an n-dimensional ecological dynamical system with n species $\mathbf{s} = (s_1, s_2, \ldots, s_n)$. The temporal variability of each variable s_i depends on the other state variables and is expressed by a set of first-order differential equations:

$$\frac{ds_i}{dt} = f_i(s_1, s_2, \ldots, s_n), \qquad (i = 1, 2, \ldots, n), \tag{4.65}$$

where $f_i(\mathbf{s})$ is a set of n functions, which can be in general nonlinear. The equilibrium states $\mathbf{s}_{st} = (s_{1,st}, s_{2,st}, \ldots, s_{n,st})$ of the system are obtained as solutions of the set of equations

$$f_i(\mathbf{s}_{st}) = 0 \qquad (i = 1, 2, \ldots, n). \tag{4.66}$$

The stability of an equilibrium state $\mathbf{s}_{st}$ is typically assessed through a linear-stability analysis, i.e., with respect to infinitesimal perturbations, which allow for a Taylor expansion of (4.65) in the neighborhood of the equilibrium point (see Box 4.3):

$$\frac{d\mathbf{s}}{dt} = \mathbf{A}\mathbf{s}, \tag{4.67}$$

where the matrix $\mathbf{A}$ is the Jacobian of (4.65), which is also known in ecology literature as the *community* or *interaction matrix* (e.g., May, 1973) because its elements $a_{i,j} = \left(\partial f_i / \partial x_j \right)_{\mathbf{s} = \mathbf{s}_{st}}$ express the interactions between the state variables s_i and s_j (i.e., the effect of s_j on the dynamics of s_i). In the absence of interactions we have that $a_{i,j} = 0$

for $i \neq j$. In this case the equilibrium point $\mathbf{s}_{st}$ is stable if all the coefficients on the diagonal of $\mathbf{A}$ are negative (i.e. $a_{i,i} < 0$, $(i = 1, 2, \ldots, n)$. We can set the time scale of convergence to equilibrium by taking all the diagonal coefficients equal to -1, as in May (1972). To study how random interactions among state variables (e.g., among the populations of different species existing in the system) may affect the stability of the system, we allow for the occurrence of random interactions between a fraction C of pairs of state variables (s_i, s_j). To this end, we set the corresponding extradiagonal elements $a_{i,j}$ of $\mathbf{A}$ equal to values drawn from a distribution of random numbers with mean zero. Thus the interactions can be either positive or negative, and their intensity is on average zero. The community matrix can be then expressed as

$$\mathbf{A} = \mathbf{B} - \mathbf{I}, \tag{4.68}$$

where $\mathbf{I}$ is the identity matrix and $\mathbf{B}$ is a random matrix. With probability C the elements of $\mathbf{B}$ are nonzero, and their value is drawn from a zero-mean probability distribution with standard deviation σ, whereas with probability $1 - C$ the elements of $\mathbf{B}$ are equal to zero. Known as *connectance* (Gardner and Ashby, 1970), the parameter C expresses the probability that any pair of state variables (e.g., populations) interacts.

Notice how, once the coefficients of $\mathbf{A}$ have been selected, the dynamics are completely deterministic and the stability of $\mathbf{s}_{st}$ can be investigated with the standard methods for deterministic systems. Thus the stability of $\mathbf{s}_{st}$ requires that all the eigenvalues of $\mathbf{A}$ have negative real parts (see Box 4.3). Using the theory of random matrices, May (1972) noticed that this condition is met when

$$\sigma < \frac{1}{\sqrt{nC}}, \tag{4.69}$$

whereas the equilibrium point $\mathbf{s}_{st}$ is unstable otherwise. The probability that the dynamics of randomly constructed communities are unstable (i.e., at least one eigenvalue has a positive real part) increases with an increasing number of species. Known as the *complexity paradox*, this result is consistent with earlier numerical simulations by Gardner and Ashby (1970) and indicates that a stable multivariate system can be destabilized by an increase either in the number of species or in the connectivity (or both). This finding provided new insights on the relation between complexity and stability in ecological systems (e.g., May, 1973; Pimm, 1984). Another interesting result appearing in condition (4.69) is associated with a the randomness of the interactions. In a system with a given connectance and number of species, an increase in the variance of interspecies interaction may induce instability. The emergence of instability with increasing values of σ is clearly a noise-induced effect, though, as noted, these dynamics are entirely deterministic once the random coefficients have been selected. The applicability of these results to actual ecosystems has been often challenged (e.g., De Angelis, 1975; Pimm, 1984) in that the randomness of the interactions is not a realistic assumption for real-food Web models. Because in nature complex communities are often found to be stable (e.g., Polis, 1991; Goldwasser and Roughgarden, 1993), it can be argued that community composition is unlikely to be random. In fact, natural selection is expected to play a role in determining how species interact within a community.

4.8 Stochastic resonance and coherence in environmental processes

The concept of stochastic resonance was first formulated (Benzi et al., 1982a, 1982b) to explain the emergence of periodic fluctuations in climate dynamics in the presence of relatively weak periodic drivers. Typical examples include the 100-kyr periodicity observed in climate-change episodes (Benzi et al., 1982b), the periodic fluctuations in the Holocene thermohaline circulation (Vélez-Belchi et al., 2001), and the periodicity in the occurrence of Dansgaard–Oeschger (DO) events (i.e., short and abrupt warming episodes) during the last glaciation (Alley et al., 2001; Ganopolski and Rahmstorf, 2002). Coherence resonance was recently invoked to explain species coexistence and biodiversity (Lai and Liu, 2005) and regular random fluctuations in predator–prey systems (Sieber et al., 2007).

4.8.1 The Benzi–Parisi–Sutera–Vulpiani climate-change model

The notion of stochastic resonance was introduced by Benzi et al. (1982a), 1982b) to study periodical changes in the Earth's climate observed in palaeoclimate records (Mason, 1976). Benzi et al. (1982a) developed the stochastic-resonance framework to show how the 100,000-yr periodicity observed in climate fluctuations may result from the combined effect of a weak (deterministic) astronomic forcing and background environmental noise in a complex, nonlinear system with internal feedbacks. The synergism between the weak periodic forcing and a noise of suitable intensity would be able to cause organized periodic fluctuations in the climate system, which would not appear in the absence of noise. Thus noise would induce order and regularity in these long-term climate dynamics.

Benzi et al. (1982b) modified the Budyko–Sellers model (Budyko, 1969; Sellers, 1969) of the Earth's energy balance:

$$c\frac{dT}{dt} = -\frac{dV(T)}{dT} = R_{in} - R_{out}, \tag{4.70}$$

where c is the active thermal inertia of the Earth system, T is the Earth's temperature, $V(T)$ is the potential function associated with the dynamics of T, R_{in} is the incoming solar radiation, and R_{out} is the outgoing (radiated and reflected) radiation. The incoming radiation is affected by periodical astronomic forcing, whereas outgoing radiation is the sum of reflected $[\alpha(T)R_{in}]$ and long-wave $[E(T)]$ radiation:

$$R_{in} = Q[1 + a\cos(\omega t)], \qquad R_{out} = \alpha(T)R_{in} + E(T), \tag{4.71}$$

where Q is the solar constant, a is the amplitude of the periodic forcing, ω is its angular frequency (with $a = 5 \times 10^{-4}$ and $2\pi/\omega = 92$ kyr), and $\alpha(T)$ is the albedo. Invoking the results of earlier studies, the authors assume that, in the absence of the periodic forcing (i.e., $a = 0$), dynamics (4.71) are bistable, with an unstable state $T = T_2$ symmetrically located between the two stable states $T = T_{1,3}$. To this end, they replaced the right-hand side of (4.70) [with R_{in} and R_{out} expressed by (4.71) with $a = 0$] with a function proportional to a third-order polynomial with roots T_1, T_2, and T_3, i.e.,

$$Q(1 - \alpha(T)) - E(T) = E(T)\beta \left(1 - \frac{T}{T_1}\right)\left(1 - \frac{T}{T_2}\right)\left(1 - \frac{T}{T_3}\right). \tag{4.72}$$

Thus, for the generic case $a \neq 0$, Eq. (4.70) becomes

$$c\frac{\mathrm{d}T}{\mathrm{d}t} = F(T) = E(T) \tag{4.73}$$

$$\times \left[\beta \left(1 - \frac{T}{T_1}\right)\left(1 - \frac{T}{T_2}\right)\left(1 - \frac{T}{T_3}\right)[a\cos(\omega t) + 1] + a\cos(\omega t)\right].$$

The dynamical properties of this system strongly depend on the bistability assumption expressed by (4.72). Although Benzi et al. (1982b) do not explain the physical rationale underlying this assumption, other studies conjectured that the climate system might be bistable (e.g., Mason, 1976). In the absence of noise, periodic forcing is unable to induce transitions across the potential barrier for realistic values of the system's parameters ($T_1 = 278.6$ K, $T_2 = 283.3$ K, $T_3 = 288$ K, $c = 4000$). The presence of additive noise may profoundly affect the dynamical properties of the system. Benzi et al. (1982b) considered the effect of environmental variability on dynamics (4.73). In particular, they concentrated on the case of additive white Gaussian noise ξ_{gn} with mean zero and intensity s_{gn}:

$$\frac{\mathrm{d}T}{\mathrm{d}t} = \frac{F(T)}{c} + \xi_{\text{gn}}. \tag{4.74}$$

Notice how this system differs from the one presented in Chapter 3. In fact in this case the periodic forcing appears as a multiplicative term in the bistable potential.

In the absence of periodic fluctuations (i.e., $a = 0$) noise induces transitions between the wells of the potential. The mean exit time from the domain of attraction of T_3 can be calculated with *Kramer's method* (see Chapter 3):

$$\langle t_c\rangle_{2,3} \approx \frac{\pi}{\sqrt{|V''(T_2)V''(T_3)|}}e^{\frac{\Delta V_{2,3}}{s_{\text{gn}}}} \tag{4.75}$$

where $V(T)$ is the potential function defined by (4.70) and $\Delta V_{2,3}$ is the height of the potential barrier from the stable state, T_3, $\Delta V_{2,3} = -\int_{T_2}^{T_3} F(T)\,\mathrm{d}T$.

When $a \neq 0$ the periodic forcing modulates the shape of the potential function and the height of the potential barrier fluctuates between a minimum $\Delta V_{2,3}^{\text{min}}$ and a

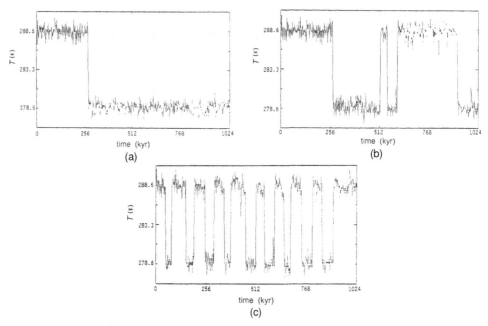

Figure 4.28. Numerical simulation of the dynamics expressed by Eq. (4.73) with (a) stochastic forcing but no periodic driver (i.e., $s_{gn} = 0.01$ and $a = 0$), and no stochastic resonance is observed; (b) weak stochastic forcing and periodic driver (i.e., $s_{gn} = 0.025$ and $a = 0.5 \times 10^{-4}$), and no stochastic resonance is observed; (c) stronger stochastic forcing and periodic driver (i.e., $s_{gn} = 0.05$ and $a = 0.5 \times 10^{-4}$), and in these conditions stochastic resonance emerges. Taken from Benzi et al. (1982b).

maximum $\Delta V_{2,3}^{\max}$ with a period $2\pi/\omega$. Stochastic resonance occurs when the noise level (i.e., the intensity s_{gn} of ξ_{gn}) is strong enough to allow for the occurrence of transitions when the potential barrier has minimum height, but not when it is maximum (otherwise the periodic forcing would not be able to regularize the timing of these occurrences). The first condition is met when the value of $\langle t_c \rangle_{2,3}$ calculated with $\Delta V_{2,3}^{\min}$ is close to the semiperiod π/ω, whereas the second condition requires that, when $\Delta V_{2,3}$ is maximum, the mean crossing time $\langle t_c \rangle_{2,3}$ be longer than the period $2\pi/\omega$. For the parameters reported in this subsection (see Benzi et al., 1982b) stochastic resonance occurs when the intensity s_{gn} of the additive noise is contained within the interval 0.03 K^2 yr^{-1} $< s_{gn} < 0.07$ K^2 yr^{-1}. Figure 4.28 shows some examples of time series generated by this model with different values of s_{gn}. In the case of Fig. 4.28(a), s_{gn} is by far below the lower end of this interval; in these conditions no transition occurs. In the case of Fig. 4.28(b), s_{gn} is slightly smaller than the lower minimum value of the interval; we observe that some transitions occur but they do not exhibit a well-defined periodicity. In this case noise is too weak to induce order in the timing of these occurrences (i.e., for stochastic resonance to occur). For larger values of s_{gn} stochastic resonance occurs, as shown in Fig. 4.28(c). We note that the periodicity of these fluctuations is imposed by the period of the external forcing. Other authors (Pelletier, 2003) argued that the observed 100,000-yr periodicity is not imposed by a periodic forcing but it is inherent in the dynamics, whereas noise enhances the occurrence of 100,000-yr fluctuations through a *coherence-resonance* mechanism (see Chapter 3).

4.8.2 Fluctuations in the glacial climate: An effect of stochastic or coherence resonance?

Stochastic resonance was recently invoked (Alley et al., 2001; Ganopolski and Rahmstorf, 2002; Braun et al., 2005b; Ditlevsen et al., 2005) to explain the occurrence of temporary and abrupt warming episodes – known as *Dansgaard–Oeschger (DO) events* – in the course of the last glacial period (100–10 kyr before the present). Alternative mechanisms based on the coherence-resonance theory have also been proposed (Timmermann et al., 2003).

Ice-core records indicate that the waiting times between two consecutive DO episodes are clustered around 1500 yr, and – with lower probability – around 3000 and 4500 yr (Alley et al., 2001), as shown in Fig. 4.29(e). The occurrence of DO events in 1500-yr cycles and in cycles with periods that are integer multiples of 1500 yr is suggestive of a stochastic-resonance mechanism (see Chapter 3). However, in this case stochastic resonance occurs in a system that is slightly different from those presented in Chapter 3. In fact, in this complex system the detection of stochastic resonance requires simulations with global-climate models with several state variables and parameters, rather than simplified models with one differential equation (Ganopolski et al., 1998; Ganopolski and Rahmstorf, 2002; Timmermann et al., 2003). Moreover, in this case the underlying deterministic dynamics are not bistable (Ganopolski and Rahmstorf, 2001). In fact, while under present-day climate conditions, the Atlantic Ocean theormohaline circulation – a major contributor to the regional heat budget – has two stable states, bistability did not exist under glacial climate conditions (Ganopolski and Rahmstorf, 2001). The glacial Atlantic Ocean circulation used to have only one stable mode, known as the cold conveyor belt (stadial conditions). Warm DO events occurred as perturbations of the "cold" climate conditions, and were presumably triggered by small salinity changes in the North Atlantic. Each event involved only a temporary shift to the unstable warm circulation state, and included three distinctive phases: abrupt warming by about 10 °C, slow cooling, and abrupt cooling to the cold stadial conditions.

Thus the glacial Atlantic behaved like an excitable system and could have exhibited either coherence-resonant or stochastic-resonant behavior, depending on whether the freshwater inflow was perturbed only with noise or also with periodic forcing. Using global-climate simulations, Ganopolski and Rahmstorf (2002) showed that, in

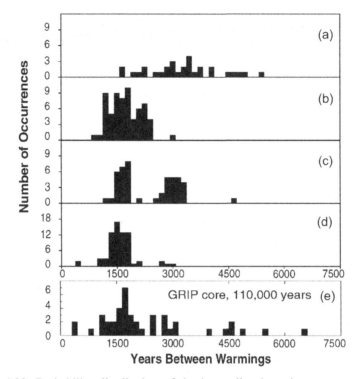

Figure 4.29. Probability distribution of the interspike times between consecutive DO events based on simulations with (a) relatively weak noise (noise intensity $s_{gn} = 6.125 \times 10^{-4}$) and no periodic forcing; (b) stronger noise ($s_{gn} = 1.25 \times 10^{-3}$) and no periodic forcing; (c) both periodic and "weak" stochastic forcing ($\sigma = 6.125 \times 10^{-4}$); (d) both periodic and "strong" stochastic forcing ($\sigma = 1.25 \times 10^{-3}$); (e) interspike time distribution obtained from the analysis of Greenland ice-core project (GRIP) data from Alley et al. (2001). Figure taken from Ganopolski and Rahmstorf (2002).

the absence of an external periodic forcing, the noise term may induce temporary transitions to the unstable circulation mode. Interestingly, when the noise levels (i.e., variance) are sufficiently high, the distribution of waiting times between DO events is clustered in the 1000–2500-yr range (Fig. 4.29). Thus noise is able to enhance and unveil an intrinsic waiting time time between DO occurrences. Moreover, this waiting time is close to the observed 1500-yr periodicity [Fig. 4.29(e)]. Thus this analysis would suggest that the periodic DO events are an example of coherence resonance (Timmermann et al., 2003). However, the analysis of ice-core data shows the existence of smaller modes at 3000 and 4500 yr in the waiting-time distribution of DO events. These modes could emerge as an effect of stochastic resonance. Ganopolski and Rahmstorf (2002) forced their global-climate model both with (additive) noise and with periodic fluctuations in freshwater inputs. They found that the three modes appear for adequate noise levels, consistent with the analysis of ice-core records [Figs. 4.29(c)–(e)]. Although, by itself, this weak periodic forcing is not able to induce DO transitions, its combined effect with noise synchronizes the transitions, inducing order and regularity in the intervals between events.

Depending on whether DO events result from coherence resonance or stochastic resonance, the underlying dynamics would exhibit different properties: (i) in the case of coherence resonance the 1500-yr periodicity would be inherent to the dynamics of the system (Timmermann et al., 2003); (ii) in the case of stochastic resonance this periodicity would result from an externally imposed periodic fluctuation in freshwater inputs (Ganopolski and Rahmstorf, 2002). Thus a stochastic-resonance theory of DO events requires an explanation of the physical processes determining a 1500-yr periodicity in freshwater inputs. These inputs are presumably associated with fluctuations in hydrologic conditions. It is not clear what could induce 1500-yr periodicity in these fluctuations (Ganopolski and Rahmstorf, 2001), though some more recent analyses have shown (Braun et al., 2005a) that the 1500-yr cycle could emerge when the periodic forcing is obtained as a suitable superposition of two harmonics corresponding to the 87-yr and 210-yr solar cycles. Thus nonlinear excitable or bistable systems forced by a periodic driver with more than one frequency may show resonance at a frequency different from those of its driver. This effect is also known as *ghost resonance*.

4.8.3 A coherence-resonance mechanism of biodiversity

In Subsection 4.2.3 we showed how noise can promote species diversity through the beneficial effect of environmental fluctuations of intermediate frequency. Lai and Liu (2005) demonstrated how species coexistence (and hence diversity) can also result from a coherence-resonance mechanism.

Recent studies have shown how environmental variability may favor the coexistence of species with different dispersal rates (e.g., Holt and McPeek, 1996). Temporal variability may either be induced by externally imposed drivers or result from endogenous chaotic dynamics (Harrison et al., 2001). A simplified framework recently used to investigate these dynamics is based on the Holt–McPeek model (Holt and McPeek, 1996), i.e., a two-species and two-patch system with four state variables, $N_{i,j}$ (i, $j = 1, 2$), representing the density of species i in patch j. Species compete within patches and disperse between patches. Each species behaves identically within each patch (i.e., with the same competitive abilities) but disperses differently between patches. Thus, in each patch, the density dependence is a function of the total species density, $N_{t,j} = N_{1,j} + N_{2,j}$ ($j = 1, 2$). The local population growth W_j in patch j ($j = 1, 2$) is species independent:

$$W_j = \exp\left\{ r_j \left[1 - \frac{N_{t,j}(t)}{K_j} \right] \right\}, \tag{4.76}$$

where K_j is the carrying capacity in patch j and r_j is the a parameter determining the growth rate in patch j. Dispersal is modeled assuming that a fraction e_i of species i migrates at each time step (generation) from its patch to the other one. This migratory

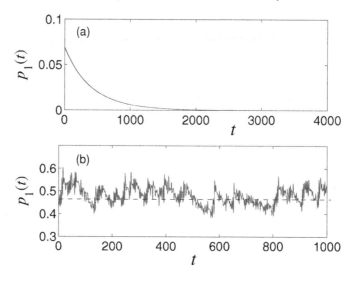

Figure 4.30. (a) Deterministic (i.e., $s_{gn} = 0$) and (b) stochastic ($s_{gn} = 8$) dynamics of the modified Holt–McPeek system (Lai and Liu, 2005) calculated with $r = 2.6$, $K_1 = 100$, $K_2 = 50$, $e_1 = 0.5$, $e_2 = 0.01$, $m = 1$ (after Lai and Liu, 2005).

fraction undergoes mortality (the *cost of dispersal*) at a rate $1 - m$ (i.e., only a fraction m of the migrators survive). The dynamics take place in such a way that density-dependent growth precedes both dispersal and cost of dispersal. The overall dynamics can be expressed as

$$N_{1,1}(t + 1) = (1 - e_1)W_1(t)N_{1,1}(t) + me_1N_{1,2}(t),$$

$$N_{1,2}(t + 1) = (1 - e_1)W_2(t)N_{1,2}(t) + me_1N_{1,1}(t),$$

$$N_{2,1}(t + 1) = (1 - e_2)W_1(t)N_{2,1}(t) + me_2N_{2,2}(t),$$

$$N_{2,2}(t + 1) = (1 - e_2)W_2(t)N_{2,2}(t) + me_2N_{2,1}(t). \tag{4.77}$$

Holt and McPeek (1996) investigated the dynamical properties of this system when one species has a much higher dispersal rate than the other ($e_1 = 0.5$ and $e_2 = 0.01$). For relatively small values of r the dynamics are stable and converge to a state in which species 1 goes extinct. Thus the frequency of the high-dispersal species,

$$p_1(t) = \frac{N_{1,1}(t) + N_{1,2}(t)}{N_{t,1} + N_{t,2}}, \tag{4.78}$$

tends to zero [Figure 4.30(a)]. This happens for $r < 3$ in the case shown in Fig. 4.30. As r tends to 3, the dynamics exhibit a cyclic behavior. For larger values of r, a transition to chaos occurs: Instead of tending to zero, the density of the high-dispersal species undergoes episodic and abrupt increases. This behavior is completely deterministic and does not depend on the existence of any external forcing (Holt and McPeek, 1996).

Harrison et al. (2001) interpreted this transition to chaos as a possible mechanism able to maintain the coexistence of two species. Lai and Liu (2005) investigated the ability of a stochastic forcing to induce species coexistence in Holt–McPeek systems in nonchaotic conditions. To this end, they accounted for the effect of environmental noise by considering the carrying capacity as a random variable K_j, modeled as a white-noise term with mean $\bar{K}_j$ and intensity s_{gn} [i.e., $K_j = \bar{K}_j + \xi_j(t)$, where ξ_j is a Gaussian-white-noise term with zero mean and intensity s_{gn}]. When the noise level (i.e., s_{gn}) exceeds a critical value, the stochastic forcing can prevent $p_1(t)$ from converging to zero [see Fig. 4.30(b)]. As s_{gn} increases, the average frequency $\langle p_1 \rangle$ of species 1 increases at the expenses of species 2. Thus coexistence occurs for intermediate noise levels. Lai and Liu (2005) interpreted this behavior as a noise-induced effect, particularly as a stochastic resonance. However, in the classification presented in Chapter 3 this behavior is suggestive of a coherence resonance. In fact, Lai and Liu (2005) did not use any periodic deterministic forcing. The stochastic forcing seems to trigger abrupt increases in the size of the population of species 1 with a temporary transition to chaotic dynamics, followed by a relaxation phase back to stable dynamics. Thus the process resembles the dynamics of excitable systems with endogenous time scales.

4.8.4 Coherence resonance in excitable predator–prey systems

Predator–prey dynamics are known for their ability to exhibit interesting nonlinear deterministic behaviors, including limit cycles (e.g., Goel et al., 1971). Thus, in the presence of a stochastic driver, these dynamics could lead to coherence resonance and behave as excitable systems. An example of coherence resonance in predator–prey systems can be found in the coupled dynamics of phytoplankton (P) and zooplankton (Z). These dynamics are often investigated with the Truscott and Brindley (1994) model:

$$\epsilon \frac{dP}{dt} = rP(1 - P) - \frac{a^2 P^2}{1 + b^2 P^2} Z,$$

$$\frac{dZ}{dt} = \frac{a^2 P^2}{1 + b^2 P^2} Z - m_3 Z; \qquad (4.79)$$

the first equation expresses the dynamics of the phytoplankton (prey) as a harvest process with a harvest (or *grazing*) rate proportional to the zooplankton biomass and dependent on P according to a Holling type III grazing model with maximum rate a^2/b^2. The dynamics of Z are expressed by a growth–death model with the growth rate determined by the grazing process and a mortality term proportional to Z with rate m_3. The parameter ϵ determines the time scale of the response of P, and r expresses the growth rate of the logistic term in the dynamics of P.

The dynamics expressed by Eqs. (4.79) may exhibit interesting cyclic behaviors (Fernandez et al., 2002), similar to those of other grazing systems (Noy-Meir, 1975). Sieber et al. (2007) used this model to investigate the viral infection of the phytoplankton population. In particular, they concentrated on the effect of lysogenic infection, i.e., on the case of viruses integrated in the genome of the host cells. Thus the reproduction of the host genome occurs concurrently to the reproduction of the viral genome. Sieber et al. (2007) modified Eqs. (4.79) to account for the fact that the phytoplankton may be either susceptible to viruses (Z_s) or infected (Z_i). Indicating the zooplankton population by Z_z, we can express the dynamics of the state variable, $\mathbf{Z}(t) = \{Z_s(t), Z_i(t), Z_z(t)\}$, as

$$\frac{dZ_s(t)}{dt} = \epsilon^{-1}\left[rZ_s(1 - Z_s - Z_i) - \frac{a^2 Z_s(Z_s + Z_i)}{1 + b^2(Z_s + Z_i)^2}Z_z - \lambda\frac{Z_s Z_i}{Z_s + Z_i}\right],$$

$$\frac{dZ_i(t)}{dt} = \epsilon^{-1}\left[rZ_i(1 - Z_s - Z_i) - \frac{a^2 Z_i(Z_s + Z_i)}{1 + b^2(Z_s + Z_i)^2}Z_z + \lambda\frac{Z_s Z_i}{Z_s + Z_i} - m_2 Z_i\right],$$

$$\frac{dZ_z(t)}{dt} = \frac{a^2(Z_s + Z_i)^2}{1 + b^2(Z_s + Z_i)^2}Z_z - m_3 Z_z, \tag{4.80}$$

where λ represents the virus transmission rate. The fraction of infected phytoplankton (Z_i) increases as an effect of virus transmission, while the susceptible population (Z_s) decreases. The infected population is then prone to virus-induced mortality at rate m_2. The zooplankton dynamics are the same as in (4.79). It can be shown that in the dynamics described by (4.80) the ratio $i = Z_i/(Z_s + Z_i)$ remains constant for $\lambda = m_2$. Sieber et al. (2007) used this property to investigate (4.80) as a 2D system. Depending on the initial value of i, the system may exhibit a variety of dynamical behaviors, including the convergence to a stable limit cycle or to a stable focus. For relatively low values of m_3 the dynamics may exhibit excitation followed by relaxation to stable conditions.

Interesting properties emerge when this system is forced with a multiplicative noise. To this end, Sieber et al. (2007) added to the right-hand side of Eqs. (4.80) a term, $Z_j \xi_j(t)$ ($j = s, i, z$), where $\xi_i(t)$ is a zero-mean Gaussian white noise with intensity s_{gn}. The effect of this stochastic forcing is to induce noise-sustained oscillations. With adequate noise intensity this system may exhibit coherence resonance, i.e., a high degree of regularity (or *coherence*) in noise-induced oscillations. An optimal value of s_{gn} exists at which these fluctuations are most coherent. This effect is shown in Fig. 4.31 for three different noise levels (with $r = 1$, $a = 4$, $b = 12$, $\lambda = m_2 = 0.01$, $m_3 = 0.0525$, and $\epsilon = 10^{-3}$): With relatively high or low noise levels (Fig. 4.31) we observe a random occurrence of excitation episodes followed by relaxation. With intermediate values of s_{gn} the dynamics exhibit coherence resonance. This effect can be assessed by calculation of the CV of the time between two consecutive

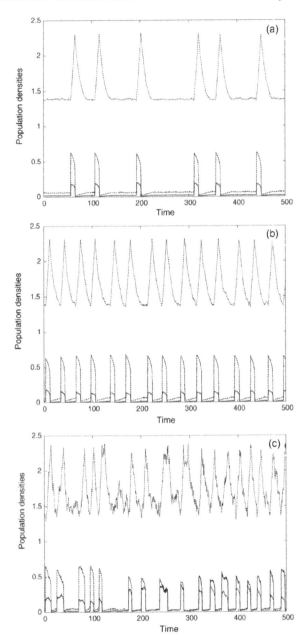

Figure 4.31. Stochastic predator–prey dynamics of phytoplankton and zooplankton populations affected by viral infection of the phytoplankton. Dashed curves, susceptible phytoplankton; solid curves, infected phytoplankton; dotted curves, zooplankton. Calculated with $r = 1$, $a = 4$, $b = 12$, $\lambda = m_2 = 0.01$, $m_3 = 0.0525$, $\epsilon = 10^{-3}$, and with (a) $s_{gn} = 1.25 \times 10^{-5}$, (b) $s_{gn} = 2 \times 10^{-4}$, (c) $s_{gn} = 1.01 \times 10^{-3}$. After Sieber et al. (2007).

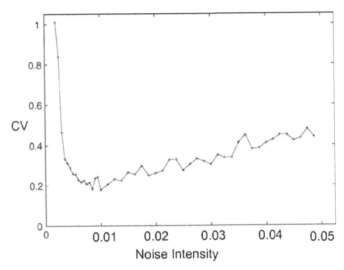

Figure 4.32. Relation between the coefficient of variation (CV) of the times between consecutive excitation spikes and the intensity of the stochastic forcing. Modified after Sieber et al. (2007).

excitation events. Figure 4.32 shows the dependence of the CV on the intensity s_{gn} of the stochastic forcing. It is observed that the CV exhibits a minimum (i.e., more regular – or *coherent* – oscillations) for intermediate noise intensity, consistent with the notion of coherence resonance.

5

Noise-induced pattern formation

5.1 Introduction

5.1.1 General aspects

A spatially extended dynamical system exhibits *geometrical patterns* when its state variables show spatial coherence, i.e., their spatial distribution is neither uniform nor random but exhibits organized spatial structures characterized by a certain regularity. The occurrence of patterns is an important feature of spatiotemporal systems, as it is a signature of the existence of order embedded in the underlying dynamics. This fact, along with the beauty of some of their geometrical features, explains why patterns have always drawn the attention of the science community.

Patterns are in fact ubiquitous in nature. In particular, a number of environmental processes are known for their ability to develop highly organized spatial features (e.g., Greig-Smith, 1979). For example, remarkable degrees of organization can be found in the spatial distribution of convective clouds (e.g., Lovejoy, 1982), dryland and riparian vegetation (e.g., Macfadyen, 1950), sand ripples and dunes (e.g., Lancaster, 1995), river channels (e.g., Allen, 1984; Ikeda and Parker, 1989; Rodriguez-Iturbe and Rinaldo, 2001), soil cracks (Lachenbruch, 1961), coastlines (Bird, 2000; Ashton et al., 2001), peatlands (Eppinga et al., 2008, e.g.), and arctic hummocks and patterned ground (Gleason et al., 1986).

These features are often undetectable on the ground, but have become visible with the advent of aerial photography (e.g., Macfadyen, 1950). Figures 5.1, 5.2, 5.3, 5.4, and 5.5 show some examples of spectacular spatially periodic natural patterns in a number of landscapes around the world. These patterns exhibit amazing regular configurations of vegetation, landforms, or clouds. In some cases patterns may spread over relatively large areas (up to several square kilometers) (White, 1971; Eddy et al., 1999; Valentin et al., 1999; Esteban and Fairen, 2006) and can be found on different soils and with a broad variety of vegetation species and life-forms

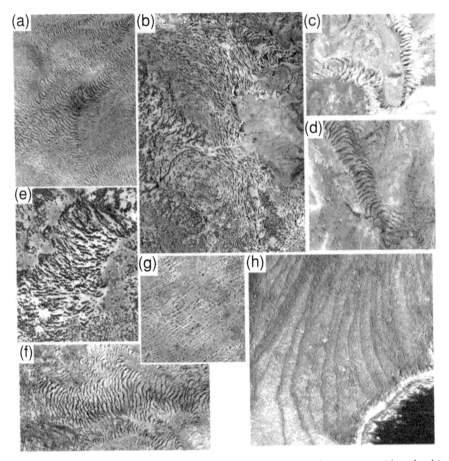

Figure 5.1. Example of aerial photographs showing vegetation patterns (tiger bush); (a) Somalia (9°20′N, 48°46′E), (b) Niger (13°21′N, 2°5′E), (c) Somalia (9°32′N, 49°19′E), (d) Somalia (9°43′N, 49°17′E), (e) Niger (13°24′N, 1°57′E), (f) Somalia (7°41′N, 48°0′E), (g) Senegal (15°6′N, 15°16′W), and (h) Argentina (54°51′S, 65°17′W) (after Borgogno et al., 2009).

(Worral, 1959, 1960; White, 1969, 1971; Bernd, 1978; Mabbutt and Fanning, 1987; Montana, 1992; Bergkamp et al., 1999; Dunkerley and Brown, 1999; Eddy et al., 1999; Valentin et al., 1999).

The study of patterns is often motivated by the possibility to infer from their occurrence and geometrical features useful information on the underlying processes. In recent years, several authors investigated the mechanisms of pattern formation in nature and their response to changes in environmental conditions or disturbance regime. For example, in the case of landscape ecology, these studies related vegetation patterns to the underlying ecohydrological processes (e.g., Schlesinger et al., 1990; Klausmeier, 1999; Barbier et al., 2006; Ridolfi et al., 2007), the nature of the interactions among plant individuals (e.g., Lefever and Lejeune, 1997;

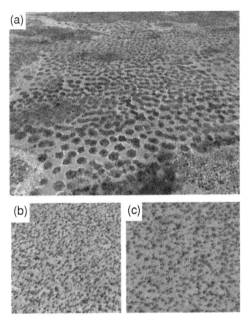

Figure 5.2. Example of aerial photographs showing vegetation patterns (spots): (a) Zambia (15°38′S, 22°46′E), (b) Australia (15°43′S, 133°10′E), (c) Australia (16°14′S, 133°10′E). Google Earth imagery © Google Inc. Used with permission.

Barbier et al., 2006), the stability and resilience of dryland ecosystems (Rietkerk et al., 2002; van de Koppel and Rietkerk, 2004), and the landscape's susceptibility to desertification under different climate drivers and management conditions (e.g., von Hardenberg et al., 2001; D'Odorico et al., 2006a).

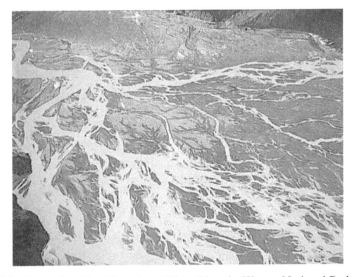

Figure 5.3. Example of braiding river: Slims River in Kluane National Park, Yukon, Canada. Used with permission.

Figure 5.4. Example of meandering river in an arid environment.

Patterns can exhibit a remarkable spatial regularity with the periodic recurrence in space of some geometrical features, such as circles, stripes, gaps, or labyrinths (e.g., see Figs. 5.1, 5.2). In these cases, the main sign of the pattern formation is the occurrence of a dominant wavelength in the field variable. This definition of pattern based on periodicity is often used in the literature on pattern formation for several reasons: First, it corresponds to a clearly defined concept of regularity (i.e., existence of periodicity); second, periodicity is a field property that is easily caught by the eyes and can be detected by spectral analysis; third, the occurrence of dominant wavelengths can be studied theoretically by use of established mathematical tools such as stability analysis by normal modes or the analysis of the structure function (see Subsection 5.1.2.3). Although these three reasons explain why pattern and spatial periodicities are often considered to be the same thing, it is important to stress that sometimes spatial fields are defined as patterned even when several length scales are present and no clear

Figure 5.5. Example of desert ripples on dunes.

wavelength emerges. In these cases, the concept of pattern is somewhat more vague and a framework for rigorous mathematical definition is lacking. The occurrence of a significant spatial coherence is generally assessed through a comparison with the homogeneous case or with disordered noisy spatial fields. Sometimes this approach can be quite subjective, and in this case the notion of pattern may depend on the specific topic under investigation. However, this broader definition of spatial pattern is not rare in the environmental sciences (Manor and Shnerb, 2008; von Hardenberg et al., 2010), in which the number of concomitant processes and forcings can hamper the occurrence of a clear wavelength, and coherence (though nonperiodical) is considered a sufficient indicator of pattern formation.

In the following discussion we refer to both types of patterns, and we distinguish *periodic* from *multiscale* patterns, depending on the presence of only one dominant length scale or several wavelengths at the same time, respectively. As a further attribute, we define *fringed* patterns as those spatial structures in which the boundaries of the coherence regions are disturbed by random local irregularities.

Depending on their behavior in time, patterns can be also classified as *steady* or *transient*, depending on whether the spatial coherence is constant in time or appears only temporarily with a tendency to fade out with time. In the case of *oscillating* patterns, the spatial coherence instead fluctuates with time and patterns periodically emerge and disappear.

5.1.2 Overview of stochastic mechanisms

A few deterministic and stochastic mechanisms were proposed to explain the formation of spatial patterns. The study of deterministic mechanisms has a relatively long history (e.g., Turing, 1952; Cross and Hohenberg, 1993) and a rich literature exists with applications to the biogeosciences (e.g., Murray, 2002; Borgogno et al., 2009). In Appendix B we review the main theories of deterministic pattern formation. Stochastic models were developed more recently (e.g., Garcia-Ojalvo and Sancho, 1999; Sagues et al., 2007). They explain pattern formation as a noise-induced effect, in the sense that patterns emerge as a result of the randomness inherent in environmental fluctuations and disturbance regime. These random drivers induce a symmetry-breaking instability: They destabilize a homogeneous (and thus symmetric) state of the system and determine a transition to an ordered phase, which exhibits a degree of spatial organization. In the thermodynamics literature, these order-forming transitions are often referred to as *nonequilibrium transitions*, to stress the fundamental difference in the role of noise with respect to the classical case of *equilibrium transitions*, which exhibit an increase in disorder as the amplitude of internal fluctuations increases. Thus nonequilibrium transitions are a typical example of the counterintuitive constructive (or order-forming) effect of noise in nonlinear dynamical systems, similar to those discussed in the third chapter for the case of nonspatial systems.

In this chapter we concentrate mainly on the case of dynamics in which the state of the system is determined by only one state variable, ϕ. These spatiotemporal dynamics can in general be modeled by a stochastic partial differential equation expressing the temporal variability of ϕ at any point, $\mathbf{r} = (x, y)$, as the sum of five terms: (i) a function $f(\phi)$ of local conditions [i.e., of the value of ϕ at (x, y)]; (ii) a multiplicative-noise term, $g(\phi)\xi_m(\mathbf{r}, t)$; (iii) a term, $D\mathcal{L}[\phi]$, accounting for the spatial interactions with the other points of the domain, (iv) a term expressing the effect of a time-dependent forcing, $F(t)$, which can in general be modulated by a function $h(\phi)$ of the local state of the system, and (v) an additive-noise component $\xi_a(\mathbf{r}, t)$. Therefore the dynamics are

$$\frac{\partial \phi}{\partial t} = f(\phi) + g(\phi)\xi_m(\mathbf{r}, t) + D\mathcal{L}[\phi] + h(\phi)F(t) + \xi_a(\mathbf{r}, t), \qquad (5.1)$$

where $\mathcal{L}$ is a (differential or integral) operator expressing the spatial coupling of the dynamics and D is a parameter expressing the strength of the spatial coupling. Depending on the specific stochastic model, some components in Eq. (5.1) can be absent. In the following discussion, we use the subscripts a and m only when essential to distinguish the dynamic role of the noise (additive or multiplicative).

The pattern-formation mechanism is sometimes called *breaking of ergodicity*. The meaning of this expression can be understood from general equation (5.1). The spatial coupling $\mathcal{L}$ prevents the dynamics of the state variable $\phi(\mathbf{r}^*, t)$, at a certain point $\mathbf{r}^*$, from spanning across the same phase space that would be explored by the corresponding zero-dimensional system (i.e., in the case of dynamics with no spatial coupling). Therefore the spatial coupling imposes a constraint on the dynamics and can modify the portion of the phase space that is explored by the spatiotemporal system. In this sense, spatially extended systems can exhibit breaking of ergodicity because the probability distribution of ϕ is different from that of the dynamics with no spatial coupling. This means that, when each point in the domain fluctuates independently of its neighbors (and of the rest of the field), the spatial average at a given time is the same as the temporal average at a certain point (i.e., the system is ergodic). Conversely, in the presence of spatial coupling, the spatial average of the field can be different from the temporal average at a given point. Thus the spatial coupling breaks the ergodicity of the system.

The role of noise can be also interpreted as an external input of energy that is dissipated by the dynamical system. This explains why noise-induced patterns are sometimes called *dissipative structures* (Manneville, 1990).

5.1.2.1 Models of the noise term

The key aspect in the dynamical systems investigated in this chapter is that patterned states are noise induced, i.e., they are induced by random fluctuations and do not occur in the deterministic counterpart of the dynamics. In fact, these symmetry-breaking states vanish as the noise intensity (i.e., the variance of the noise) drops below a critical value depending on the specific spatiotemporal stochastic model considered. At the same time, in some cases these noise-induced transitions have been found to be *reentrant*, in that the ordered phase is destroyed when the noise intensity exceeds another threshold value. In these cases, the noise has a constructive effect only when the variance is within a certain interval of values. Smaller or larger values of the variance correspond to conditions in which noise is either too weak or too strong to induce ordered states.

We concentrate on the case of noise with an isotropic covariance structure given by

$$\langle \xi(\mathbf{r}, t)\xi(\mathbf{r}', t') \rangle = s C \left(\frac{|\mathbf{r} - \mathbf{r}'|}{d}, \frac{|t - t'|}{\tau_c} \right), \tag{5.2}$$

where s is the noise intensity, $C(|\mathbf{r} - \mathbf{r}'|, |t - t'|)$ is the correlation function, and d and τ_c are the noise-correlation length and time, respectively. When $d \rightarrow 0$ and $\tau \rightarrow 0$, the case of spatially and temporally uncorrelated (i.e., white) noise is recovered. Most of the research on noise-induced pattern formation considered cases in which ξ_m and ξ_a are modeled as white-Gaussian-noise terms. Only a few papers investigated the impact of colored Gaussian noises, and even fewer studies explored the case of spatiotemporal dynamics driven by dichotomous noises and white shot noises. For this reason, unlike in the previous part of the book, in this chapter we place more emphasis on dynamics driven by Gaussian noise.

5.1.2.2 Models of spatial coupling

Several mathematical models are frequently used to express spatial coupling among different points in a field. Because the main focus of this chapter is on pattern formation, we propose to classify the mathematical operators that represent the spatial coupling as *pattern-forming* and *non-pattern-forming* operators. We define *pattern forming* as those operators that are able to generate periodic patterns in a deterministic univariate system when suitable parameter values (e.g., the strength D of the spatial coupling) and shape of the function $f(\phi)$ are used (see Appendix B). In contrast, non-pattern-forming operators may induce spatial coherence, but in a deterministic setting they do not produce patterns characterized by a clear dominant length scale. Multiscale patterns (see Subsection 5.1.1) may emerge under suitable conditions, but these are typically transient or oscillatory in noise-free conditions.

A typical example of a non-pattern-forming operator is the Laplacian operator,

$$\mathcal{L}[\phi] = \nabla^2 \phi = \frac{\partial^2 \phi}{\partial x^2} + \frac{\partial^2 \phi}{\partial y^2}, \tag{5.3}$$

which is often used to represent the effect of a diffusive process on a random field. A possible schematic representation of the effect of ∇^2 on one-dimensional (1D) spatial dynamics is reported in Fig. 5.6(a); the initial condition is a field where the values of

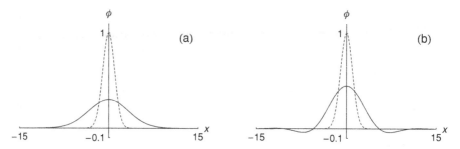

Figure 5.6. Example of the effect of (a) the Laplacian term and (b) the fourth-derivative term. Panel (a) refers to equation $\partial\phi/\partial t = \partial^2\phi/\partial x^2$ and panel (b) to the model $\partial\phi/\partial t = -\partial^4\phi/\partial x^4$. In both panels the initial condition (a bell-shaped distribution, $\phi = \exp[-x^2/2]$), is shown with the dashed curve, and the solid curves refer to the solution after 5 time units.

ϕ have a bell-shaped distribution, $\phi[x, t = 0] = \exp[-x^2/2]$. The field is subjected to a diffusive term $\nabla^2\phi$, i.e., $\partial\phi/\partial t = \partial^2\phi/\partial x^2$, and the distribution of ϕ is investigated at $t = 5$ time units. As expected, the Laplacian operator has the effect of producing a redistribution of ϕ along the x axis, with a smoothing of the peak and an increase of spatial coherence.

A mathematical structure that is useful in the description of pattern-forming couplings is

$$\mathcal{L}[\phi] = -a_0\nabla^2\phi - \nabla^4\phi, \tag{5.4}$$

where a_0 is a parameter and ∇^4 is the so-called biharmonic term, equal to

$$\nabla^4\phi = \frac{\partial^4\phi}{\partial x^4} + 2\frac{\partial^4\phi}{\partial x^2\partial y^2} + \frac{\partial^4\phi}{\partial y^4}. \tag{5.5}$$

If we reconsider the same example as before, but with $\partial\phi/\partial t = -\nabla^4\phi$, the result in Fig. 5.6(b) is obtained. It is evident that the spatial coupling in this case produces coherent structures with a clear periodicity, i.e., periodic patterns (see Appendix B).

Note that all the spatial-coupling operators considered so far have the property that they do not affect a spatially homogeneous field. In fact, $\mathcal{L}[\phi] = 0$ if the field is homogeneous. In contrast, the Swift–Hohenberg operator,

$$\mathcal{L}[\phi] = -(\nabla^2 + k_0^2)^2\phi \tag{5.6}$$

(where k_0 is a parameter), does not have this property, but can be decomposed into a pattern-forming term of the form of Eq. (5.4) (with $a_0 = 2k_0^2$) plus a drift term $f(\phi) = -k_0^4\phi$.

Other examples of mathematical operators that can be used to express the spatial coupling is provided by the integral operator

$$\mathcal{L}[\phi(\mathbf{r})] = \int \phi(\mathbf{r}')\omega(\mathbf{r} - \mathbf{r}')d\mathbf{r}', \tag{5.7}$$

where $\omega(\mathbf{r} - \mathbf{r}')$ is a kernel function. Depending on the shape of the kernel function, this integral spatial coupling may belong to the pattern-forming or non-pattern-forming category. Further details on these integral operators are provided in Chapter 6 and in Appendix B, along with a description of the linkage between operators based on derivatives and integrals.

Note that the presence of a pattern-forming coupling by itself does not mean that a spatial pattern will necessarily emerge. In fact, in many cases the shape of the function $f(\phi)$ or the intensity of the coupling may not allow the occurrence of steady patterns. In these cases, noise may have the key role of creating the dynamical conditions for which the spatial coupling exerts its pattern-forming action even though the deterministic counterpart of the dynamics is unable to generate stationary patterns. Several examples are provided in this chapter.

5.1.2.3 *Prognostic tools for pattern detection*

Unfortunately, the mathematical complexity of the spatiotemporal models expressed by Eq. (5.1) hampers general analytical solutions. For this reason, several approximate analytical techniques have been developed for obtaining useful indications about pattern formation. When the occurrence of a dominant wavelength is the main symptom of pattern formation, the first available prognostic tool is provided by stability analysis by normal modes. This analysis is based on the idea of disturbing the basic state (i.e., the homogeneous equilibrium state) of the system with an infinitesimal perturbation and to assess whether the perturbation grows in time (in that case patterns may emerge). The stability of the basic state is assessed with the dispersion equation, which is the condition relating the growth factor γ of the perturbation to its wave numbers $\mathbf{k} = (k_x, k_y)$. We refer the interested reader to Box 5.1 for details on the way the dispersion relation can be obtained.

Dispersion relation (B5.1-4) can exhibit different scenarios: (i) the growth factor γ is negative for all wave numbers. In this case no wavelength is unstable and no pattern may emerge; (ii) there is a range of wave numbers with positive growth factors, with γ exhibiting a maximum for a wave number k_{max} that is finite and different from zero. In this case, periodic patterns occur with a wavelength close to $\lambda = 2\pi/k_{max}$ (the exact wavelength will depend also on the boundary conditions); (iii) $\gamma(k_{max}) > 0$ but the most unstable mode is $k_{max} = 0$. In this case multiscale patterns may occur.

The dependence of the dispersion relation on the noise intensity s_m is generally studied to determine the transition from the first scenario (i.e., the case with no unstable modes) and the other two (i.e., with unstable modes). In particular, from Eq. (B5.1-4) it is possible to obtain the threshold value of the noise intensity

$$s_c = -\frac{f'(\phi_0) + Dh_{\mathcal{L}}(k)}{g'_s(\phi_0)}, \tag{5.8}$$

Box 5.1: Stability analysis by normal modes

The use of stability analysis in stochastic models follows the same conceptual steps described in Appendix B for deterministic systems, but here it is applied to the spatiotemporal dynamics of the ensemble average of the field variable ϕ. We consider the case of systems whose dynamics are expressed by Eq. (5.1) with no time-dependent forcing [i.e., $h(\phi)F(t) = 0$]. The linear-stability analysis includes three steps. First, a deterministic equation for the spatiotemporal dynamics of the ensemble average of the field variable $\langle\phi\rangle$ is obtained from (5.1) and the homogeneous basic state (i.e., the homogeneous steady state) is found. In this step, the most complex aspect is the evaluation of the ensemble average of the multiplicative random component $\langle g(\phi)\xi_m\rangle$. As shown in Box 3.2, this term may be expressed as the product of the noise intensity, s_m, by a function $g_S(\phi)$ of the state variable, $\langle g(\phi)\xi_m\rangle = s_m\langle g_S(\phi)\rangle$. The specific function $g_S(\phi)$ varies depending on the type of noise and its interpretation. $g_S(\phi) = 0$ when $g(\phi) = $ const or the noise term in the Langevin equation is interpreted in Ito's sense. In the case of Langevin equations with Gaussian white noise interpreted in the Stratonovich sense, the Novikov theorem provides the result $\langle g(\phi)\xi_{gn}\rangle = s_{gn}\langle g(\phi)g'(\phi)\rangle$ [see Eq. B3.2-2)], i.e., $g_S(\phi) = g(\phi)g'(\phi)$. Thus, taking the ensemble average of Eq. (5.1) [with $h(\phi)F(t) = 0$], we obtain

$$\frac{\partial\langle\phi\rangle}{\partial t} = \langle f(\phi)\rangle + s_m\langle g_S(\phi)\rangle + D\mathcal{L}[\langle\phi\rangle]. \tag{B5.1-1}$$

The basic state, $\langle\phi\rangle = \phi_0 = $ const, is obtained as the zero of Eq. (B5.1-1) at steady state, i.e., $f(\phi_0) + s_m g_S(\phi_0) = 0$.

   In the second step of the stability analysis we look at the effect of small perturbations of the basic state. To this end, Eq. (B5.1-1) is linearized, in that the effect of nonlinearities becomes negligible in the case of infinitesimal perturbations. The Taylor's expansion of Eq. (B5.1-1) around $\phi = \phi_0$, truncated to first order, leads to

$$\frac{\partial\langle\phi\rangle}{\partial t} = f'(\phi_0)\langle\phi\rangle + s_m g_S'(\phi_0)\langle\phi\rangle + D\mathcal{L}[\langle\phi\rangle], \tag{B5.1-2}$$

where $f'(\phi_0) = \frac{\mathrm{d}f(\phi)}{\mathrm{d}\phi}|_{\phi=\phi_0}$ and $g_S'(\phi_0) = \frac{\mathrm{d}g_S(\phi)}{\mathrm{d}\phi}|_{\phi=\phi_0}$. The basic state $\langle\phi\rangle = \phi_0$ is now disturbed (third step) by adding an infinitesimal harmonic perturbation

$$\langle\phi\rangle = \phi_0 + \hat{\phi}e^{\gamma t + i\mathbf{k}\cdot\mathbf{r}}, \tag{B5.1-3}$$

where $\hat{\phi}$ is the perturbation amplitude, γ is the growth factor, $i = \sqrt{-1}$ is the imaginary unit, $\mathbf{k} = (k_x, k_y)$ is the wave-number vector of the perturbation, and $\mathbf{r} = (x, y)$ is the coordinate vector. If (B5.1-3) is inserted into Eq. (B5.1-2), we obtain the so-called dispersion relation:

$$\gamma(k) = f'(\phi_0) + s_m g_S'(\phi_0) + Dh_{\mathcal{L}}(k), \tag{B5.1-4}$$

where $h_{\mathcal{L}}(k)$, the Fourier transform of $\mathcal{L}(\phi)$, is a function of the wave number $k = |\mathbf{k}|$ and depends on the specific form of spatial coupling $\mathcal{L}$ considered. For example, if $\mathcal{L}(\phi) = \nabla^2(\phi)$, $h_{\mathcal{L}}(k) = -k^2$; if $\mathcal{L}(\phi) = \nabla^4(\phi)$, $h_{\mathcal{L}}(k) = k^4$.

corresponding to the occurrence of a wave number with $\gamma = 0$, i.e., of conditions of marginal stability. When $s_m > s_c$, positive growth factors emerge, and spatial patterns (periodic or multiscale, depending on the value of k_{max}) are expected to occur.

Stability analysis may be a useful tool for pattern recognition, but it is not without faults: Its effectiveness depends on the specific stochastic model under consideration. We can also use other more complicated, prognostic tools based on the structure function. The basic idea behind this method is that the presence of patterns will eventually modify the correlation structure of the field. Instead of considering the correlation function, this method analyzes its Fourier transform in space, which is known as the *structure function* or power spectrum (see Appendix A):

$$S(\mathbf{k}, t) = \int_{\mathcal{D}} e^{-i\mathbf{k}\cdot\mathbf{r}} G(\mathbf{r}, t) d\mathbf{r}, \tag{5.9}$$

where $G(\mathbf{r}, t)$ is the correlation function,

$$G(\mathbf{r}, t) = \frac{1}{A} \int_{\mathcal{D}} \phi(\mathbf{r}', t)\phi(\mathbf{r}' + \mathbf{r}, t) d\mathbf{r}' - \frac{M(t)^2}{A^2}, \tag{5.10}$$

$\mathcal{D}$ is the 2D spatial domain, A is its area, and

$$M(t) = \int_{\mathcal{D}} \phi(\mathbf{r}, t) d\mathbf{r}. \tag{5.11}$$

The structure function is useful in describing the spatial periodicity of the system, similar to the way the power spectrum describes the temporal periodicity of a temporal signal. The use of the structure function as a prognostic tool requires some methods to determine the dynamics of $S(\mathbf{k}, t)$ from Eq. (5.1). These methods are explained in Box 5.2.

Once the equation describing the time evolution of the structure function [i.e., Eq. (B5.2-5)] is obtained, we can determine the corresponding steady-state expression [Eq. (B5.2-7)] and use it to assess whether periodic patterns – corresponding to the existence of a maximum of the structure function for wave numbers k different from zero – are expected to appear. Equation (B5.2-7) clearly shows that the additive noise is fundamental to having a nonnull steady-state structure function. The role of the multiplicative noise is instead fundamental in determining the wavelength of the emerging patterns: In fact, the dominant wavelength k_{max} is found where the denominator of Eq. (B5.2-7) is maximum.[1]

The third prognostic tool to foresee pattern formation is based on the generalized mean-field theory, and it is presented in Box 5.3.

In Boxes 5.1, 5.2 and 5.3 we have described three analytical prognostic tools that may be useful for assessing the occurrence of spatial patterns when the equation

---

[1] Note that the denominator is rather similar to dispersion relation (B5.1-4): In particular, when $g(\phi) = \phi$ and the multiplicative noise is white and Gaussian, we have that $g'_s(\phi) = 1$. Thus $\gamma(k) = f'(\phi_0) + Dh_{\mathcal{L}}(k) + s_m$, as in the denominator of Eq. (B5.2-7).

Box 5.2: Structure function

The structure function is defined in Fourier space as

$$S(\mathbf{k}, t) = \langle \hat{\phi}(\mathbf{k}, t)\hat{\phi}(-\mathbf{k}, t)\rangle, \tag{B5.2-1}$$

where $\hat{\phi}(-\mathbf{k}, t)$ is the Fourier transform of $\phi(\mathbf{r}, t)$.

The first-order temporal derivative of the structure function is

$$\frac{\partial S(\mathbf{k}, t)}{\partial t} = \frac{\partial}{\partial t}\langle \hat{\phi}(\mathbf{k}, t)\hat{\phi}(-\mathbf{k}, t)\rangle$$

$$= \left\langle \frac{\partial \hat{\phi}(\mathbf{k}, t)}{\partial t}\hat{\phi}(-\mathbf{k}, t)\right\rangle + \left\langle \frac{\partial \hat{\phi}(-\mathbf{k}, t)}{\partial t}\hat{\phi}(\mathbf{k}, t)\right\rangle. \tag{B5.2-2}$$

The terms on the right-hand side of Eq. (B5.2-2) depend on the first-order temporal derivative of the Fourier transform of $\phi(\mathbf{r}, t)$, which we can calculate by taking the Fourier transform of the linearized version of Eq. (5.1) [with $F(t) = 0$], applying a procedure similar to the one used for Eq. (B5.1-2):

$$\frac{\partial \hat{\phi}(\mathbf{k}, t)}{\partial t} = \int_0^\infty \left\{ f'(\phi_0)\phi(\mathbf{r}, t) + g'(\phi_0)\phi(\mathbf{r}, t)\xi_m(\mathbf{r}, t) + \xi_a(\mathbf{r}, t) \right.$$

$$\left. + D\mathcal{L}[\phi(\mathbf{r}, t)] \right\} e^{-i\mathbf{k}\cdot\mathbf{r}}d\mathbf{r}. \tag{B5.2-3}$$

Equation (B5.2-3) can be substantially simplified if the noise terms are uncorrelated (or white) in space. In this case Eq. (B5.2-3) can be rewritten as

$$\frac{\partial \hat{\phi}(\mathbf{k}, t)}{\partial t} = f'(\phi_0)\hat{\phi}(\mathbf{k}, t) + g'(\phi_0)\hat{\phi}(\mathbf{k}, t)\xi_m(t) + \xi_a(t) + Dh_\mathcal{L}(k)\hat{\phi}(\mathbf{k}, t), \tag{B5.2-4}$$

where $h_\mathcal{L}(k)$ is the same operator already defined in Box 5.1. Using Eq. (B5.2-4) in (B5.2-2) we obtain

$$\frac{\partial}{\partial t}\langle \hat{\phi}(\mathbf{k}, t)\hat{\phi}(-\mathbf{k}, t)\rangle = 2\left[f'(\phi_0) + Dh_\mathcal{L}(k) \right] \langle \hat{\phi}(\mathbf{k}, t)\hat{\phi}(-\mathbf{k}, t)\rangle$$

$$+ 2g'(\phi_0)\langle \hat{\phi}(\mathbf{k}, t)\hat{\phi}(-\mathbf{k}, t)\xi_m(t)\rangle$$

$$+ \langle \hat{\phi}(\mathbf{k}, t)\xi_a(t)\rangle + \langle \hat{\phi}(-\mathbf{k}, t)\xi_a(t)\rangle. \tag{B5.2-5}$$

We can now recognize that $\langle \hat{\phi}(\mathbf{k}, t)\hat{\phi}(-\mathbf{k}, t)\rangle = S(\mathbf{k}, t)$ by definition; the remaining terms involve products of $\hat{\phi}$ (or $\hat{\phi}^2$) and noise, and can be treated as follows. If we define $z = \hat{\phi}(\mathbf{k}, t)\hat{\phi}(-\mathbf{k}, t)$, i.e., $S(\mathbf{k}, t) = \langle z\rangle$, the second and third addenda in Eq. (B5.2-5) can be written in the form $\langle u(z)\xi\rangle$, with $u(z) = z$ in the second addendum and $u(z) = \sqrt{z}$ in the third addendum. The expected value of the product of a function of a stochastic variable, and a noise term assumes different values depending on the considered type of noise. In the case of Gaussian white noise in a Langevin equation interpreted in the Stratonovich sense we can apply Novikov's theorem (see Box 3.2), $\langle u(z)\xi_{gn}\rangle = s_{gn}\langle u(z)u'(z)\rangle$, i.e., $\langle \hat{\phi}(\mathbf{k}, t)\hat{\phi}(-\mathbf{k}, t)\xi_{gn,m}(t)\rangle = s_{gn,m}S(\mathbf{k}, t)$ and $\langle \hat{\phi}(\pm\mathbf{k}, t)\xi_{gn,a}(t)\rangle = \frac{s_{gn,a}}{2}$, where $s_{gn,a}$ is the intensity of the additive white Gaussian noise.

When the various terms are recomposed, Eq. (B5.2-5) becomes

$$\frac{\partial S(\mathbf{k}, t)}{\partial t} = 2\left[f'(\phi_0) + Dh_{\mathcal{L}}(k) + g'(\phi_0)s_{gn,m}\right]S(\mathbf{k}, t) + s_{gn,a}. \tag{B5.2-6}$$

Therefore at steady state the structure function is

$$S_{st}(\mathbf{k}) = -\frac{s_{gn,a}}{2\left[f'(\phi_0) + Dh_{\mathcal{L}}(k) + g'(\phi_0)s_{gn,m}\right]}. \tag{B5.2-7}$$

expressing the spatiotemporal evolution of ϕ is known. In the following subsections each of these methods is applied to specific examples and the skills and limitations of each approach are discussed. It is worth mentioning that in general the wavelength $2\pi/k_{max}$, resulting from stability analysis, structure function, or generalized mean field, will actually emerge only in an infinite domain. In this case, possible (small) differences in wavelength – with respect to $2\pi/k_{max}$ – observable in the numerical simulations of the complete model are due to approximations associated with the mathematical techniques used in each of these prognostic methods. An example is the linear approximation involved in stability analysis. In contrast, when finite domains are considered (e.g., no-flux boundary conditions), wave numbers significatively different from k_{max} may emerge because the dominant wavelength has to be compatible with the size of the domain. Generally the most unstable wavelength compatible with these boundary conditions is selected in the process of pattern formation (see Murray (2002)).

When the expected spatial fields do not have a clear periodicity, i.e., in the case of multiscale patterns, it is difficult to foresee the pattern shape by using these theoretical analyses. Dispersion relation, steady-state structure function, and generalized mean-field analysis can give some insight into pattern formation, but the definitive way to assess noise-induced pattern formation is to numerically simulate the dynamics and visually inspect the results. Such qualitative analysis can then be complemented by quantitative evaluations, such as the analysis of transects and the investigation of the statistical properties of coherence regions (e.g., Dale, 1999; Kent et al., 2006).

5.1.2.4 Numerical simulation of random fields

Along with mathematical prognostic tools, numerical simulations of stochastic models are fundamental to verifying the analytical findings and characterizing patterns. The typical approach is to discretize the continuous spatial domain by use of a regular Cartesian lattice with spacing $\Delta x = \Delta y = \Delta$. Original stochastic partial differential

Box 5.3: Generalized mean-field theory

The mean-field theory is typically used to provide an approximated solution of stochastic partial differential equations for spatially extended systems. The method is valuable mainly for a qualitative analysis of (stochastic) spatiotemporal dynamics (van den Broeck et al., 1994; van den Broeck, 1997; Buceta and Lindenberg, 2003; Porporato and D'Odorico, 2004).

The mean-field technique adopts a finite-difference representation of stochastic spatiotemporal dynamics (5.1):

$$\frac{d\phi_i}{dt} = f(\phi_i) + g(\phi_i)\xi_{m,i} + D\, l(\phi_i, \phi_j) + \xi_{a,i}, \qquad (B5.3\text{-}1)$$

where for sake of simplicity we have set $F(t) = 0$. In Eq. (B5.3-1), ϕ_i, $\xi_{m,i}$, and $\xi_{a,i}$ are the values of ϕ, ξ_m, and ξ_a at site i, respectively; i runs across all the discretization cells, $j \in nn(i)$ refers to the neighbors of the ith cell, and $l(\phi_i, \phi_j)$ expresses the spatial coupling between cell i and its neighbors (see Subsection 5.1.2.4). A general expression for $l(\phi_i, \phi_j)$ is

$$l(\phi_i, \phi_j) = w_i\phi_i + \sum_{j \in nn(i)} w_j\phi_j, \qquad (B5.3\text{-}2)$$

where w_i and w_j are weighting factors. Expressions for these weighting factors are specified in Subsection 5.1.2.4 for some commonly used spatial couplings.

The analytical solution of Eq. (B5.3-1) is hampered by the fact that the dynamics of ϕ_i are coupled to those of the neighboring points. In fact, the spatial interaction term in (B5.3-1) depends on the values ϕ_j of ϕ in the neighborhood of i. To avoid this obstacle, the mean-field approach assumes that (i) the variables ϕ_j can be approximated by the local ensemble mean $\langle \phi_j \rangle$, and (ii) there is a link between $\langle \phi_j \rangle$ and the ensemble average $\langle \phi_i \rangle$ in the ith cell.

Typically we are interested in the stability and instability conditions of the homogeneous basic state with respect to periodic perturbations. For this reason, the ith site is placed on a local maximum of the field, and the pattern is approximated by a harmonic function,

$$\langle \phi_j \rangle = \langle \phi_i \rangle \cos[\mathbf{k} \cdot (\mathbf{r}_i - \mathbf{r}_j)], \qquad (B5.3\text{-}3)$$

where $\mathbf{k} = (k_x, k_y)$ and k_x and k_y are the two wave numbers along the x and y axes, respectively. It follows that the function $l(\phi_i, \phi_j)$ in Eq. (B5.3-1) is approximated as

$$l(\phi_i, \phi_j) \approx l_h(\phi_i, \langle \phi_i \rangle, k_x, k_y), \qquad (B5.3\text{-}4)$$

where $l_h(\cdot)$ is a function whose structure depends on the spatial coupling considered.

Under assumption (B5.3-4), dynamics (B5.3-1) of ϕ_i do not depend anymore on those of the neighboring points, and it is possible to determine exact expressions for the steady-state probability distributions, $p_{st}(\phi; \langle \phi_i \rangle, k_x, k_y)$, of ϕ by the methods described in Chapter 2. $p_{st}(\phi; \langle \phi_i \rangle, k_x, k_y)$ will necessarily depend on a number of parameters of the dynamics, on k_x and k_y, and on $\langle \phi_i \rangle$, which remains unknown. To determine $\langle \phi_i \rangle$, we observe that the mean of $p_{st}(\phi; \langle \phi_i \rangle, k_x, k_y)$ must coincide with $\langle \phi_i \rangle$. Therefore the self-consistency condition,

$$\langle \phi_i \rangle = \int_{-\infty}^{+\infty} \phi \, p_{st}(\phi; \langle \phi_i \rangle, k_x, k_y) \, d\phi = F(\langle \phi_i \rangle, k_x, k_y), \qquad \text{(B5.3-5)}$$

can be used to obtain the unknown $\langle \phi_i \rangle$ as a function of the wave numbers k_x and k_y. Notice that when $k_x = k_y = 0$, Eq. (B5.3-3) becomes $\langle \phi_j \rangle = \langle \phi_i \rangle$. This corresponds to the classic mean-field analysis, which is illustrated in Box 5.4.

The occurrence of solutions of Eq. (B5.3-5) different from $\langle \phi_i \rangle = \phi_0$ (where ϕ_0 is the basic state of the system) corresponds to the loss of stability of the uniform basic state with respect to periodic disturbances. It is expected that this loss of stability takes place for only some specific value of the wave numbers k_x and k_y. More in detail, for each form of the spatial coupling and of the equation governing the dynamics of ϕ, a specific wave number $\mathbf{k}_{max}$ will have the maximum destabilizing effect on ϕ_0. Spatial patterns will likely evolve in these cases, with characteristic wavelength $2\pi/k_{max}$.

equation (5.1) is then transformed into a system of coupled stochastic ordinary differential equations:

$$\frac{d\phi_i}{dt} = f(\phi_i) + g(\phi_i)\,\xi_{m,i}(t) + D\,l(\phi_i, \phi_j) + h(\phi_i)\,F(t) + \xi_{a,i}(t), \qquad (5.12)$$

where $\phi_i(t)$ is the value of the discretized state variable in the ith cell of the lattice, $l(\phi_i, \phi_j)$ is the discretized counterpart of the specific spatial coupling considered, and j runs over a suitable set of neighbors of cell i. As mentioned in Box 5.3, a general expression for $l(\phi_i, \phi_j)$ is

$$l(\phi_i, \phi_j) = w_i\phi_i + \sum_{j \in nn(i)} w_j\phi_j, \qquad (5.13)$$

where the number of neighbors $nn(i)$ and weighting factors w_i and w_j depend on the specific finite-difference scheme adopted to numerically approximate the spatial operator (Strikwerda, 2004). For example, in the case of the Laplacian operator the simplest (and most-used) scheme is

$$\mathcal{L}[\phi] = \nabla^2\phi \approx l(\phi_i, \phi_j) = \frac{1}{\Delta^2} \sum_{j \in nn(i)} (\phi_j - \phi_i), \qquad (5.14)$$

where the four nearest neighbors are involved, $w_i = -4$, and $w_j = 1$. This corresponds to the set of weighting factors schematically reported in Fig. 5.7. In the case of the ∇^4 operator, the standard numerical approaches adopt the weighting factors reported in Fig. 5.8.

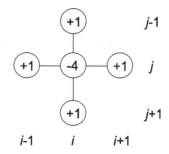

Figure 5.7. Weighting factors for the discretization of the Laplacian operator $\nabla^2\phi$ in the cell $\{i, j\}$.

Infinitely extended random fields are generally suitably simulated by periodic boundary conditions. Numerical simulation of the stochastic equations is not trivial, and attention has to be paid to the interplay between the temporal and the spatial discretizations and the simulation of the noise component. In this chapter we adopt the methods described by Carrillo et al. (2003) and Garcia-Ojalvo and Sancho (1999). We refer the readers to these publications for further details.

5.1.3 Other noise-induced phenomena in spatiotemporal systems

Formation of spatial patterns is not the only noise-induced phenomenon that might be of interest in the environmental sciences. Indeed, even though there exists no univocal definition of noise-induced phenomena in a spatiotemporal dynamical system, a broad definition could encompass all phenomena in which the presence of (noisy) stochastic

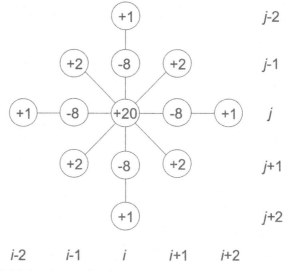

Figure 5.8. Weighting factors for the discretization of the biharmonic operator $\nabla^4\phi$ in the cell $\{i, j\}$.

forcing significantly modifies some prominent statistical feature of the system. In the case of spatial patterns the signature of noise-induced phenomenon is a modification of the spatial-correlation function (or the structure function), but noise-induced modifications of other statistical descriptors of the random field (e.g., mean, modes, or variance) can also be very relevant.

Usually modifications of *order parameters* (which are global quantities used to describe general aspects of the spatial distribution of the state variable ϕ) are known as *phase transitions*. In zero-dimensional systems (see Chapter 3), we considered the mode of the pdf as the order parameter. In fact, in pure temporal systems the mean can be poorly representative of transitions, whereas the modes indicate the most frequently visited states of the dynamics and are therefore particularly suitable to represent structural changes of the system's behavior. Moreover, analytical expressions for the modes can be determined (Chapter 3), and this allows a theoretical analysis of transitions. Conversely, in spatiotemporal systems exact analytical results are very rare, and this hampers the analysis of the modes. For this reason, the simplest order parameter – i.e., the spatiotemporal average of the state variable at steady state – is more frequently used[2] because analytical approximated techniques have been developed for the analysis of this order parameter (see Box 5.4). In this case, the order parameter is

$$m = \frac{\overline{M(t)}}{A},\qquad(5.15)$$

where the overbar indicates the temporal mean, and a noise-induced phase transition occurs when the random component is able to change the value of m with respect to the basic homogeneous steady state ϕ_0.[3]

Notice that the occurrence of a nonequilibrium (i.e., order-forming) phase transition is neither a necessary nor a sufficient condition for noise-induced pattern formation. In fact, a nonequilibrium phase transition implies that noise is able to change the value of the order parameter, but not that ordered geometrical structures necessarily appear. Conversely, a number of cases exist in which patterns occur but m remains unchanged with respect to the disordered case. In the thermodynamics literature, fields with $m \neq \phi_0$ are sometimes called ordered phases (hence the use of the term *order parameter* to indicate m) and nonequilibrium phase transitions often seem to be used, implying pattern occurrence. For the sake of clarity, in the following discussion we keep the occurrence of patterns distinct from the existence of a phase transition in the mean.

---

[2] Another possible order parameter that is (less frequently) used is

$$J_{st} = \frac{\bar{J}(t)}{A} \quad \text{with} \quad J(t) = \int_{\mathcal{D}} \phi^2(\mathbf{x}, t)\mathrm{d}\mathbf{x},$$

where J_{st} recalls the flux of convective heat (Garcia-Ojalvo and Sancho, 1999).

[3] m is often used as an order parameter because of its correspondence to the so-called *magnetization* used in the physics literature in the study of equilibrium systems.

Box 5.4: Classic mean-field analysis

The classic mean-field theory can be presented as a simplified version of the generalized mean field described in Box 5.3. In this case it is assumed that all cells have the same mean, which coincides with the spatiotemporal mean of the field. This corresponds to taking $\mathbf{k} = 0$ in Eq. (B5.3-3), i.e., $\langle \phi_j \rangle = \langle \phi_i \rangle = m$, with m defined in Eq. (5.15). In this case, Eq. (B5.3-5) in Box 5.3 becomes

$$m = \int_{-\infty}^{+\infty} \phi \, p_{st}(\phi; m) \, d\phi = F(m). \tag{B5.4-1}$$

Multiple solutions of Eq. (B5.4-1) indicate the existence of multiple phase transitions of the system. In this case the mean-field analysis is not used to investigate the emergence of periodic patterns but the occurrence of phase transitions. This zero-wave-number version of the mean-field approximation is the standard mean-field method, and the mean-field analysis that involves $k_x \neq 0$ (or $k_y \neq 0$) is called generalized mean-field theory (Sagues et al., 2007).

The effectiveness of the standard mean-field approximation can be improved by expressing the values of ϕ_j in the neighborhood of point i as the average between the spatiotemporal mean and the local value of ϕ at point i, namely

$$\phi_j \approx \frac{1}{2}(m + \phi_i). \tag{B5.4-2}$$

This correction of the mean-field approximation accounts for the dependence of ϕ_j on the local conditions (see Sagues et al., 2007). On the basis of this analysis we conclude that the standard mean-field assumption can overestimate the strength of the spatial coupling, and the diffusivity D in (B5.4-1) could be replaced with a new diffusivity $D_n = D/2$ (Sagues et al., 2007). For sake of simplicity in the examples presented in this chapter we apply the classical form of the mean-field approximation without using this correction.

The mean-field framework is also used to investigate different mechanisms of noise-induced pattern formation. For example, Zhonghuai et al. (1998) used this method to study noise-induced phase transitions in generic two-variable systems exhibiting Turing instability. When a control parameter of the kinetics is perturbed by noise, new kinds of patterns arise (transition from single-spiral to double-spiral waves). A similar study was developed by Carrillo et al. (2004) for the analysis of pattern formation in chemical reactions and fluid convection. Further, this method was used in Chapter 4 to investigate the emergence of phase transitions in multivariate (temporal) population dynamics.

Section	Ingredients	Prototype	Pattern shape (field pdf)	Prognostic tool	Role of interpretation of the Langevin equation
5.2	additive noise + pattern-forming spatial coupling	Eq. (5.16)	periodic (unimodal)	structure function (peak in $k \neq 0$)	patterns with Stratonovich and Ito interpretations
5.3	additive noise + no pattern-forming spatial coupling	Eq. (5.29)	multiscale (unimodal)	structure function (peak in $k = 0$)	patterns with Stratonovich and Ito interpretations
5.4	multiplicative noise + pattern-forming spatial coupling	Eq. (5.34)	periodic (unimodal)	stability analysis structure function	pattern only with Stratonovich interpretation
5.5	multiplicative noise + no pattern-forming spatial coupling	Eq. (5.58)	multiscale (unimodal)	stability analysis structure function	pattern only with Stratonovich interpretation
5.7	temporal phase transition	Eq. (5.73) Eq. (5.81)	periodic or multiscale (unimodal or bimodal)	mean-field analysis (when the spatial coupling is pattern forming)	patterns with Stratonovich and Ito interpretations

Figure 5.9. Models of noise-induced pattern formation resulting from the interaction among the local dynamics, $f(\phi)$, the noise component, and spatial coupling.

Finally, it is worth mentioning other interesting noise-induced phenomena in spatiotemporal systems, as self-organized criticality (Bak, 1996), noise-induced synchronization and control (e.g., Sagues et al., 2007), and the dynamics of fronts (e.g., Garcia-Ojalvo and Sancho, 1999). However, these phenomena are not discussed here as they are beyond the scope of this book. Thus, to avoid dispersion in too broad a research field, in this chapter we concentrate only on noise-induced pattern formation.

5.1.4 Chapter organization

In this chapter the description of the models of noise-induced pattern formation is organized as follows: We classify the models based on (i) whether the noise is additive or multiplicative, and (ii) the nature of the spatial coupling, which may or may not be able to form patterns in the deterministic counterpart of the dynamics. We refer to *pattern-forming spatial coupling* when the deterministic counterpart of the dynamics exhibits pattern formation within a certain range of parameter values. This approach forms the skeleton of the chapter and allows us to elucidate the key points of the interplay among the local dynamics, $f(\phi)$, the noise component, and the spatial coupling (the table shown in Figure 5.9 synthesizes the main characteristics of the models presented in the following subsections). Finally, the sections at the end of the chapter are devoted to the description of two peculiar mechanisms of noise-induced patterning that need the cooperation of a temporal periodicity, namely spatiotemporal stochastic resonance and spatiotemporal coherence resonance.

5.2 Additive noise and pattern-forming spatial couplings

Consider the stochastic model

$$\frac{\partial \phi}{\partial t} = a\phi + D\,\mathcal{L}[\phi] + \xi_{\mathrm{gn}}, \qquad (5.16)$$

where $\phi(\mathbf{r}, t)$ is the scalar field, a is a parameter, and ξ_{gn} is zero-mean Gaussian white (in space and time) noise with intensity s_{gn}. Equation (5.16) is the prototype model to illustrate how patterns may occur when both multiplicative noise and time-dependent components are not present in general model (5.1) [i.e., $F(t) = 0$ and $g(\phi) = 0$]. In this section we study the case in which $\mathcal{L}[\phi]$ is a pattern-forming spatial coupling; the case with non-pattern-forming coupling is considered in the next section.

In more detail, $\mathcal{L}[\phi]$ is modeled here as a Swift–Hohenberg coupling [see Eq. (5.6)]. As mentioned in Subsection 5.1.2.2, the Swift–Hohenberg coupling can be decomposed into a drift term $-k_0^4\phi$ and a pattern-forming spatial coupling given by Eq. (5.4). The Swift–Hohenberg coupling is one of the simplest couplings able to form patterns in a deterministic setting, and for this reason we adopt it as an exemplifying case. In the sixth chapter we discuss in detail the physical meaning of the biharmonic operator and the link between the mathematical structure $-D(\nabla^2 + k_0^2)^2$ and the integral form of the spatial coupling $\mathcal{L}[\phi]$. This will help us understand why such coupling is frequently used in a number of applications (Cross and Hohenberg, 1993). Here we recall only that structure (5.6) was introduced by Swift and Hohenberg (1977) to study the effect of hydrodynamic fluctuations in systems exhibiting Rayleigh–Benard convection (Chandrasekhar, 1981). This phenomenon leads to the emergence of organization in atmospheric convection, which is often evidenced by well-organized cloud patterns. The organization results from (symmetry-breaking) thermoconvective instability typically observed when a fluid overlies a hot surface (Chandrasekhar, 1981).

5.2.1 Analysis of the deterministic dynamics

We first study the following deterministic dynamics associated with model (5.16):

$$\frac{\partial \phi}{\partial t} = f(\phi) + D\,\mathcal{L}[\phi] = a\phi - D(\nabla^2 + k_0^2)^2\phi. \qquad (5.17)$$

When $a < 0$, the asymptotic steady state is the homogeneous field $\phi = 0$, regardless of the initial conditions (i.e., $\phi = 0$ is a stable state). However, if the transient between the initial condition and the steady state is sufficiently long and D/a sufficiently large, the spatial coupling may be able to cause the emergence of transient spatial patterns. Figure 5.10 shows an example of this behavior: If weak spatial gradients existing in the initial condition activate spatial interactions, during the transient the spatial coupling term [i.e., $-D(\nabla^2 + k_0^2)^2\phi$] is able to form well-developed patterns that

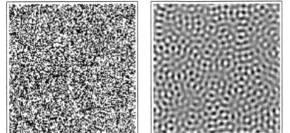

Figure 5.10. Dynamics described by model (5.17) when a is negative ($a = -0.1$, $D = 10, k_0 = 1$). The three panels correspond to t equal to 0, 5, and 35 time units, the field has 128×128 pixels, and periodic boundary conditions are set. The gray-tone scale spans the interval $[10^{-3}, 10^3]$. The initial condition is $\phi(\mathbf{r}, t = 0) = \xi_u$, where ξ_u is a noise uniformly distributed in the interval $[10^{-2}, 10^2]$.

tend to vanish as the system tends to the homogeneous stable state $\phi = 0$. Notice that the transient pattern exhibits the periodicity of about $2\pi/k_0$ pixels imposed by the spatial coupling (see Appendix B).

Conversely, when a is positive no steady states exist and the dynamics of ϕ diverge. However, even in this case the spatial terms in (5.17) are able to induce patterns that become more and more pronounced (i.e., with stronger spatial gradients) as the dynamics diverge (see Fig. 5.11). To stabilize the dynamics to a steady state, it is necessary to introduce a nonlinear term that hampers the dynamics in diverging. A relatively simple nonlinear term is $-\phi^3$; in this case the dynamics turn into the celebrated Ginzburg–Landau model, with local dynamics expressed as $f(\phi) = a\phi - \phi^3$,

$$\frac{\partial \phi}{\partial t} = a\phi - \phi^3 - D(\nabla^2 + k_0^2)^2 \phi. \tag{5.18}$$

For small values of ϕ – in the interval $[-\sqrt{a}, \sqrt{a}]$ – the diverging effect of the linear term $a\phi$ prevails (with $a > 0$), and as ϕ increases the stabilizing effect of the

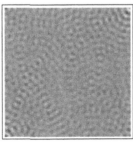
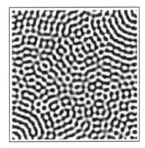

Figure 5.11. Example of the dynamics described by deterministic model (5.17) when a is positive ($a = +0.1$, $D = 10$, $k_0 = 1$). The three panels correspond to t equal to 0, 30, and 60 time units, and the gray-tone scale spans the interval $[-0.1, 0.1]$. The other conditions are as in Fig. 5.10.

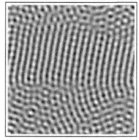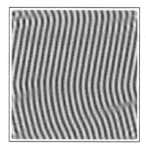

Figure 5.12. Dynamics described by Ginzburg–Landau model (5.18) when a is positive ($a = +0.1$, $D = 10$, $k_0 = 1$). The three panels correspond to t equal to 0, 150, and 300 time units, and the gray-tone scale spans the interval $[-0.5, 0.5]$. Other conditions are as in Fig. 5.10.

nonlinear term comes into play. In this way, the linear term maintains the dynamics away from the uniform state, $\phi = 0$, thereby allowing the spatial terms to remain active (i.e., different from zero) and to generate patterns. Because of the nonlinear term, the dynamics do not diverge but tend to a statistically steady state. Figure 5.12 shows an example of this type of dynamic. Notice that the dominant length scale is again $\lambda = 2\pi/k_0$, but the competition among the unstable modes close to k_0 gives rise to a transition between a spots-and-gaps pattern to quasi-cylindrical waves.

The occurrence of a periodic pattern when $a > 0$ can be easily assessed by the analysis by normal modes of the homogeneous state $\phi = 0$, as described in Box 5.1. Using Eq. (B5.1-4) we obtain the dispersion relation

$$\gamma(k) = a - D(k_0^2 - k^2)^2, \tag{5.19}$$

where $k = |\mathbf{k}|$. The most unstable mode is $k = k_0$, with $\gamma(k_0) = a$. Therefore when $a > 0$ the homogeneous state $\phi = 0$ is unstable and periodic patterns emerge if $k_0 \neq 0$. Notice that only the linear component of $f(\phi)$ appears in the linear-stability analysis (i.e., stability analysis with respect to infinitesimal harmonic perturbations) in that the nonlinear terms contribute with higher-order infinitesimals. Therefore the linear-stability analysis is unable to assess whether the nonlinear terms can prevent divergence. Moreover, the linear-stability analysis by normal modes is not able to detect the transient occurrence of patterns when $a < 0$, because such a technique investigates only the long-term behavior of the system (Trefethen et al., 1993). For $a < 0$, dispersion relation (5.19) always gives $\gamma < 0$, indicating that no patterns exist in the steady state of the system (i.e., for $t \rightarrow \infty$).

This analysis of deterministic model (5.17) shows that in this system periodic patterns may always emerge because of the presence of pattern-forming spatial coupling. If $a < 0$ the local dynamics $f(\phi)$ tend to relax to a stable state. In this case patterns are transient and disappear after some time. If in contrast $a > 0$, the local component

$f(\phi)$ is able to sustain the dynamics, the patterns remain stable in time, and a steady state is reached if a suitable nonlinear component is also present. However, it is important to stress that such a nonlinear term is not crucial to pattern formation, in that its role is only to stabilize the field ϕ that would otherwise diverge.

This example suggests a general rule: When a spatial coupling able to induce periodic patterns is present in a spatiotemporal dynamical system, patterns may emerge, provided that the local dynamics maintain the system away from the homogeneous steady state. This happens when the homogeneous steady state is unstable, i.e., when

$$\frac{\mathrm{d}f(\phi)}{\mathrm{d}\phi}\bigg|_{\phi_0} > 0. \tag{5.20}$$

On the contrary, when the derivative in (5.20) is negative, patterns can occur, but only transiently.

5.2.2 The role of the additive noise

These analyses of the deterministic dynamics underlying stochastic model (5.16) are fundamental to understanding the role of additive noise and its ability to induce pattern formation. In fact, the key idea is that additive noise ξ_a is able to keep the dynamics away from their homogeneous steady state even when this state is stable. Although in this case the deterministic system can exhibit only transient patterns that vanish as $t \to \infty$, in the presence of additive noise, patterns may persist, sustained by the noise. To demonstrate this pattern-inducing role of the additive noise ξ_a, Fig. 5.13 shows the results of numerical simulations of stochastic model (5.16). The results confirm that a very clear and stable pattern occurs in spite of a being negative: The noise component maintains the dynamics away from the deterministic steady state $\phi_0 = 0$, thereby allowing the spatial differential terms to drive the system into the patterned state with wavelength $2\pi/k_0$. In this sense, the pattern is noise induced; in fact, if the noise variance is set to zero, patterns disappear, and the system converges to its homogeneous stable state $\phi = 0$. Finally, we notice that the patterns in Fig. 5.13 are similar to those shown in Figs. 5.10 and 5.11, because the main geometric characteristic of these patterns (i.e., the dominant periodicity) is dictated by spatial coupling $\mathcal{L}[\phi]$, which remains the same in all cases. The most obvious difference between deterministic and stochastic patterns induced by additive noise (beside the fact that deterministic patterns are transient or unsteady) is the irregularity of the contours in the stochastic case (i.e., patterns are *fringed*), which is due to the local disturbance caused by the noise.

This pattern-inducing role of the additive noise can be detected by the structure function, as defined in Box 5.2. Using Eq. (B5.2-7) we obtain

$$S_{\mathrm{st}}(k) = \frac{s_{\mathrm{gn}}}{2\left[D(-k^2 + k_0^2)^2 - a\right]}. \tag{5.21}$$

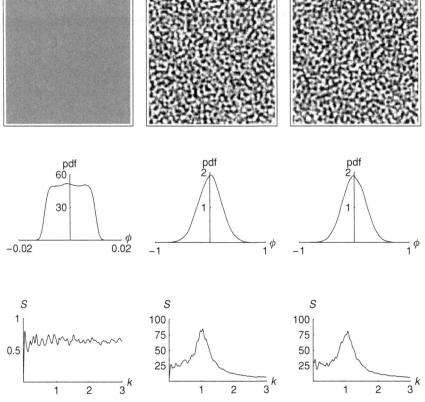

Figure 5.13. Numerical simulation of stochastic model (5.16) with $a = -1$, $D = 10$, $k_0 = 1$, and $s_{\mathrm{gn}} = 0.5$ (the other conditions are as in Fig. 5.10). The columns refer to 0, 50, and 100 time units. The first row shows the field (the gray-tone scale spans the interval $[-0.2, 0.2]$), the second row shows the pdf of the field variable ϕ, and the third row shows the azimuth-averaged spectrum of the field. The boundary conditions are periodic.

Thus the steady-state structure function has a maximum at $k = k_0$ even for $a < 0$ [see Fig. 5.14(a)]. This result confirms that additive random fluctuations are able to induce a stable pattern. Conversely, when the noise is absent ($s_{\mathrm{gn}} = 0$) the steady-state structure function is uniformly null and no pattern exists in the long-term steady state of the system.

It is worth noticing that patterns become more evident as the magnitude of the peak in the structure function increases. The intensity of the peak, i_p, can be evaluated as the ratio between the value of S_{st} in $k = k_0$ and the area subtended by the structure function, namely $i_p = S_{\mathrm{st}}(k_0)/\int_0^\infty S_{\mathrm{st}}(k)\mathrm{d}k$. Using Eq. (5.21), we find that i_p grows with $D/|a|$ [see Fig. 5.14(b)]. This is consistent with the fact that D controls the intensity of the pattern-forming spatial coupling while a modulates the strength of the local deterministic dynamics $f(\phi)$, which tends to drive the system toward the

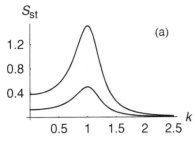
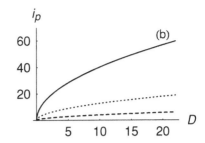

Figure 5.14. (a) Shape of the structure function at steady state for model (5.16) with $\mathcal{L}[\phi]$ given by expression (5.6); the upper curve corresponds to $s_{gn} = 3$, the lower one to $s_{gn} = 1$ ($k_0 = 1, a = -1, D = 3$). (b) Dependence of the peak intensity on strength of spatial coupling D for $k_0 = 2$; the solid, dotted, and dashed curves correspond to $a = -0.01, -0.1$, and -1, respectively.

(homogeneous) basic state $\phi_0 = 0$. Notice that i_p depends on a, D, and k_0 but not on the noise intensity s_{gn}; therefore, although noise is necessary for the emergence of patterns, it does not affect pattern intensity.

We observe that, in this case, pattern occurrence cannot be detected through the analysis of the dynamics of the ensemble mean $\langle\phi\rangle$. In fact, by taking the ensemble average of Eq. (5.16), in the case of a zero-mean additive noise we obtain (see Box 5.1)

$$\frac{\partial\langle\phi\rangle}{\partial t} = a\langle\phi\rangle - D(\nabla^2 + k_0^2)^2\langle\phi\rangle, \tag{5.22}$$

which is an equation formally identical to the deterministic part of original model (5.16), in which the effect of the noise is completely hidden. Therefore the conditions controlling the stability and instability of the zero-mean (i.e., $\langle\phi\rangle = 0$) homogeneous state are the same as those controlling the stability of the homogeneous state $\phi = 0$ in (5.16). Thus, using the ensemble mean to detect noise-induced changes, the stochastic dynamical system would not seem to be able to generate stable patterns when $a < 0$.

The generalized mean-field technique is also unable to detect the constructive (i.e., pattern-forming) role of additive noise. In fact, the application of this technique (see Box 5.3) leads to

$$\frac{d\phi_i}{dt} = a\phi_i - D\omega(\mathbf{k})\phi_i - D_{eff}(\phi_i - \langle\phi_i\rangle) + \xi_{gn,i}, \tag{5.23}$$

where

$$\omega(\mathbf{k}) = \left\{k_0^2 - \frac{2}{\Delta^2}[2 - \cos(\Delta k_x) - \cos(\Delta k_y)]\right\}^2, \tag{5.24}$$

$$D_{eff} = D\left[\left(\frac{4}{\Delta^2} - k_0^2\right)^2 + \frac{4}{\Delta^4} - \omega(\mathbf{k})\right]. \tag{5.25}$$

As the most unstable (or least stable) modes correspond to the condition $\omega(\mathbf{k}) = 0$, it is sufficient to investigate the self-consistency equation for the Langevin equation,

$$\frac{d\phi_i}{dt} = a\phi_i - D_{\text{eff}}(\phi_i - \langle\phi_i\rangle) + \xi_{\text{gn},i}, \qquad (5.26)$$

where

$$D_{\text{eff}} = D\left[\left(\frac{4}{\Delta^2} - k_0^2\right)^2 + \frac{4}{\Delta^4}\right]. \qquad (5.27)$$

Equation (5.26) does not provide multiple solutions even when well-defined patterns emerge (see Fig. 5.13).

Equation (5.23) shows that the most unstable modes correspond to the condition $\omega(\mathbf{k}) = 0$ [in fact, $\omega(\mathbf{k}) \geq 0$; see Eq. (5.24)], which can be satisfied by infinite pairs of wave-number components, k_x, k_y, different from zero. Thus stable patterns can emerge from the interactions among several unstable modes. Notice that the condition with zero wave number (i.e., $k_x = k_y = 0$) gives a more restrictive condition of instability:

$$\frac{d\phi_i}{dt} = f(\phi_i) + g(\phi_i)\xi_i - Dk_0^4\phi_i - D\left(\frac{20}{\Delta^4} - \frac{8k_0^2}{\Delta^2}\right)(\phi_i - m) + \xi_{\text{gn},i}. \qquad (5.28)$$

In other words, phase transitions (i.e., $m \neq 0$) can occur only when many wave numbers different from zero have already become unstable.

The classical mean-field analysis (see Box 5.4) correctly detects that the order parameter m does not change, and then the periodic patterns sustained by additive noise do not entail phase transitions. The pdf's of the field variable ϕ, shown in Fig. 5.13, confirm this theoretical finding. In fact, the pdf's of ϕ remain unimodal, symmetrical, and with zero mean even after the appearance of patterns.

Some additional comments should complete the picture drawn in this section on the interplay between additive noise and pattern-forming spatial coupling:

1. The nonlinear component of $f(\phi)$ does not play any fundamental role in the mechanism's inducing and sustaining spatial patterns. In fact, close to the deterministic stable state the deterministic behavior is determined by the linear component of $f(\phi)$, as demonstrated by the analysis of the structure function (and stability analysis) for the linearized models. Thus the addition of the nonlinear term $-\phi^3$ to model (5.16) does not substantially change any of the previous results. The fine details of the patterns can change, but neither their stable occurrence nor their dominant wavelength changes.

2. Pattern formation induced by additive noise is usually introduced in the science literature as an example of a noisy precursor to deterministic pattern-forming bifurcation (Sagues et al., 2007). According to this point of view, additive noise acts on a deterministic system that exhibits a bifurcation point between a homogeneous stable state and a stable patterned state [an example is Ginzburg–Landau model (5.18)]. In this case, the role of additive noise is to anticipate the transition through the bifurcation point, inducing patterns even when the deterministic system is in subcritical conditions. In other words, additive random

fluctuations unveil the intrinsic periodicity of the deterministic system even before reaching the pattern-forming bifurcation. In this sense, noise provides a premonitory sign of the proximity to marginal stability condition even if the system is subcritical. Thus these noise-induced precursory behaviors near order-forming bifurcations are interpreted as the spatial counterpart of the emergence of coherence resonance in the temporal dynamics of subthreshold systems close to a Hopf bifurcation (see Chapter 4).

This point of view is formally correct, but we think that it can be limiting and potentially misleading. In fact, it suggests that a deterministic bifurcation is necessary for additive noise to generate patterns. The example with $f(\phi) = a\phi$ presented in this section demonstrates that this is not necessarily the case. In fact, in this case no bifurcation is present because the dynamical system diverges when $a > 0$. Therefore a deterministic bifurcation is not a necessary condition. More generally, the additive noise unveils the capability of the deterministic component of the dynamical system to induce transient periodic patterns even when the asymptotic stable state is homogenous. Additive noise exploits this capability of generating transient patterns (which is due to the pattern-forming spatial coupling) and hampers their disappearance.

3. The pattern-forming process described in this section is general and is not limited to the case of the Swift–Hohenberg coupling. Qualitatively similar results can be extended to other types of pattern-forming spatial coupling. For example, the integral coupling $\mathcal{L}[\phi]$ in Eq. (5.7) can have similar interactions with additive noise. Moreover, additive noise can also activate Turing instability and pattern formation (see Appendix B) in subcritical conditions (Hutt et al., 2008).

5.3 Additive noise and non-pattern-forming spatial couplings

In this section, we consider the same interplay between the local deterministic term $f(\phi)$ and additive noise, but we consider a spatial coupling that is now unable to select a specific wavelength. To this end, we adopt the Laplacian operator, which is commonly used to express spatial coupling in the spatiotemporal modeling of environmental dynamics. Thus, in this case, the prototype model is

$$\frac{\partial \phi}{\partial t} = a\phi + D\nabla^2\phi + \xi_{\mathrm{gn}}. \tag{5.29}$$

The deterministic part of the dynamics – i.e., $\partial\phi/\partial t = a\phi + D\nabla^2\phi$ – is unable to generate patterns for any value of a. However, the introduction of an additive random component leads to the emergence of interesting patterns. In this case the effect of additive noise is even more astonishing than in the case of systems with pattern-forming couplings (see Section 5.2). In fact, in that case, (unsteady) patterns were already potentially present in the deterministic dynamics, whereas the deterministic counterpart of Eq. (5.29) does not exhibit pattern formation. Notice that when $a = 0$ Eq. (5.29) coincides with the Edwards–Wilkinson equation (Edwards and Wilkinson, 1982), adopted to describe surface-growth and roughening processes (Krug, 1997).

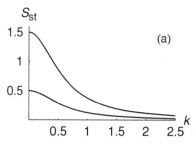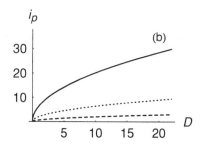

Figure 5.15. (a) Shape of the structure function at steady state for model (5.29); the upper curve corresponds to $s_{gn} = 3$, the lower one to $s_{gn} = 1$ ($a = -1$, $D = 3$). (b) Dependence of the peak intensity on the strength of the spatial coupling D; the solid, dotted, and dashed curves correspond to $a = -0.01, -0.1$, and -1, respectively.

If we consider again the prognostic tools to predict pattern occurrence, it is possible to have a clue about the role of additive noise; in fact, the steady-state structure function corresponding to stochastic model (5.29),

$$S(k, t) = \frac{s_{gn}}{2(Dk^2 - a)}, \qquad (5.30)$$

exhibits a maximum at $k = 0$ (see Fig. 5.15). It follows that no strong periodicity is selected, but a range of wave numbers close to zero competes to produce patterns more complex than those discussed in the previous section. However, it is difficult to anticipate the geometrical characteristics of these patterns from only the analysis of the structure function. Therefore numerical simulations are also needed. An example is shown in Fig. 5.16, which shows patterns substantially different from those observed when pattern-forming spatial couplings are present (see Fig. 5.13). As expected from the analysis of the structure function, no clear periodicity is detectable, and many wavelengths are present; moreover, the boundaries of the coherence regions are irregular. These spatial structures fall then in the class of multiscale fringed patterns, which are especially relevant in the environmental sciences, in which there are several instances of variables that exhibit a spatial distribution very similar to the one shown in Fig. 5.16; a typical example is the distribution of vegetated sites in semiarid environments; see Chapter 6.

Notice that in this case the other prognostic tools used to assess pattern occurrence – i.e., the dispersion equation and the generalized mean-field analysis – fail to provide useful indications. Even though these patterns are very different from those encountered in the previous section, the pdf of the field exhibits similar characteristics (compare the second row of Fig. 5.16 with Fig. 5.13): The pdf is unimodal and its mean coincides with the basic homogeneous stable state (i.e., $m = \phi_0 = 0$). Therefore, in this case, no phase transitions in the mean occur, as indicated by the classical mean-field analysis. This behavior follows the general notion that Gaussian additive

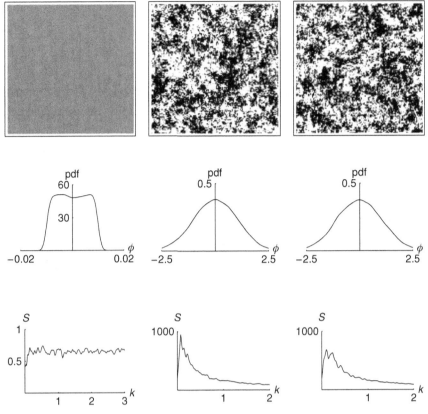

Figure 5.16. Numerical simulation of stochastic model (5.29) with $a = -0.1$, $D = 2.5$, and $s_{\mathrm{gn}} = 5$ (the other conditions are as in Fig. 5.10). The columns refer to 0, 200, and 400 time units. The first row shows the field (the gray-tone scale spans the interval $[-0.5, 0.5]$), the second row shows the pdf of the field variable ϕ, and the third row shows the azimuth-averaged spectrum of the field.

noise is unable to give rise to phase transitions (i.e., changes of m), regardless of the type of spatial coupling.

5.4 Multiplicative noise and pattern-forming spatial couplings

Similar to the case of additive noise, multiplicative noise and pattern-forming spatial coupling can also cooperate to generate steady patterns. Such cooperation is, however, more subtle than in the case of additive noise in that it is based on two key actions: (i) the multiplicative random driver temporarily destabilizes the homogeneous stable state ϕ_0 of the underlying deterministic dynamics, and (ii) the spatial coupling acts during this instability, generating and stabilizing a pattern. Notice that the instability is induced by the multiplicative-noise term and not by the spatial coupling, as in the case of the deterministic mechanism of Turing instability presented in Appendix B.

To describe the interplay between short-term temporal instability and spatial dynamics, we refer to the model

$$\frac{\partial \phi}{\partial t} = f(\phi) + g(\phi)\,\xi_m(\mathbf{r}, t) + D\mathcal{L}[\phi], \qquad (5.31)$$

obtained from general equation (5.1), with $F(t)$ set to zero because this mechanism does not require a time-dependent forcing, and without additive noise, ξ_a, in order to isolate the role of multiplicative noise. ξ_m is a zero-average noise with intensity s_m. We indicate by ϕ_0 the stable homogeneous state of the system in the deterministic case. Namely, $\phi(\mathbf{r}, t) = \phi_0$ is a homogeneous solution of (5.31) when $s_m = 0$ [i.e., $f(\phi_0) = 0$ because $\mathcal{L}[\phi] = 0$ in homogeneous states]. Moreover, we consider cases in which $g(\phi_0) = 0$, so that the noise does not have the possibility of destabilizing the homogeneous steady state (otherwise, the role of multiplicative noise would be similar to that of additive noise).

The key features of pattern formation induced by multiplicative noise are that for values of s_m lower than a critical value s_c, the state variable $\phi(\mathbf{x}, t)$ experiences fluctuations about ϕ_0 but noise does not play any constructive role. At any point in space, the evolution of the ensemble average $\langle\phi\rangle$, starting from an initial condition $[\phi(t = 0) \neq \phi_0]$, is similar to the one indicated in Fig. 5.17 by the dotted curve: $\langle\phi\rangle$ decreases monotonically to the statistically steady state ϕ_0, i.e.,

$$\frac{d\langle\phi\rangle}{dt} < 0 \qquad (5.32)$$

at any time (we use the total derivative because any point exhibits the same behavior). Thus the system remains locked in the disordered phase and no pattern occurs. In this case the spatial coupling does not play any significant role at steady state. Only transiently, during the descent from $\phi(t = 0)$ to ϕ_0, would the spatial coupling be able to induce a pattern. This transient occurrence is most probable when the descent is slow and D is large, but patterns fade out as the system converges to its steady state. The temporal behavior of $\langle\phi\rangle$ depicted in Fig. 5.17 remains qualitatively the same regardless of the value of D. This means that the evolution of $\langle\phi\rangle$ coincides qualitatively with that of the correspondent zero-dimensional system, in which $D = 0$, which evolves according to $d\phi/dt = f(\phi) + g(\phi)\xi_m$.

The dynamics change when the noise intensity increases above a critical level (i.e., $s_m > s_c$). At first, let us consider only the zero-dimensional dynamics. When $s_m > s_c$, the multiplicative noise is able to induce a short-term instability of the equilibrium state ϕ_0, i.e., with

$$\lim_{t \to 0} \frac{d\langle\phi\rangle}{dt} > 0, \qquad \lim_{t \to \infty} \frac{d\langle\phi\rangle}{dt} \leq 0. \qquad (5.33)$$

The first condition determines an initial instability: The dynamics initially tend to move away from the basic state $\langle\phi\rangle = \phi_0$; however, the second condition establishes

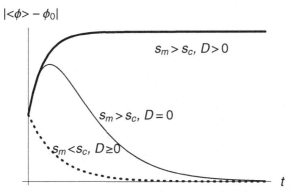

Figure 5.17. Scheme of the possible evolutions of small initial displacement depending on the noise intensity and the strength of the spatial coupling.

that noise-induced instability is transient, in the sense that the state ϕ_0 is recovered in the long term. This behavior is shown by the thin continuous curve in Fig. 5.17: Noise initially induces a growth (i.e., an instability) of $\langle \phi \rangle$, but for $t \to \infty$ the difference $|\langle \phi \rangle - \phi_0|$ tends to zero. This behavior (i) explains why this mechanism is called *short-term instability* and (ii) suggests that the steady-state pdf of ϕ is expected to be unimodal, with the mode at ϕ_0, even when s_m exceeds the threshold s_c.

We now consider spatiotemporal dynamics (i.e., $D \neq 0$). Differently from the subcritical case, when $s_m > s_c$, the spatial coupling introduces qualitative changes with respect to the zero-dimensional case. In fact, for suitable values of the strength D of the coupling, the spatial term in Eq. (5.31) takes advantage of the noise-induced short-term instability that also exists in the spatiotemporal dynamics (in fact, $\partial \langle \phi \rangle / \partial t \simeq \mathrm{d} \langle \phi \rangle / \mathrm{d}t$ for $t \to 0$) and prevents the displacement $|\langle \phi \rangle - \phi_0|$ from decaying to zero. As a consequence, the spatial coupling locks the system in a new ordered state, with $\langle \phi \rangle$ different from ϕ_0 and variable in space. In this case the temporal trajectory of $\langle \phi \rangle$ does not exhibit a convergence to ϕ_0 for $t \to \infty$, as shown by the thick curve in Fig. 5.17. Thus the spatial coupling is responsible for maintaining the dynamics away from the state ϕ_0. Before investigating the behavior of some prototype models, we describe in Box 5.5 an analytical tool used to detect the possible presence of short-term instability in stochastic dynamical systems.

Once the short-term instability has been detected, the ability of spatiotemporal stochastic dynamics (B5.5-1) to give rise to periodic patterns is typically investigated by the prognostic tools described in Boxes 5.1, 5.2, and 5.3.

5.4.1 Prototype model

To illustrate how pattern formation can be induced by multiplicative noise, we consider the prototype model:

$$\frac{\partial \phi}{\partial t} = a\phi - \phi^3 + \phi \xi_{\mathrm{gn}} - D(k_0^2 + \nabla^2)^2 \phi, \qquad (5.34)$$

Box 5.5: Short-term instability

We consider the model

$$\frac{\partial \phi}{\partial t} = f(\phi) + g(\phi)\xi_m(\mathbf{r}, t) + D\mathcal{L}[\phi]. \tag{B5.5-1}$$

The first steps of the stability analysis are the same as those described in Box 5.1 and lead to the following equation for the evolution of the ensemble average

$$\frac{\partial \langle \phi \rangle}{\partial t} = \langle f(\phi) \rangle + s_m \langle g_S(\phi) \rangle + D\mathcal{L}[\langle \phi \rangle]. \tag{B5.5-2}$$

If f and g are nonlinear functions of ϕ, their ensemble average involves all moments of the pdf of $\langle \phi \rangle$. For example, for $f(\phi)$ the expansion is (van Kampen, 1992)

$$\langle f(\phi) \rangle = f(\langle \phi \rangle) + \frac{1}{2} \langle (\phi - \langle \phi \rangle)^2 \rangle \frac{d^2 f(\langle \phi \rangle)}{d\langle \phi \rangle^2} + \cdots. \tag{B5.5-3}$$

Inserting Eq. (B5.5-3) into (B5.5-2), we find that the dynamics of $\langle \phi \rangle$ do not depend on only $\langle \phi \rangle$ itself but also on the fluctuations of ϕ around $\langle \phi \rangle$. However, as we are interested here in the initial evolution of small displacements from ϕ_0 – i.e., we are focusing on $t \to 0$ – the nonlinear terms of the expansion can be neglected and the zero-order Taylor's expansion,

$$\langle f(\phi) \rangle \approx f(\langle \phi \rangle), \qquad \langle g_S(\phi) \rangle \approx g_S(\langle \phi \rangle), \tag{B5.5-4}$$

can be used. Similarly, we can also neglect the spatial gradients of the fluctuations and assume that they are small in the short term. It follows that, for $t \to 0$, Eq. (B5.5-2) can be approximated at any point in space as (Sagues et al., 2007)

$$\frac{d\langle \phi \rangle}{dt} \approx f(\langle \phi \rangle) + s_m g_S(\langle \phi \rangle) = f_{\text{eff}}(\langle \phi \rangle) = -\frac{d\mathcal{V}(\langle \phi \rangle)}{d\langle \phi \rangle}, \tag{B5.5-5}$$

where f_{eff} and $\mathcal{V}$ are often indicated as the *effective kinetics* and the *effective* or *stochastic potential* (see Box 3.1), respectively. Notice that Eq. (B5.5-5) is identical to the macroscopic equation obtained in the short term in a zero-dimensional dynamical system with the same functions f and g. Therefore the short term stability analysis of the state ϕ_0 expressed by Eq. (B5.5-5) deals only with the temporal dynamics at any given point of the domain.

Equation (B5.5-5) clearly shows that noise can destabilize the stable state ϕ_0 of the underlying deterministic dynamics. We consider two cases. In the first case the roots of $f_{\text{eff}}(\langle \phi \rangle) = 0$ do not coincide with those of $f(\langle \phi \rangle) = 0$, and in the second case the state ϕ_0 remains a zero of the right-hand side of Eq. (B5.5-5) – i.e., $f_{\text{eff}}(\phi_0) = 0$ also for $s_m > 0$ – but sufficiently high noise intensities cause the instability of ϕ_0. In the first case, ϕ_0 is a stable state of the deterministic dynamics [i.e., for $s_m = 0$ $f(\phi_0) = 0$ and $f'(\phi_0) < 0$], whereas it is not an equilibrium point of the stochastic dynamics, i.e., for $s_m > 0$, $f_{\text{eff}}(\phi_0) \neq 0$. Multiplicative noise destabilizes the state ϕ_0 if

$$\frac{d\langle \phi \rangle}{dt} = f_{\text{eff}}(\phi_0) = s_m g_S(\langle \phi_0 \rangle) > 0. \tag{B5.5-6}$$

Notice how this instability arises as an effect of the spurious drift term in the macroscopic equation, which would not exist if the noise were additive or Ito's interpretation were used. In the second case $f_{\text{eff}}(\phi_0) = 0$ also in the stochastic dynamics (i.e., when $s_m > 0$). Thus, using in Eq. (B5.5-5) the first-order truncated Taylor's expansion of $f_{\text{eff}}(\phi)$ around ϕ_0, we find that instability emerges when

$$\frac{\mathrm{d} f_{\text{eff}}}{\mathrm{d}\langle\phi\rangle}\bigg|_{\phi_0} = -\frac{\mathrm{d}^2 \mathcal{V}(\langle\phi\rangle)}{\mathrm{d}\langle\phi\rangle^2}\bigg|_{\phi_0} \geq 0. \tag{B5.5-7}$$

In fact, the first-order truncated Taylor expansion of the function f_{eff} around ϕ_0 yields

$$\frac{\mathrm{d}\langle\phi\rangle}{\mathrm{d}t} \approx \frac{\mathrm{d} f_{\text{eff}}}{\mathrm{d}\langle\phi\rangle}\bigg|_{\phi_0}\langle\phi\rangle. \tag{B5.5-8}$$

Thus the sign of $\mathrm{d} f_{\text{eff}}(\phi_0)/\mathrm{d}\langle\phi\rangle$ determines the stability or instability of the dynamics with respect to a (small) perturbation around ϕ_0. The noise threshold can therefore be calculated when condition (B5.5-7) is equal to zero.

In both cases noise has to be multiplicative [i.e., $g(\phi)$ should not be a constant] for a transition to unstable dynamics to emerge. When the noise is additive the effective kinetics are $f_{\text{eff}}(\langle\phi\rangle) = f(\langle\phi\rangle)$ and the stable states of (B5.5-5) are the same as those of the deterministic counterpart of the process.

It is also worth stressing the impact of the type of noise and noise interpretation. When ξ_m is white Gaussian noise and we interpret Langevin equation (B5.5-1) by using the Stratonovich rule, $g_S(\phi) = s_{\text{gn}}g(\phi)g'(\phi)$. In contrast, under Ito's interpretation, $g_S(\phi) = 0$ (see Box 3.2). These differences play an important role in the short-term instability presented in this section. In fact, such instability is triggered by spurious drift [see Eq. (B5.5-5)]. Therefore noise-induced instability is possible only under the Stratonovich interpretation of Langevin's equation, and it cannot occur when Ito's framework is adopted.

where a is a negative-valued number, the random component is modulated by a function $g(\phi) = \phi$, and ξ_{gn} is zero-mean white Gaussian noise with intensity s_{gn}; Eq. (5.34) is interpreted in the Stratonovich sense. The local dynamics, $f(\phi) = a\phi - \phi^3$, and the spatial coupling à la Swift–Hohenberg are the same as those already used to discuss patterns induced by additive noise. Thus the deterministic dynamics underlying Eq. (5.34) exhibit the behavior described in Section 5.2. In particular, the homogeneous stable state, obtained as a solution of $f(\phi_0) = 0$, is $\phi(\mathbf{r}, t) = \phi_0 = 0$.

The short-term stability of the basic state is investigated as described in Box 5.5. For model (5.34), Eq. (B5.5-5) yields

$$\frac{\mathrm{d}\langle\phi\rangle}{\mathrm{d}t} \approx a\langle\phi\rangle - \langle\phi\rangle^3 + s_{\text{gn}}\langle\phi\rangle = f_{\text{eff}}(\langle\phi\rangle). \tag{5.35}$$

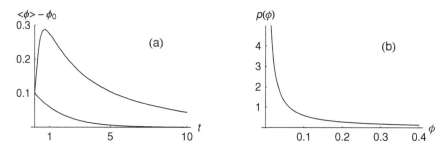

Figure 5.18. (a) Behavior of $\langle\phi\rangle - \phi_0$ obtained as an ensemble average of 10^6 realizations of model (5.37) with $a = -1$. The initial condition is $\phi = 0.1$, and $s_{gn} = 0.5$ and $s_{gn} = 5$ for the lower and the upper curve, respectively. (b) Example of the steady-state pdf of model (5.38) for $a = -1$, $s_{gn} = 5$, and positive initial condition.

In this case $f_{eff}(\phi_0) = 0$ for any value of s_{gn}. The condition for marginal stability is then

$$\frac{d f_{eff}}{d\langle\phi\rangle}\bigg|_{\phi_0} = a + s_{gn} = 0, \tag{5.36}$$

which provides the critical noise intensity $s_c = -a$. From condition (B5.5-7), it follows that ϕ_0 is stable for $s_{gn} < -a$ and becomes unstable for $s_{gn} > -a$. In the former case, small disturbances around the steady state ϕ_0 are dimmed whereas in the latter they tend to grow exponentially for $t \rightarrow 0$. However, this amplification is transient if the spatial coupling is absent (see Fig. 5.17): Fig. 5.18(a) shows the temporal behavior of the ensemble average of a number of numerically evaluated realizations of the zero-dimensional stochastic model obtained when the spatial component is eliminated from Eq. (5.34):

$$\frac{d\phi}{dt} = f(\phi) + g(\phi)\,\xi_{gn}(t) = a\phi - \phi^3 + \phi\xi_{gn}(t). \tag{5.37}$$

The results in Fig. 5.18(a) show that the growth phase appears only when $s_{gn} > s_c$ and at the beginning of simulations (i.e., at short term), whereas the effect of the initial perturbation disappears in the long term.

The steady-state pdf corresponding to model (5.37) can be found with the methods described in Chapter 2:

$$p(\phi) = C e^{-\frac{\phi^2}{2s_{gn}}} \phi^{\frac{a}{s_{gn}}-1}, \tag{5.38}$$

where C is a normalization constant and the domain of $p(\phi)$ is either the positive or the negative semiaxis of ϕ, depending on whether the initial condition is positive or negative valued. In fact, the condition $g(\phi = 0) = 0$ entails that $\phi = 0$ be a natural boundary of the dynamics [an example of pdf is shown in Fig. 5.18(b)]. Equation (5.38) shows that the mode of the steady-state distribution is always at $\phi = 0$ for any noise intensity. This confirms that in the long run perturbations of the deterministic

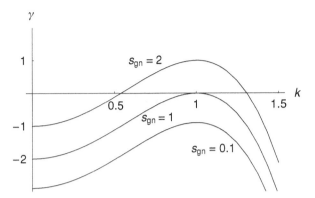

Figure 5.19. Dispersion relation for prototype model (5.34), with spatial coupling à la Swift–Hohenberg for three values of noise intensity ($a = -1$, $D = 2$, and $k_0 = 1$).

stable state ϕ_0 tend to disappear in spite of their initial amplification when $s_{gn} > s_c$. Mathematically, we can understand the short-term instability by observing that when ϕ is close to zero the disturbance effect that is due to the (multiplicative) noise tends to prevail on the restoring effect of f. Conversely, as ϕ grows, the leading term ϕ^3 of $f(\phi)$ prevails over g and the deterministic local kinetics f tends to restore the state ϕ_0.

Once the presence of a short-term instability has been detected, the capability of spatiotemporal stochastic model (5.34) to generate patterns can be investigated, for example, by use of the stability analysis by normal modes (see Box 5.1). In the case of prototype model (5.34) we obtain from Eq. (B5.1-4) a dispersion relation

$$\gamma(k) = a + s_{gn} - D(k_0^2 - k^2)^2, \tag{5.39}$$

which provides the same threshold s_c as Eq. (5.36) for neutral stability; the maximum amplification is for the wave number $k = k_0$ (see Fig. 5.19). It follows that statistically steady periodic patterns, with wavelength $\lambda = 2\pi/k_0$, emerge when the noise intensity exceeds the threshold $s_c = -a$.

The analysis of the structure function can be used as a prognostic tool to confirm these results. The linear evolution of the structure function from Eq. (B5.2-6) is

$$\frac{\partial S(k, t)}{\partial t} = 2\left[a + s_{gn} - D(k_0^2 - k^2)^2\right] S(k, t). \tag{5.40}$$

Thus neutral stability corresponds to a critical value of the noise intensity:

$$s_c = -a + D(k_0^2 - k^2)^2. \tag{5.41}$$

When $s_{gn} < s_c$, the structure function tends to zero and no patterns occur. Conversely, when threshold (5.41) is exceeded, the linear evolution of the structure function is divergent and a range of wave numbers becomes unstable. Notice that the wave number most prone to instability is $k = k_0$, which corresponds to $s_c = -a$. Therefore these results are consistent with those obtained from the stability analysis.

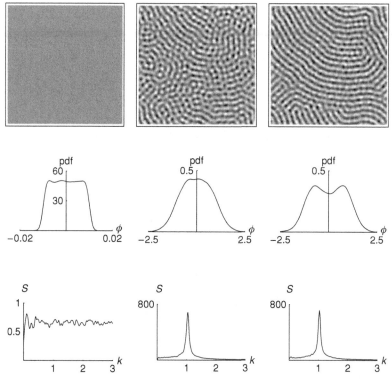

Figure 5.20. Results of the numerical simulation of prototype model (5.34). The parameters are $a = -1$, $D = 15$, $k_0 = 1$, and $s_{\mathrm{gn}} = 2.5$ (the other conditions are as in Fig. 5.10). The three panels correspond to t equal to 0, 10, and 100 time units. The gray-tone scale spans the interval $[-2, 2]$.

The numerical simulation of stochastic model (5.34) confirms these theoretical findings. Figure 5.20 shows an example of patterns emerging in the simulation. They have the same basic characteristics as those observed in the case of additive noise (see Fig. 5.13): They are statistically stable and exhibit a clear dominant wavelength corresponding to k_0. Moreover, in this case the pdf of the field is unimodal with the mean at $\phi = 0$, demonstrating that no phase transitions occur (i.e., m remains the same as in the disordered case). However, there are some differences with respect to the case with additive noise. First, the edges of the stripes or other geometrical features existing in the pattern are more regular when patterns are induced by multiplicative noise. This regularity is consistent with the fact that the power spectrum of the field exhibits a sharper peak at k_0 in the case of multiplicative noise (compare the third rows of Figs. 5.13 and 5.20). This difference is due to the fact that multiplicative noise is modulated by the local value of ϕ, which has the effect of enhancing the spatial coherence of the pattern because ϕ is spatially correlated (from the definition itself of a patterned state). Second, patterns induced by additive noise exhibit more stable shapes than those induced by multiplicative noise. For example, the pattern shown

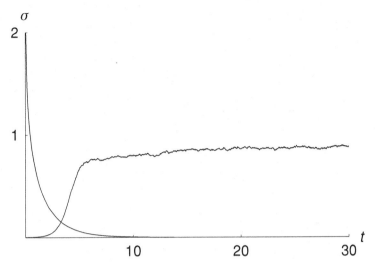

Figure 5.21. Temporal behavior of the field standard deviation σ for $s_{gn} = 0.5$ and $s_{gn} = 2.5$. The other parameters are $a = -1$, $D = 15$, and $k_0 = 1$. The decreasing curve corresponding to $s_{gn} = 1$ is amplified by a factor of 10^3.

in Fig. 5.20 seems to evolve from a labyrinthine to a striped shape. Third, a weak bimodality may appear in the pdf of ϕ in the dynamics driven by multiplicative noise (see Fig. 5.20). The presence of such bimodality depends on the model structure, parameter values, and field size.

We conclude this section stressing that the presence of pattern-forming spatial couplings is not sufficient for stable patterns to occur. The concomitant presence of short-term instability is in fact necessary. For example, if prototype model (5.34) is simulated with a noise intensity lower than s_c, no pattern emerges. Figure 5.21 shows the temporal evolution of the standard deviation of the field for two noise intensities: When $s_{gn} > s_c$, the spatial variance increases and tends to the asymptotic value corresponding to the patterned state shown in Fig. 5.20, whereas when $s_m < s_c$ the spatial variance decreases to zero, namely the field tends to the homogeneous state.

5.4.2 *The effect of nonlinear $g(\phi)$ terms*

We now consider the effect of nonlinear terms on the multiplicative factor $g(\phi)$. To this end, we consider the same mathematical structure as in Eq. (5.34), but with $g(\phi) = \phi^2$, i.e.,

$$\frac{\partial \phi}{\partial t} = a\phi - \phi^3 + \phi^2 \times \xi_{gn} - D(k_0^2 + \nabla^2)^2\phi. \tag{5.42}$$

Figure 5.22. Numerical simulation of model (5.42). The parameters are $a = -1$, $D = 10$, $k_0 = 1$, and $s_{gn} = 5$. The three panels correspond to t equal to 0, 1, and 5 time units. The gray-tone scale is in the interval $[-0.01, 0.01]$.

Thus the steady uniform case is still $\phi_0 = 0$. However, in this case, the analysis of the short-term behavior of the temporal component of the model yields

$$\frac{d\langle\phi\rangle}{dt} \approx a\langle\phi\rangle - \langle\phi\rangle^3 + 2s_{gn}\langle\phi\rangle^3 = f_{\text{eff}}(\langle\phi\rangle). \tag{5.43}$$

Thus ϕ_0 is stable when

$$\left.\frac{d f_{\text{eff}}}{d\langle\phi\rangle}\right|_{\phi_0} = a < 0, \tag{5.44}$$

indicating that the contribution of the multiplicative-noise term is negligible with respect to the linear component of $f(\phi)$. Thus Eq. (5.44) shows that in these dynamics [Eq. (5.42)] noise is not able to have an impact on the short-term behavior of the system and no short-term instability occurs. We explain this result by interpreting the short-term behavior as a result of the interplay between the restoring action that is due to the tendency of $f(\phi)$ to drive the system toward the homogeneous state $\phi = \phi_0 = 0$ and the diverging action that is due to multiplicative noise, $g(\phi)\xi_{gn}$. When ϕ is close to $\phi_0 = 0$ (i.e., for small displacements from ϕ_0, such as those studied in the linear-stability analysis), the power two in the function $g(\phi)$ strongly reduces the effect of noise, which becomes unable to hinder the action of the leading term $a\phi$ of $f(\phi)$.

This result is confirmed by the stability analysis of the dynamics of the ensemble average. In this case the dispersion relation is

$$\gamma(k) = a - D(k_0^2 - k^2)^2. \tag{5.45}$$

Thus the growth factor γ is always negative for any wave number and no patterns emerge. An example of numerical simulations of model (5.42) is shown in Fig. 5.22. Patterns occur only transiently, and the field then rapidly decays to the homogeneous state ϕ_0.

Notice that short-term instability can also be present with nonlinear forms of $g(\phi)$, provided that the multiplicative component overcomes the action of the deterministic

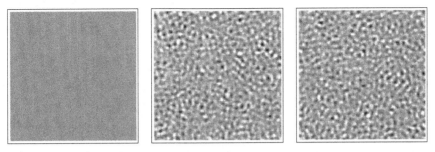

Figure 5.23. Numerical simulation of model (5.46). The parameters are $a = -1$, $D = 10$, $k_0 = 1$, and $s_{\text{gn}} = 0.5$. The three panels correspond to t equal to 0, 10, and 100 time units. The gray-tone scale is in the interval $[-0.3, 0.3]$.

term $f(\phi)$ close to ϕ_0. For example, consider the model

$$\frac{\partial \phi}{\partial t} = a\phi - \phi^3 + \phi^{1/3} \times \xi_{\text{gn}} - D(k_0^2 + \nabla^2)^2 \phi. \tag{5.46}$$

In this case the short-term analysis yields

$$\frac{d\langle \phi \rangle}{dt} \approx a\langle \phi \rangle - \langle \phi \rangle^3 + \frac{s_{\text{gn}}}{3\langle \phi \rangle^{1/3}} = f_{\text{eff}}(\langle \phi \rangle). \tag{5.47}$$

It follows that

$$\lim_{\phi \to \phi_0} \frac{d f_{\text{eff}}}{d\langle \phi \rangle} = +\infty, \tag{5.48}$$

and the short-term behavior is then unstable for any noise intensity (i.e., even when $s_c = 0$), indicating that the noise component always overwhelms $f(\phi)$ when ϕ is close to zero. Notice that the balance between $f(\phi)$ and $g(\phi)\xi$ is reversed when ϕ moves away from $\phi_0 = 0$. This fact hampers the divergence of the system. The numerical simulation shown in Fig. 5.23 confirms that model (5.46) generates stable patterns with the same dominant wavelength as those shown in Fig. 5.20.

5.4.3 Case with $g(\phi_0) \neq 0$: The van den Broeck–Parrondo–Toral model

In this subsection, we study a more complex model with respect to the one investigated in Subsection 5.4.1. It is shown that more complicated forms of the functions $f(\phi)$ and $g(\phi)$ do not qualitatively change the picture drawn in the previous subsections, provided that the interplay between short-term instability and spatial coupling remains the same. To this end, we can consider the stochastic model

$$\frac{\partial \phi}{\partial t} = -\phi(1 + \phi^2)^2 + (1 + \phi^2) \times \xi_{\text{gn}} - D(k_0^2 + \nabla^2)^2. \tag{5.49}$$

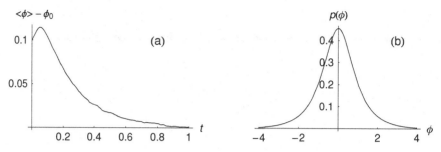

Figure 5.24. (a) Behavior of $\langle\phi\rangle - \phi_0$ obtained as ensemble average of 10^6 realizations of model (5.52). The initial condition is $\phi = 0.1$, and $s_{gn} = 4$. (b) Example of the steady-state pdf of model (5.52) for $s_{gn} = 4$.

This model – though with spatial coupling expressed by diffusion (i.e., $\mathcal{L}[\phi] = D\nabla^2\phi$) – was proposed by van den Broeck et al. (1994) as an example of a system able to exhibit noise-induced phase transitions (see Section 5.5) and is indicated here as the VPT model. Subsequently the same authors investigated the properties of (5.49) to show how multiplicative noise can induce patterns in the presence of a suitable spatial coupling (Parrondo et al., 1996; Garcia-Ojalvo and Sancho, 1999). We note that Eq. (5.49) can be also interpreted as dynamics driven by both additive and multiplicative noise; in fact, $g(\phi)\xi_{gn} = (1 + \phi^2)\xi_{gn}$ can be split into an additive term, $\xi_a \equiv \xi_{gn}$, and a multiplicative term, $\xi_m \equiv \phi^2\xi_{gn}$.

The stable homogeneous state of the underlying deterministic dynamics (which is found with $s_{gn} = 0$) is $\phi(\mathbf{r}, t) = \phi_0 = 0$, and the short-term behavior close to ϕ_0 is given by

$$\frac{d\langle\phi\rangle}{dt} \approx -\langle\phi\rangle \left(1 + \langle\phi\rangle^2\right)^2 + 2s_{gn}\langle\phi\rangle \left(1 + \langle\phi\rangle^2\right) = f_{\text{eff}}(\langle\phi\rangle). \tag{5.50}$$

The condition for the marginal stability is then

$$\left.\frac{d f_{\text{eff}}}{d\langle\phi\rangle}\right|_{\phi_0} = -1 + 2s_{gn} = 0, \tag{5.51}$$

which entails $s_c = 0.5$. ϕ_0 is then stable for $s_{gn} < 0.5$ and becomes unstable for $s_{gn} > 0.5$. However, the amplification of the initial perturbation is transient in the absence of spatial coupling. This fact is demonstrated in Fig. 5.24(a), which shows the temporal behavior of the ensemble average of the zero-dimensional stochastic model we obtain by eliminating the spatial component from Eq. (5.49):

$$\frac{d\phi}{dt} = -\phi(1 + \phi^2)^2 + (1 + \phi^2)\xi_{gn}(t). \tag{5.52}$$

Figure 5.24(a) shows that the growth phase appears only at the beginning (i.e., in the short term), whereas in the long run the effect of the initial perturbation disappears.

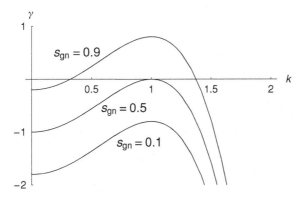

Figure 5.25. Dispersion relation of the VPT model with spatial coupling à la Swift–Hohenberg for three values of noise standard deviation ($D = 1$, $k_0 = 1$).

Figure 5.24(b) shows an example of the steady-state pdf corresponding to model (5.52):

$$p(\phi) = \frac{e^{-\frac{1+\phi^2}{2s_{gn}}}}{\pi(1 + \phi^2)\mathrm{erfc}[\frac{1}{\sqrt{2s_{gn}}}]}, \tag{5.53}$$

where erfc[·] is the complementary error function. The pdf has the mode at $\phi = 0$ for any noise intensity, which confirms that perturbations of the deterministic stable state ϕ_0 tend to disappear in spite of their initial amplification when $s_{gn} > s_c$.

To investigate the ability of dynamics (5.49) to generate periodic patterns, we focus on the deterministic differential equation describing the spatiotemporal dynamics of $\langle\phi\rangle$:

$$\frac{\partial\langle\phi\rangle}{\partial t} = -\langle\phi\rangle\left(1 + \langle\phi\rangle^2\right)^2 + 2s_{gn}\langle\phi\rangle\left(1 + \langle\phi\rangle^2\right) - D(k_0^2 + \nabla^2)^2\langle\phi\rangle. \tag{5.54}$$

The dispersion relation obtained by use of the techniques described in Box 5.1 is

$$\gamma(k) = -1 + 2s_{gn} - D(k_0^2 - k^2)^2. \tag{5.55}$$

If we focus on positive wave numbers, the maximum is localized at $k_{max} = k_0$, where $\gamma(k = k_{max}) = 2s_{gn} - 1$ (see Fig. 5.25). It follows that (i) unstable wave numbers occur only if $s_{gn} > 0.5$ and (ii) the selected pattern is periodic with a wavelength approximatively equal to $\lambda = 2\pi/k_{max}$. The first condition is the same as the one obtained from the short-term analysis and confirms that pattern formation needs short-term instability, namely patterns are absent if noise intensity is subthreshold (i.e., $s_{gn} < s_c = 0.5$).

The generalized mean-field technique is useful for understanding the role of the intensity both of noise s_{gn} and of the spatial coupling D (Parrondo et al., 1996). If this technique is applied to VPT model (5.49) and the most unstable modes are considered

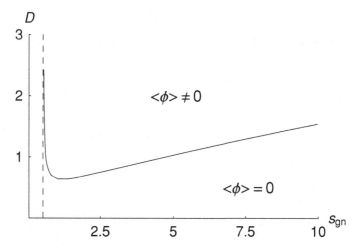

Figure 5.26. Curve of marginal instability for the VPT model according to the generalized mean-field technique ($k_0 = 1$ and $\Delta = 1$). The dashed vertical line marks the threshold $s_c = 0.5$.

[i.e., the condition $\omega(\mathbf{k}) = 0$ is imposed], we obtain

$$\frac{\mathrm{d}\phi_i}{\mathrm{d}t} = -\phi_i(1 + \phi_i^2)^2 + (1 + \phi_i)^2\xi_{\mathrm{gn}} - D_{\mathrm{eff}}(\phi_i - \langle\phi_i\rangle). \tag{5.56}$$

We can determine the steady-state pdf, $p_{\mathrm{st}}(\phi_i; \langle\phi_i\rangle)$, for zero-dimensional model (5.56) by applying the methods described in the second chapter. Using the self-consistency condition, we obtain the curve shown in Fig. 5.26, which marks the boundary between the region of parameter space in which the self-consistency equation has solutions different from zero (region inside or above the curve) and the region outside the curve, where only the solution $\langle\phi\rangle = 0$ exists. Therefore, under the simplifying assumptions of the mean-field approach (see Box 5.3), patterns can occur only for the pairs $\{s_{\mathrm{gn}}, D\}$ that fall above the region bounded by the curve. In particular, points of this curve correspond to pairs of values of $\{s_{\mathrm{gn}}, D\}$ associated with wave numbers $\{k_x, k_y\}$ that satisfy the condition $\omega(\mathbf{k}) = 0$, i.e., pairs of wave numbers $\{k_x, k_y\}$ that satisfy

$$\left\{k_0^2 - \frac{2}{\Delta^2}[2 - \cos(\Delta k_x) - \cos(\Delta k_y)]\right\}^2 = 0. \tag{5.57}$$

There are clearly infinite possible pairs of wave numbers satisfying condition (5.57), as well as many other unstable modes corresponding to the region above or inside the curve. The specific shape of the pattern emerging from the numerical simulations of the model depends on the initial and boundary conditions. However, the threshold $s_{\mathrm{gn}} = s_c = 0.5$ determined by short-term marginal stability has to be exceeded for patterns to emerge (see dashed line in Fig. 5.26). Moreover, the transition to a patterned state is reentrant, in that the stable state of the system switches back to the disordered state, $\langle\phi_i\rangle = 0$, for large values of s_{gn}. In fact, Fig. 5.26 shows that the emergence of

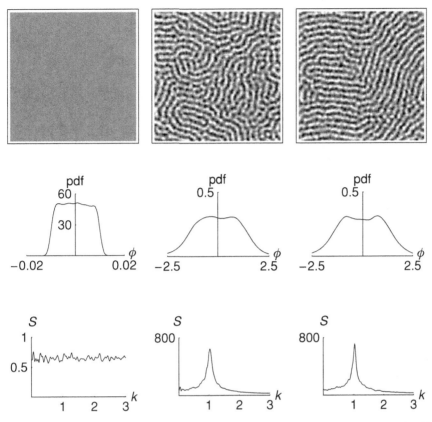

Figure 5.27. Patterns obtained from numerical simulation of the VPT model with spatial coupling à la Swift–Hohenberg. The columns correspond to $t = 0, 10$, and 100 time units. The rows show the field, the pdf of the variable ϕ, and the azimuth-averaged spectrum, respectively. The parameters are $k_0 = 1$, $s_{gn} = 2.5$, $D = 15$, and $\Delta = 1$ (the other conditions are as in Fig. 5.10). The gray-tone scale spans the interval $[-2, 2]$.

an ordered (i.e., patterned) state requires both spatial coupling (i.e., D needs to exceed a noise-dependent critical value) and intermediate noise intensities.

It is also interesting to notice that the classical form of the mean-field technique (Box 5.4) indicates that model (5.49) is unable to exhibit phase transitions in the sense defined in Subsection 5.1.3. In fact, using Eq. (5.28) with $f(\phi_i)$ and $g(\phi_i)$ given by Eq. (5.49), we obtain the fact that the order parameter m remains equal to zero for any value of s_{gn} and D.

Figure 5.27 shows some numerical simulations of Eq. (5.49). Both a visual inspection of the field and an analysis of its power spectrum averaged over the azimuthal angle clearly show the occurrence of a distinct periodic pattern. The wavelength $2\pi / k_{max} = 6.28$ pixels corresponds to the one identified by linear-stability analysis and remains stable in time. The pattern is stochastically steady (i.e., it maintains its average characteristics), although the fronts of the coherence regions continuously

Figure 5.28. Example of pattern obtained from numerical simulation of VPT model (5.49) with $s < s_c$. The parameters are $k_0 = 1$, $s_{gn} = 0.25$, $D = 2$, and $\Delta = 1$.

evolve in time. Moreover, the pdfs of the variable ϕ confirm the absence of phase transitions, consistent with the results of the mean-field analysis.

If the patterns shown in Fig. 5.27 are compared with those resulting from prototype model (5.34) (see Fig. 5.20), it is evident that the main characteristics of the steady-state field are the same. This is because the core of the dynamics is the same in both models, and it is dictated by two components: (i) the existence of a noise-induced short-term instability that triggers an initial departure from the homogeneous state $\phi_0 = 0$, and (ii) a suitable spatial coupling that is able to freeze the dynamics away from uniform steady-state conditions and to impose a dominant wavelength. Because the spatial coupling and the type of initial and boundary conditions are the same in both models, the resulting pattern is then statistically the same, provided that the threshold of noise intensity is exceeded (i.e., $s_{gn} > s_c$). Therefore the specific mathematical structure of functions $f(\phi)$ and $g(\phi)$ plays a marginal role: These functions determine secondary aspects of pattern formation, including, for example, the duration of the initial transient and some details of the pdf and power spectrum of ϕ.

Finally, it is worth noticing that the pattern-forming role played by the additive component of noise allows the emergence of patterns even for parameter values corresponding to points below or outside the curve of marginal stability displayed in Fig. 5.26. Thus patterns can also emerge for pairs of values $\{s_{gn}, D\}$, for which the multiplicative-noise component is unable to induce pattern formation. In this case, patterns result from the same mechanism as those described in Section 5.2. An example is shown in Fig. 5.28 that refers to a pattern emerging in subthreshold condition (i.e., $s_{gn} < s_c$). Therefore the additive- and multiplicative-noise components cooperate in forming patterns when the system exhibits a short-term instability, whereas only the additive component of noise is responsible of noise-induced patterns when this instability condition is not satisfied.

5.5 Multiplicative noise and non-pattern-forming spatial couplings

In this section, we explore the ability of spatiotemporal models driven by multiplicative noise to generate patterns when the spatial coupling does not fall into the class of

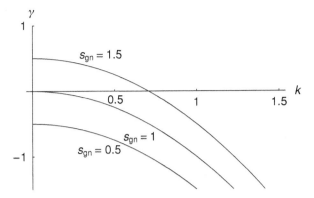

Figure 5.29. Dispersion relation for prototype model (5.58) for three values of the noise intensity s_{gn} ($a = -1$, $D = 1$).

pattern-forming spatial interactions. In other words, the spatial coupling would not be able to induce patterns in the deterministic counterpart of the system. Thus we consider models with the same temporal component, $f(\phi) + g(\phi)\xi_m$, as those studied in Section 5.4. This allows us to compare the results with those presented in Section 5.4 and to stress the role of the type of spatial coupling. In particular, we choose the diffusive differential operator, $\mathcal{L}[\phi] = D\nabla^2\phi$, as the exemplifying case of non-pattern-forming coupling. Similar to the case investigated in Subsection 5.4.1, the prototype model is then

$$\frac{\partial\phi}{\partial t} = a\phi - \phi^3 + \phi\xi_{gn} + D\nabla^2\phi, \tag{5.58}$$

where a is a parameter and ξ_{gn} is zero-mean white Gaussian noise, with intensity s_{gn}. Equation (5.58) is interpreted in the Stratonovich sense. The homogeneous deterministic stable state of Eq. (5.58) is $\phi_0 = 0$, and short-term instability occurs when $s_{gn} > s_c = -a$ (see Subsection 5.4.1).

We first study the linear stability of the basic state ϕ_0 in the dynamics of the ensemble average. Because of the linearity of the spatial coupling, such dynamics are described by

$$\frac{\partial\langle\phi\rangle}{\partial t} = a\langle\phi\rangle - \langle\phi\rangle^3 + s_{gn}\langle\phi\rangle + D\nabla^2\langle\phi\rangle. \tag{5.59}$$

The stability analysis of (5.59) leads to the dispersion relation

$$\gamma(k) = a + s_{gn} - Dk^2. \tag{5.60}$$

Figure 5.29 shows some examples of dispersion relation (5.60) calculated for different noise intensities s_{gn}. Three properties are evident. First, the value $s_{gn} = -a$ of the noise intensity marks the condition of marginal stability: No unstable wave numbers occur when $s_{gn} < -a$, whereas the wave numbers lower than $\sqrt{(s_{gn} + a)/D}$

become unstable if $s_{gn} > -a$. The threshold $s_{gn} = -a$ coincides with the one obtained in the short-term analysis. Second, the strength D of the spatial diffusive coupling affects the range of unstable wave numbers. In particular, the unstable wave numbers decrease when D increases, consistent with the fact that the diffusive coupling introduces spatial coherence in the random field. However, D does not have an impact on the occurrence of the instability, which depends only on the noise intensity. Therefore these nonequilibrium transitions are induced only by noise. Third, the most (linearly) unstable mode is always $k_{max} = 0$ (provided that $s_{gn} > -a$) for any value of D. This means that the linear component of the spatiotemporal dynamics tends to select an infinite wavelength, which corresponds to a homogeneous state.

In the study of the interplay between additive noise and diffusive spatial coupling presented in Section 5.3, the analysis of the steady-state structure function led to a result very similar to that shown in Fig. 5.29 (see Fig. 5.15). In fact, also in that case the most unstable mode corresponded to $k_{max} = 0$. However, in spite of the presence of this well-defined peak at $k = 0$, numerical simulations showed the emergence of steady multiscale patterns (see Fig. 5.16). Because of the existence of a range of unstable wave numbers (i.e., with positive growth factors γ), many competing scales are present in the pattern. The same behavior is also expected to emerge in the case of multiplicative noise discussed in this section. However, numerical simulations of (5.58) lead to a different result. In fact, Fig. 5.30 shows that a pattern without a clear periodicity occurs, but it evolves in time and tends to disappear in the long term. The pdf of the field ϕ shows that patterns occur during a phase transition from the initial basic state with order parameter $m = \phi_0 = 0$ to a new, substantially homogeneous state with $m \neq 0$. In particular, in the case shown in Fig. 5.30, numerical simulations give $m = 0.8$ for $t > 100$ time units.

The substantial difference with respect to the spatiotemporal model driven by additive noise is that patterns are now transient. The explanation for this different behavior can be found in the fact that the spatial coupling is unable to maintain the system far from the homogeneous condition, in spite of the initial instability. Similar to the case shown in Fig. 5.17, the emergence of patterns sustained by multiplicative noise requires a suitable spatial coupling, which is able to take advantage of the short-term instability to maintain the system far from the homogeneous basic condition, thereby creating a patterned state. However, this happens only when a pattern-forming type of coupling is included in the model. In contrast, when other types of spatial coupling are considered, they are unable to lock the system far from the homogeneous state. In these cases, the spatial coupling interacts with the short-term instability (which can be investigated through the dispersion relation[4]), but this interaction lasts only as long as the temporal dynamics are able to sustain the instability. As the initial

---

[4] Recall that the stability analysis is performed on an equation that approximates only the first stages of the dynamics of the ensemble average in the neighborhood of ϕ_0.

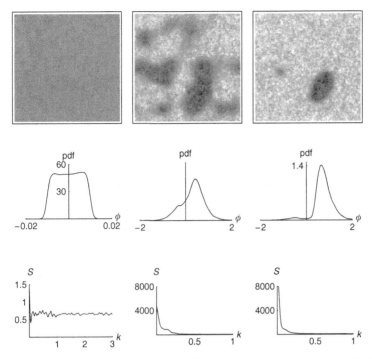

Figure 5.30. Example of numerical simulations of the prototype model (5.58). The columns refer to 0, 10, and 40 time units ($s_{gn} = 2$, $a = -1$, $D = 5$, and the other conditions are as in Fig. 5.10). The gray-tone scale spans the interval $[-2, 2]$.

instability tends to disappear, the spatial patterns follow the same fate and tend to fade out. In the long term the only legacy of the effect of the spatial coupling on the dynamics can be found in the phase transition (i.e., $m \neq 0$), though it is associated with a homogeneous field.

It is also worth noticing two other features of the patterns shown in Fig. 5.30. First, even though they are multiscale patterns, their main wave numbers are always close to zero. Thus the dominant wavelengths observable in the simulated fields are always of the same order of magnitude as the domain size. This is consistent with dispersion relation (5.60), which exhibits positive growth rates close to the maximum at $k = 0$. Therefore these noise-induced multiscale patterns fall into the category of spatial structures (see Subsection 5.1.1) with no clear dominant wavelengths. These stochastic models can provide a suitable representation of some real transient patterns found in the environmental sciences. The second interesting aspect concerns the boundaries of the geometrical features existing in patterns induced by multiplicative noise. Similar to the case discussed in Subsection 5.4.1, the coherence regions are much sharper and less fringed than those observed in the case of additive noise (compare Figs. 5.16 and 5.30). Also in this case, the difference is due to the multiplicative nature of the noise, which means that the $g(\phi)\xi_{gn}$ term is spatially correlated.

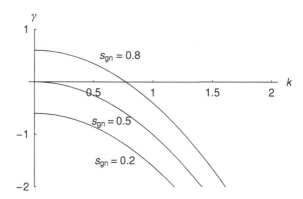

Figure 5.31. Dispersion relation of the VPT model with diffusive spatial coupling for three values of noise standard deviation, s_{gn} ($D = 1$).

5.5.1 The VPT model with diffusive spatial coupling

The VPT model with diffusive spatial coupling (see also Subsection 5.4.3),

$$\frac{\partial \phi}{\partial t} = -\phi(1 + \phi^2)^2 + (1 + \phi^2)\xi_{gn} + D\nabla^2\phi, \qquad (5.61)$$

is the original form of the VPT model (van den Broeck et al., 1994), which was proposed as one of the first examples of a spatiotemporal system exhibiting noise-induced (nonequilibrium) phase transitions.

In Subsection 5.4.3, we demonstrated that the temporal component of the VPT model shows a short-term instability when the noise intensity s_{gn} exceeds the threshold $s_c = 0.5$. The dispersion relation is very similar to the one determined for prototype model (5.58), in that both models have about the same linearized structure. In fact, the dispersion relation for model (5.61) is

$$\gamma(k) = 2s_{gn} - 1 - Dk^2 \qquad (5.62)$$

(Fig. 5.31 shows some examples). Apart from an inessential factor of 2 in the noise intensity, the stability analysis of the basic state $\phi_0 = 0$ of the ensemble leads to the same conclusions as those discussed for prototype model (5.58): (i) the threshold $s_c = 0.5$ detected by the short-term instability is confirmed, (ii) a range of wave numbers $[0, \sqrt{(2s_{gn} - 1)/D}]$ becomes unstable when $s_{gn} > s_c$, and (iii) the dispersion relation has a peak at $k_{max} = 0$.

It is instructive to explore the behavior of the diffusive VPT model by means of the generalized mean-field technique. When the spatial coupling is expressed by the diffusion term, the most unstable modes are $k_x = k_y = 0$ and correspond to the phase transition of the order parameter m. In fact, in the case of Laplacian spatial coupling, mean-field assumption (B5.3-4) is

$$l_h(\phi_i, \langle\phi_i\rangle, k_x, k_y) = \frac{2}{\Delta^2}\{[\cos(\Delta k_x) + \cos(\Delta k_y)]\langle\phi_i\rangle - 2\phi_i\}. \tag{5.63}$$

It follows that discretized equation (B5.3-1) is

$$\frac{d\phi_i}{dt} = f(\phi_i) + g(\phi_i)\xi_{m,i} + \frac{2D}{\Delta^2}\left[\cos(\Delta k_x) + \cos(\Delta k_y)\right]\langle\phi_i\rangle - \frac{4D}{\Delta^2}\phi_i + \xi_{a,i}. \tag{5.64}$$

Because this analysis assumes the ith cell to be at a maximum of the (possible) wavy pattern – i.e., $\langle\phi_i\rangle \geq 0$ – the term proportional to the ensemble average in Eq. (5.64) (i.e. the third addendum) has the maximum destabilizing effect on the dynamics of ϕ_i when $[\cos(\Delta k_x) + \cos(\Delta k_y)]$ is maximum, namely when $k_x = k_y = 0$. Thus, in the case of Laplacian spatial coupling, the most unstable modes correspond to a phase transition with no stable patterns. This phase transition can be explored with the standard mean-field assumption (B5.4-1), which for a generic diffusive model leads to

$$\frac{d\phi_i}{dt} = f(\phi_i) + g(\phi_i)\xi_{m,i} + \frac{4D}{\Delta^2}(m - \phi_i) + \xi_{a,i}. \tag{5.65}$$

Transitions can therefore be investigated by use of the discretized version of model (5.61), namely

$$\frac{d\phi_i}{dt} = -\phi_i(1 + \phi_i^2)^2 + (1 + \phi_i^2)\xi_{gn} + \frac{4D}{\Delta^2}(m - \phi_i). \tag{5.66}$$

The corresponding steady-state pdf is (see Chapter 2)

$$p(\phi_i, m) = \frac{\exp\left[-\frac{\phi_i^2}{s_{gn}} + \frac{D}{s_{gn}}\left(\frac{m\phi_i - \phi_i^2}{1 + \phi_i^2}\right) + \frac{mD\arctan(\phi_i)}{s_{gn}}\right]}{2(1 + \phi_i^2)}, \tag{5.67}$$

which allows us to impose the self-consistency condition, $m = \int_{-\infty}^{\infty} p(\phi_i, m)d\phi_i$, and to explore the existence of solutions with m values different from $\phi_0 = 0$ (van den Broeck et al., 1994). Figure 5.32(a) shows the line marking the transition between the region of the parameter space $\{s_{gn}, D\}$ where $m = 0$ is the only solution of Eq. (5.67) and the region where new solutions appear. Figure 5.32(b) shows the emergence of the phase transition as a function of s_{gn} for a given D. It can be observed that the diffusive VPT model exhibits multiple solutions of the order parameter for suitable values of the noise intensity s_{gn} and strength of the spatial coupling D. Moreover, this phase transition is clearly a second-order (nonequilibrium) phase transition because it occurs with no discontinuity in the order parameter m. Finally, the figure shows that the transition is reentrant, in that the stable state of the system switches back to the initial (disordered) state $m = 0$ for large values of s_{gn}. Thus the emergence of a new phase requires both spatial coupling (i.e., D needs to exceed a noise-dependent critical value) and intermediate noise intensities.

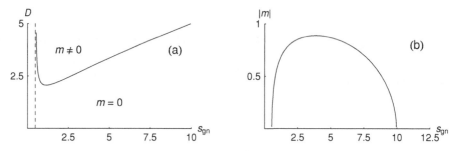

Figure 5.32. Results of the mean-field analysis of the VPT model with diffusive spatial coupling: (a) Curve delimiting the phase transition (the vertical dashed line marks the threshold $s_c = 1$), (b) behavior of the order parameter m as a function of the noise intensity for $D = 5$.

It is worth recalling that the existence of phase transitions does not necessarily imply the occurrence of patterns. Pattern formation needs to be investigated through numerical simulations of Eq. (5.61). Using periodic boundary conditions and random initial conditions, these simulations show (Fig. 5.33) the emergence of patterns with features that are qualitatively similar to those already discussed for prototype model (5.58). Regions with strong spatial coherence occur, but do not exhibit any obvious periodicity. Such regions tend to fade out after some time. Thus, after a transient (whose duration increases with the domain size), the field appears to be practically homogeneous with average $m = 0.63$. This value is close to the one calculated with the mean-field analysis for the same noise intensity (see Fig. 5.32). The final value of the order parameter m is then different from $\phi_0 = 0$ and can be either positive or negative, depending on the initial conditions. The numerical simulations therefore confirm that the diffusive VPT model exhibits noise-induced phase transitions and irregular transient patterns and that no stable periodic structures emerge.

Figure 5.33 shows the probability density distributions corresponding to the simulated fields. Such pdf's are substantially unimodal (only a weak bimodality occurs in some realizations). Therefore coherence regions are not due to a temporal noise-induced transition in the local stochastic dynamics (like those described in Chapter 3), which change the shape of the local pdf from unimodal to bimodal.

5.6 The role of the temporal autocorrelation of noise

This section investigates the role of colored noise in spatially extended systems. In particular, we focus on the same prototype models as those described in the previous sections, but we model the random fluctuations by using DMN instead of Gaussian white noise. As remarked in Chapter 2, DMN is a simple form of colored noise, characterized by the autocovariance function

$$\langle \xi_{dn}(t)\xi_{dn}(t')\rangle = \frac{k_1 k_2(\Delta_2 - \Delta_1)^2}{(k_1 + k_2)^2}e^{-|t-t'|(k_1+k_2)}. \tag{5.68}$$

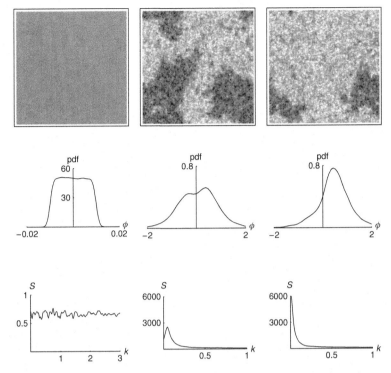

Figure 5.33. Numerical simulations of the diffusive VPT model. The columns refer to 0, 50, and 200 time units ($s_{gn} = 2$, $D = 5$, and the other conditions are as in Fig. 5.10). The gray-tone scale spans the interval $[-2, 2]$.

For simplicity, in this section we consider processes driven by symmetrical DMN, i.e., processes with $k_1 = k_2 = k$ and $\Delta_1 = -\Delta_2 = \Delta$; the corresponding noise intensity is $s_{dn} = \Delta^2$. Other forms of colored noise could of course be used to represent the temporal correlation of the random forcing, but the results would be analogous to those found with DMN.

The four models considered are

$$\frac{\partial \phi}{\partial t} = a\phi - D(k_0^2 + \nabla^2)^2\phi + \xi_{dn}, \tag{5.69}$$

$$\frac{\partial \phi}{\partial t} = a\phi + D\nabla^2\phi + \xi_{dn}, \tag{5.70}$$

$$\frac{\partial \phi}{\partial t} = a\phi - \phi^3 + \phi\,\xi_{dn} - D(k_0^2 + \nabla^2)^2\phi, \tag{5.71}$$

$$\frac{\partial \phi}{\partial t} = -a\phi - \phi^3 + \phi\,\xi_{dn} + D\nabla^2\phi, \tag{5.72}$$

which correspond to prototype models (5.16), (5.29), (5.34), and (5.58), respectively, analyzed in previous sections.

(a)   (b)

(c)   (d)

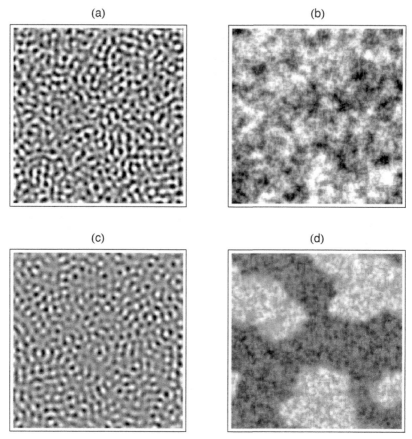

Figure 5.34. Numerical simulation of stochastic models forced by dichotomous noise: (a) corresponds to the model (5.69); the parameters are $a = -1$, $D = 10$, $k_0 = 1$, $k = 0.5$ ($\tau_c = 1$), $s_{dn} = 2$, $t = 50$ time units. (b) refers to model (5.70) with parameters $a = -1$, $D = 2.5$, $k = 0.5$ ($\tau_c = 1$), $s_{dn} = 20$, $t = 50$ time units. (c) shows the outcome of model (5.71) with parameters $a = -1$, $D = 15$, $k_0 = 1$, $k = 0.5$ ($\tau_c = 1$), $s_{dn} = 200$, $t = 50$ time units. (d) corresponds to model (5.72) with parameters $a = -1$, $D = 5$, $k = 0.5$ ($\tau_c = 1$), $s_{dn} = 200$, $t = 50$ time units.

Figure 5.34 shows an example of four fields resulting from the simulation of models (5.69)–(5.72). We observe that patterns are qualitatively very similar to those described in the previous sections, when the stochastic forcing was a white Gaussian noise. This testifies that the temporal-noise correlation generally does not introduce new pattern morphologies, which remain essentially dictated by the deterministic components. However, the autocorrelation of the noise term can have two important effects. The first effect concerns the pattern intensity: The noise correlation tends to make patterns sharper and better defined and – in the case of periodic patterns – with a more evident dominant wavelength. To quantify this effect, we use the ratio $i_p / i_{p,0}$ between the peak value of the power spectrum and the area subtended by the spectrum.

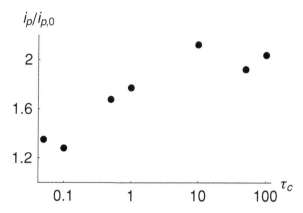

Figure 5.35. Impact of the noise correlation τ_c on the intensity of the pattern i_p (this latter quantity is normalized to the value $i_{p,0}$ it takes when $\tau_c = 0$). The correlation scale is expressed in time units.

Figure 5.35 shows how – for the case of Eq. (5.69) – this ratio tends to increase as a function of the noise correlation $\tau_c = 0.5/k$ apart from some fluctuations that are due to the limited number of realizations.

The second effect of autocorrelation is to change – with respect to the white-noise case – the boundaries of the region in the parameter space $\{s_{dn}, D\}$ in which patterns occur. Unfortunately, correlation makes analytical results much more difficult to obtain than in the case of white noise. However, some studies have shown that temporal correlation can induce both an expansion and a contraction of the region of the parameter space corresponding to pattern formation. For example, Mangioni et al. (1997) investigated the emergence of noise-induced transitions in the VPT model driven by autocorrelated Gaussian noise (i.e., O-U process) and they found that temporal-noise autocorrelation reduces the range of parameter values in which patterns emerge.

5.7 Patterns accompanied by temporal phase transitions

In the third chapter we described noise-induced transitions in zero-dimensional dynamical systems. Such transitions correspond to the occurrence of steady-state probability distributions whose modes are different from the equilibrium states of the corresponding deterministic system. A relevant case is represented by systems exhibiting noise-induced bistability. In this case, suitable noise intensities are able to generate pdf's with two modes even though the deterministic dynamics have only one stable steady state. Two key ingredients are needed to activate this type of stochastic dynamics, when no constraints are imposed by the boundaries of the domain. First, a deterministic local kinetics $f(\phi)$, tending to drive the dynamical system toward the steady state, $\phi = \phi_0$, is needed. The local potential is then monostable with a minimum at $\phi = \phi_0$: If the system is moved away from the state ϕ_0, the local kinetics tends to restore this state. For this reason, the deterministic local dynamics are also called *relaxational*. The second ingredient is a multiplicative random component that tends to drive the state of the system away from $\phi = \phi_0$. To this end, the multiplicative function $g(\phi)$ of the noise term generally has a maximum at $\phi = \phi_0$ and decreases as ϕ moves away from ϕ_0. As a result of the balance between deterministic and stochastic components, bimodal probability distributions of ϕ may emerge at steady state. In fact, when the noise is sufficiently intense the random component is able to maintain the system far from $\phi = \phi_0$ in spite of the fact that the local deterministic dynamics tend to drive the system toward the neighborhood of ϕ_0.

In the case of spatiotemporal dynamical systems, an ordered state may emerge if the spatial coupling cooperates with the stochastic component to prevent the relaxation imposed by the local dynamics, thereby maintaining the system away from the uniform state $\langle \phi \rangle = \phi_0$. Moreover, the spatial coupling needs to give spatial coherence to the field, creating a patterned state in which the coherent regions correspond to the two modes existing in the underlying temporal dynamics. In these conditions, the off-center (i.e., for $\phi \neq \phi_0$) modes of the underlying temporal dynamics are crucial to the emergence of spatially ordered structures. This mechanism of pattern formation can then be considered the direct extension to spatially extended systems of noise-induced transitions in zero-dimensional systems. This mechanism is sometimes called *entropy-driven pattern formation* (Sagues et al., 2007) in that the dynamical system escapes from the minimum of the potential (i.e., $\phi = \phi_0$) because of the strength of noise (which is an entropy source).

It is worth noticing that the mechanism of noise-induced pattern formation discussed in this section has two key differences with respect to those presented in Sections 5.4 and 5.5, in that (i) it does not emerge from a short-term instability, and (ii) patterns also emerge if the Langevin equation is interpreted according to Ito's rule. The following subsection is devoted to presenting this mechanism of pattern formation through a simple example.

5.7.1 Prototype model

Consider the following spatiotemporal stochastic model:

$$\frac{\partial \phi}{\partial t} = -a\phi + \sqrt{\frac{1}{1+c\phi^2}}\, \xi_{gn}(t) - D(k_0^2 + \nabla^2)^2 \phi, \qquad (5.73)$$

where a and c are two positive-valued parameters, ξ_{gn} is a white zero-mean Gaussian noise with intensity s_{gn}, and the spatial coupling is expressed by the Swift–Hohenberg operator, where D modulates the coupling intensity and k_0 determines the dominant wavelength. A similar model was proposed by Buceta et al. (2003).

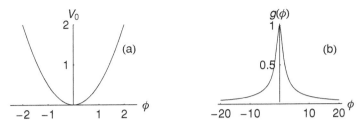

Figure 5.36. (a) Potential corresponding to the local kinetics in model (5.73) and (b) behavior of the function $g(\phi)$. In both cases $a = 1$ and $c = 1$.

The local kinetics $f(\phi) = -a\phi$ corresponds to the monostable potential

$$V(\phi) = -\int_\phi f(\phi')\mathrm{d}\phi' = \frac{a\phi^2}{2}, \tag{5.74}$$

shown in Fig. 5.36(a). It follows that $\phi = \phi_0 = 0$ is the only deterministically stable homogeneous state. Figure 5.36(b) shows the behavior of the function $g(\phi) = \sqrt{1/(1 + c\phi^2)}$. It has a maximum at $\phi = \phi_0$ [where $g(0) = 1$] and decays symmetrically for increasing values of $|\phi|$. We can observe that model (5.73) does not exhibit short-term instability. When Eq. (5.73) is interpreted according to Ito, we have $\mathrm{d}\langle\phi\rangle/\mathrm{d}t = f(\langle\phi\rangle) = -a\langle\phi\rangle$, which denotes stability of the basic state $\langle\phi\rangle = 0$ (recall that $a > 0$). Also, when the Stratonovich interpretation is adopted, following the methods in Box 5.4, we obtain

$$\frac{\mathrm{d}\langle\phi\rangle}{\mathrm{d}t} = -a\langle\phi\rangle - s_{\mathrm{gn}}\frac{c\langle\phi\rangle}{(1 + c\langle\phi\rangle^2)^2}; \tag{5.75}$$

because the spurious drift term is also negative, it is unable to destabilize the basic state.

The purely temporal component of model (5.73), namely

$$\frac{\mathrm{d}\phi}{\mathrm{d}t} = -a\phi + \sqrt{\frac{1}{1 + c\phi^2}}\xi_{\mathrm{gn}}(t), \tag{5.76}$$

exhibits a noise-induced bistability. In fact, the steady-state pdf is

$$p(\phi) = C(1 + c\phi^2)^\nu \exp\left[-\frac{a\phi^2(2 + c\phi^2)}{4s_{\mathrm{gn}}}\right], \tag{5.77}$$

where C is the normalization constant and $\nu = 1/2$ or $\nu = 1$, depending on the use of the Stratonovich or Ito interpretation, respectively. The modes ϕ_m correspond to the zeros of

$$-a\phi_m + \frac{2\nu c s_{\mathrm{gn}}\phi_m}{(1 + c\phi_m^2)^2} = 0. \tag{5.78}$$

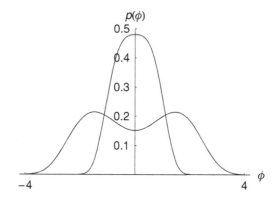

Figure 5.37. Examples of pdf (5.77) in the case of Stratonovich interpretation of the noise (i.e., $\nu = 1/2$). The unimodal pdf corresponds to $s_{gn} = 0.8$, and the bimodal one refers to $s_{gn} = 10$ (in both cases $a = 1$ and $c = 1$).

When $s_{gn} < a/2\nu c$ the pdf has only one mode at $\phi = 0$, whereas when the noise intensity exceeds the threshold $s_{c,t} = a/(2\nu c)$ (the subscript t indicates that such a threshold refers to the temporal dynamics), two noise-induced maxima occur at $\pm\sqrt{(\sqrt{2\nu s_{gn} c/a} - 1)/c}$ (see Fig. 5.37).

The generalized mean-field analysis allows us to explore whether patterns may emerge from the dynamics of Eq. (5.73). If the most unstable wave number is considered – i.e., the condition $\omega(\mathbf{k}) = 0$ is assumed (see Box 5.3) – according to self-consistency equation (B5.3-5) patterns are expected to emerge when $s_{gn} > a/c$, i.e., when the noise is able to induce bimodality in the corresponding temporal dynamics, under both the Stratonovich and the Ito interpretation. On the other hand, the classical mean-field analysis indicates that the order parameter m remains equal to zero for any pair of values $\{a, c\}$; thus, in this case, pattern formation is not associated with the occurrence of a phase transition.

It is worth noticing that stability analysis by normal modes is unable to detect pattern occurrence. In fact, the equation describing the linear spatiotemporal dynamics of the ensemble average is

$$\frac{\partial \langle \phi \rangle}{\partial t} = -\left(a + s_{gn}c\right)\langle \phi \rangle - D(k_0^2 - \nabla^2)^2\langle \phi \rangle, \tag{5.79}$$

which gives the dispersion relation,

$$\gamma(k) = -\left(a + s_{gn}c\right) - D(k_0^2 - k^2)^2, \tag{5.80}$$

which shows that the growth factor γ is always negative for all values of noise-intensity wave numbers.

Figures 5.38 and 5.39 show some examples of patterns resulting from the numerical simulation of Eq. (5.73), under the Stratonovich or the Ito interpretation, respectively. Well-defined patterns emerge with the dominant wave number dictated by the spatial

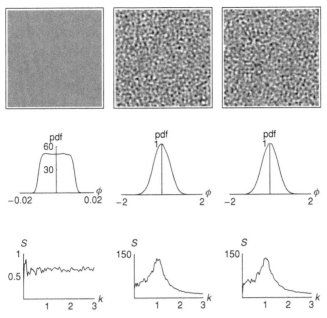

Figure 5.38. Example of patterns resulting from the numerical simulation of model (5.73) under the Stratonovich interpretation. The boundary conditions are periodic. The columns refer to 0, 50, and 100 time units. The parameters are $a = 1$, $c = 1$, $k_0 = 1$, $D = 40$, and $s_{gn} = 10$.

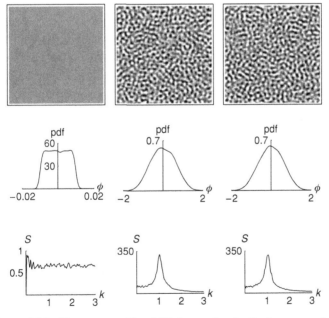

Figure 5.39. The same as Fig. 5.38, but under the Ito interpretation.

coupling, consistent with the results of the mean-field analysis. It is interesting to notice that the pdf of the field is unimodal. This demonstrates that the spatial coupling has also the effect of eliminating the bimodality intrinsic to the purely temporal dynamics. No substantial differences emerge between the case with the Ito or the Stratonovich interpretation of the noise term in that in both cases the power spectra exhibit a clear maximum at $k = k_0$) however, in the case of the Ito interpretation, patterns appear to be more regular than in the Stratonovich case, consistent with similar observations reported for other spatiotemporal models (Sagues et al., 2007).

5.7.2 Prototype model with non-pattern-forming coupling

It is interesting to analyze what happens when the Swift–Hohenberg coupling in model (5.73) is substituted with a non-pattern-forming coupling, e.g., the classical diffusion $D\nabla^2\phi$. In this case the model is

$$\frac{\partial \phi}{\partial t} = -a\phi + \sqrt{\frac{1}{1 + c\phi^2}}\, \xi_{\mathrm{gn}}(t) + D\nabla^2\phi, \qquad (5.81)$$

and the dispersion relation is

$$\gamma(k) = -\left(a + s_{\mathrm{gn}}c\right) - Dk^2. \qquad (5.82)$$

Thus the stability analysis does not detect unstable modes for any noise intensity. Moreover, the classical mean-field analysis shows that the order parameter m remains equal to zero, indicating that no phase transitions occur. In spite of these analytical results, numerical simulations of model (5.81) display some interesting features. An example is shown in Fig. 5.40 for the case of the Ito interpretation (very similar patterns emerge also in the case of the Stratonovich interpretation). Although the patterns are less defined than in the other cases shown in this chapter, a fringed multiscale pattern emerges. Similar to the case with Swift–Hohenberg spatial coupling, the pdf of the field is unimodal.

5.7.3 A case with $g(\phi_0) = 0$

In the previous subsection, we discussed the occurrence of (periodic or multiscale) patterns as the result of the cooperation between a noise-induced temporal transition and a spatial coupling. However, the specific form of $g(\phi)$ used in model (5.73) suggests another possible interpretation. In fact, $g(\phi)$ is formed by a constant component (corresponding to additive noise) plus a ϕ-dependent component (corresponding to purely multiplicative noise). Such two components are evident in the series expansion

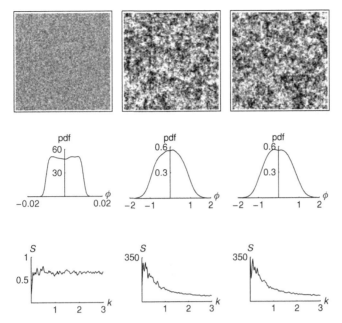

Figure 5.40. Example of patterns resulting from the numerical simulation of model (5.81). The columns refer to 0, 50, and 100 time units. The parameters are $a = 1$, $c = 1$, $D = 10$, and $s_{\text{gn}} = 10$.

of $g(\phi)$ around zero:

$$g(\phi) = \sqrt{\frac{1}{1 + c\phi^2}} = 1 - \frac{c\phi^2}{2} + \frac{3c^2\phi^4}{8} - \frac{5c^3\phi^6}{16} + \cdots . \qquad (5.83)$$

Therefore, when ϕ is close to zero, the ϕ-dependent component tends to zero, but the additive component remains [i.e. $g(\phi_0) \neq 0$] and is able to unlock the system from the deterministic stable state ϕ_0, thereby maintaining the dynamics away from equilibrium and allowing the spatial coupling to exert its pattern-inducing effect. From this point of view, the mechanism of noise-induced pattern formation described in this section is very similar to the one described in Sections 5.2 and 5.3. However, the purely multiplicative component of g is not secondary; in fact, it has the role of dimming the noise effect far from ϕ_0, thereby hampering the divergence of the dynamical system away from $\phi = \phi_0$. In other words, when $|\phi|$ is high, the function $g(\phi)$ tends to zero, and the effect of noise on the dynamics of ϕ tends to disappear; consequently the deterministic component prevails and the system goes back to values of ϕ close to ϕ_0. In this sense the dynamical role of the ϕ-dependent component of $g(\phi)$ resembles the role of the term $-\phi^3$ in the deterministic Ginzburg-Landau model (see Subsection 5.2.1).

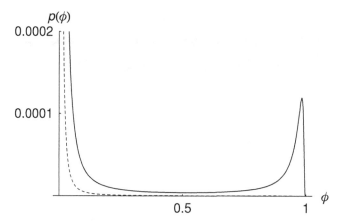

Figure 5.41. Occurrence of a noise-induced transition in the purely temporal component of model (5.84) (i.e., $D = 0$) when the noise strength s_{gn} crosses the threshold
$s_c = 4a$. The dotted and continuous pdf's refer to $s_{gn} = 3$ and $s_{gn} = 30$, respectively
($a = 1$ in both cases). The areas of the pdf's appear to be different because of the
truncation of the left peak.

To show what happens when $g(\phi_0) = 0$, we consider the model

$$\frac{\partial \phi}{\partial t} = -a\phi + \phi(1 - \phi)\xi_{gn}(t) - D(k_0^2 + \nabla^2)^2\phi, \tag{5.84}$$

interpreted according to Ito. The purely temporal version of model (5.84) [i.e.,
$d\phi/dt = -a\phi + \phi(1 - \phi)\xi_{gn}$] shows a noise-induced transition for $s_c = 4a$. In fact,
the steady-state pdf is

$$p(\phi) = \frac{(1 - \phi)^{\frac{a}{s_{gn}} - 2}}{\phi^{\frac{a}{s_{gn}} + 2}} \exp\left[-\frac{a}{s_{gn}(1 - \phi)} \right], \qquad \phi \in]0, 1[, \tag{5.85}$$

and always has a mode for $\phi \to 0$, whereas a new mode occurs at

$$\phi = \frac{3}{4} + \sqrt{\frac{s_{gn} - 4a}{16s_{gn}}} \tag{5.86}$$

when $s_{gn} > s_c$ (Fig. 5.41 shows an example of this noise-induced bimodality). It is
instructive to take a closer look at the causes of this bimodality. In fact, its emergence
is different from the one observed in model (5.76). In that case, the function $g(\phi)$
exhibited a decrease as ϕ moved away from $\phi = 0$. Thus a sufficiently strong noise
was able to unlock the deterministic stable state, and two modes emerged at values of
ϕ where the actions of noise and $f(\phi)$ tended to compensate. Differently, in model
(5.84) bimodality is due to cooperation between the noise and the natural boundaries
at $\phi = 0$ and $\phi = 1$. When the noise is sufficiently strong, the system tends to move
away from $\phi = 0$, but the boundary at $\phi = 1$ prevents the system from visiting the

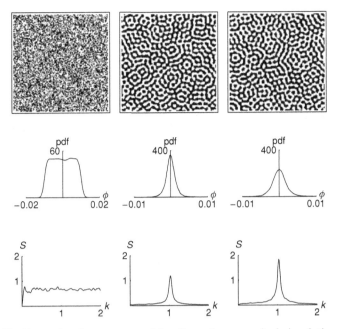

Figure 5.42. Example of patterns resulting from the numerical simulation of model (5.84) under the Ito interpretation. The columns refer to 0, 50, and 100 time units. The parameters are $a = 1 \times 10^{-3}$, $k_0 = 1$, $D = 10$, and $s_{gn} = 1$.

whole real axis, thereby causing an accumulation of probability close to the upper limit of the domain.

Equation (5.84) does not exhibit any short-term instability because the equation is interpreted according to Ito [hence, $\langle g(\phi)\xi_{gn} \rangle = 0$]. However, the spatial coupling is able to induce patterns exploiting the temporal-noise-induced transition. Figure 5.42 shows an example: Similar to the case of the spatiotemporal dynamics (5.73), in this case the pdf of the field is also unimodal even though pdf (5.85) of the temporal dynamics exhibits a strong bimodality for the same values of noise strength.

It is worth noticing that in model (5.84) patterns can emerge also when the diffusive spatial coupling is used in place of the Swift–Hohenberg operator,

$$\frac{\partial \phi}{\partial t} = -a\phi + \phi(1 - \phi)\xi_{gn} + D\nabla^2\phi. \tag{5.87}$$

Figure 5.43 shows an example of these patterns. Notice that high noise strengths s_{gn} and coupling coefficients D are necessary to generate well-defined patterns.

5.7.4 A particular subclass of processes

We conclude this section by mentioning a particular subclass of processes that attracted a certain interest in this field (Buceta and Lindenberg, 2004; Sagues et al., 2007). This

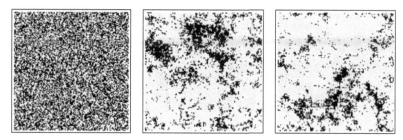

Figure 5.43. Example of pattern resulting from the numerical simulation of model (5.87) under the Ito interpretation. The columns refer to 0, 50, and 100 time units. The parameters are $a = 0.001$, $k_0 = 1$, $D = 25$, and $s_{gn} = 40$.

class of models has a peculiar mathematical structure that reads

$$\frac{\partial \phi}{\partial t} = g^2(\phi) \left[-\frac{dV_0(\phi)}{d\phi} + \mathcal{L}[\phi] \right] + g(\phi)\xi_{gn}(\mathbf{r}, t), \tag{5.88}$$

where V_0 is the local deterministic monostable potential with a minimum at ϕ_0, $\mathcal{L}[\phi]$ is the spatial coupling, and $\xi(\mathbf{r}, t)$ is the usual white (in space and time) Gaussian noise with zero mean and strength s_{gn}. The key characteristic of these models [Eq. (5.88)] is the function $g(\phi)$ that modulates, with a different exponent, the three components of the model: the local deterministic dynamics, $f(\phi) = -g^2(\phi)dV_0(\phi)/d\phi$; the noise component, $g(\phi)\xi_{gn}$; and the spatial coupling, $g^2(\phi)\mathcal{L}[\phi]$. If functions $V_0(\phi)$ and $g(\phi)$ are suitably chosen, the zero-dimensional component of model (5.88) exhibits noise-induced temporal transitions similar to those of model (5.73).

A case investigated in detail (see Carrillo et al., 2003; Buceta and Lindenberg, 2004) refers to the monostable potential

$$V(\phi) = \frac{a}{2}\phi^2, \tag{5.89}$$

with $a > 0$,

$$g(\phi) = \sqrt{\frac{1}{1 + c\phi^2}} \tag{5.90}$$

[notice that this function is the same as the one in Eq. (5.73)], and with Swift–Hohenberg coupling, $-D(k_0^2 + \nabla^2)^2\phi$. The steady-state pdf of the corresponding temporal dynamics is

$$p(\phi) = C(1 + c\phi^2)^{\nu}\exp\left[-\frac{a\phi^2}{2s_{gn}}\right], \tag{5.91}$$

where C is the normalization constant and ν depends on the interpretation of the noise term. The pdf (5.91) undergoes a noise-induced transition at $s_{c,t} = a/(2c\nu)$.

The short-term stability analysis of model (5.88) with (5.89) and (5.90) shows that the homogeneous solution $\phi_0 = 0$ is stable for any noise intensity, whereas the stability

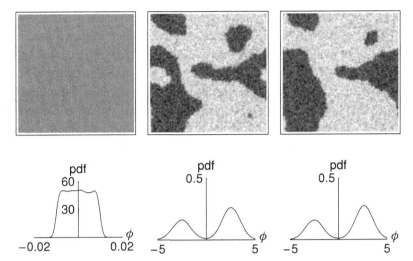

Figure 5.44. Example of pattern resulting from the numerical simulation of model (5.88) with potential $V(\phi)$ and function g given by (5.89) and (5.90), respectively, under the Ito interpretation. The columns refer to 0, 500, and 1000 time units. The parameters are $a = 1$, $D = 4$, $c = 3$, and $s_{\mathrm{gn}} = 3.5$.

analysis by normal modes exhibits a dispersion relation that is always negative for any wave number. In contrast, the generalized mean-field analysis is able to diagnose the occurrence of patterns (not shown). Figure 5.44 shows an example of the numerical simulations obtained with these models. Notice that in this case the bimodality of the pdf's remains also in the spatially extended system.

Buceta and Lindenberg (2004) made a comprehensive analysis of the class of the models expressed by (5.88). They showed that this type of system may exhibit a remarkable variety of transition scenarios, depending on the properties of the local potential and the kinetic coefficient. Four simple forms of local potential and kinetic coefficients allow us to describe all the major cases. These forms of $V(\phi)$ and $g(\phi)$ are

$$V_1(\phi) = \frac{\phi}{2}, \qquad\qquad V_2(\phi) = \frac{\phi^4}{4} - \frac{\phi^2}{2}, \qquad\qquad (5.92)$$

$$g_1(\phi) = \sqrt{\frac{1 + \phi^2}{1 + \phi^4}}, \qquad\qquad g_2(\phi) = \sqrt{\frac{1}{1 + \phi^2}}. \qquad\qquad (5.93)$$

5.8 Patterns induced by periodic or random switching between deterministic dynamics

In the mechanisms of noise-induced pattern formation described in the previous sections, ordered states of the system emerge always as a result of local, random fluctuations of the state variable, which can be either spatially correlated or uncorrelated. In another major class of stochastic models, noise-induced patterns result from the random switching between dynamics that simultaneously occurs at all points of the spatial domain (Buceta and Lindenberg, 2002a, 2002b; Buceta et al., 2002a, 2002b). This mechanism is based on the idea that if the random switching between the dynamics is global (i.e., it simultaneously occurs across the domain) and rapid, the system behaves as if it was undergoing the average dynamics obtained as a weighted mean of the two states. In this context, by rapid switching we mean that the average residence time of the dynamics in each of its two dynamical states is much shorter than the time needed by the system to reach the equilibrium state.

Thus if we take as an example two Turing models[5] and we randomly and rapidly switch between them, the system experiences only the average Turing dynamics. We can envision cases in which, separately, neither one of the two dynamics is able to lead to pattern formation, and their average exhibits diffusion-induced symmetry-breaking instability. In these conditions patterns emerge from the nonequilibrium random (global) alternation between the dynamics. Similar models can be constructed with two suitable biharmonic (or neural) models (Buceta and Lindenberg, 2002a) that by themselves are unable to exhibit symmetry-breaking instability. The random switching between these two models may lead to mean dynamics that are capable of generating patterns.

We consider as an example a system in which Turing's instability is induced by noise. To this end, consider a two-variable dynamical system described by reaction–diffusion equations, namely

$$\frac{\partial \phi_1}{\partial t} = f_{1,2}(\phi_1, \phi_2) + \nabla^2 \phi_1, \tag{5.95a}$$

$$\frac{\partial \phi_2}{\partial t} = g_{1,2}(\phi_1, \phi_2) + d_{1,2}\nabla^2 \phi_2, \tag{5.94b}$$

where $f_{1,2}(\phi_1, \phi_2)$ and $g_{1,2}(\phi_1, \phi_2)$ are the two pairs of functions describing the local dynamics of states 1 and 2. The system switches between state 1, in which the local kinetics is expressed by f_1 and g_1 and the diffusion ratio is d_1 (these three quantities are hereafter simply indicated as A_1), and state 2, in which the kinetics is modeled by the functions f_2 and g_2 and the diffusion ratio is d_2 (indicated as A_2). Neither of Eqs. (5.95) for state 1 or 2 meets all the conditions for pattern formation in the Turing models [see Appendix B, Eqs. (B.5) and (B.9)]. Thus, separately, the two dynamics do not lead to pattern formation. Each control parameter (f, g, or d) alternates

---

[5] Turing models (Turing, 1952) involve a system of two or more coupled deterministic reaction–diffusion equations in which a homogeneous equilibrium state is destabilized by the spatial coupling (i.e., by diffusion). This instability leads to pattern formation when the most unstable mode has a wave number different from zero. More details on the Turing model can be found in Appendix B.

dichotomously in a way that the temporal evolution $A(t)$ of each parameter can be expressed as

$$A(t) = A_1 \Lambda(t) + A_2[1 - \Lambda(t)], \qquad (5.95)$$

where $\Lambda(t)$ is a dichotomous variable assuming values 0 and 1. Notice that in this case dichotomous noise is used for its ability to provide a mechanism of random alternation between two deterministic dynamics. This use of dichotomous noise is consistent with the mechanistic interpretation discussed in Chapters 2 and 3.

When the switching is fast in the sense discussed before, $\Lambda(t)$ can be replaced with its average value $\Lambda(t) \simeq \langle \Lambda(t) \rangle$, and in this case

$$A(t) \simeq \bar{A} = A_1 P_1 + A_2 P_2, \quad \text{with} \quad P_1 = \langle \Lambda \rangle, \quad P_2 = 1 - \langle \Lambda \rangle. \qquad (5.96)$$

The dynamics resulting from the fast switching between the two states are then

$$\frac{\partial \phi_1}{\partial t} = \bar{f}(\phi_1, \phi_2) + \nabla^2 \phi_1, \qquad (5.97a)$$

$$\frac{\partial \phi_2}{\partial t} = \bar{g}(\phi_1, \phi_2) + \bar{d}\nabla^2 \phi_2, \qquad (5.97b)$$

where $\bar{f} = f_1 P_1 + f_2 P_2$, $\bar{g} = g_1 P_1 + g_2 P_2$, and $\bar{d} = d_1 P_1 + d_2 P_2$. Patterns emerge if the average dynamics meet the conditions of Turing's instability [Eqs. (B.5) and (B.9)] presented in Appendix B.

The emergence of switching-induced patterns depends on the velocity of the alternation between the two dynamics. Over a relatively long time both dynamics would lead to spatially homogeneous configurations. However, if the switching is sufficiently fast the system can experience the average dynamics (5.97). In fact, in these conditions, the homogeneous steady state can never be reached and the system always remains in a nonequilibrium configuration, which can be described by the mean of the two states. The separation between slow and fast switching can be defined with a control parameter r, representing the ratio between an external time scale t_{ext}, associated with the random switching (i.e., the average time the system spends in each configuration), and an internal time scale, t_{int}, associated with the time needed by the system to reach equilibrium in each of the two states. If $r = t_{ext}/t_{int} \to \infty$, no switching-induced instability emerges. On the other hand, when $r < 1$ the dynamics can be described in average terms as in Eq. (5.97), and patterns may emerge if conditions (B.5) and (B.9) are met. We note the following facts:

1. Patterns may emerge from random alternation of dynamics even when both dynamics have the same homogeneous steady state. In this case, it has been found that the random switching leads to spotted structures, whereas, when the two dynamics have different steady states, the random alternation may lead to labyrinthine-striped configurations (Buceta and Lindenberg, 2002a).

2. A similar model of pattern formation can be developed with a random switching between two biharmonic (Buceta and Lindenberg, 2002b) or two neural models (D'Odorico et al., 2006b), as discussed in Chapter 6.

3. Although the random switching can play a crucial role in this process of pattern formation, similar patterns emerge when the switching mechanism is deterministic. In fact, Buceta and Lindenberg (2002b) showed that periodic alternation of dynamics can also lead to pattern formation. Thus patterns emerging in nonequilibrium systems from the global alternation of dynamics are not noise-induced *sensu strictu*. We refer the interested reader to Bena (2006) for a more detailed discussion of ordered states induced by periodic and random drivers.

We can consider as an example the case of system (5.95) that switches between the following two states: state 1, expressed by

$$f_1(u, v) = avu^2, \qquad g_1(u, v) = bv, \qquad (5.98)$$

with $a = 27.5$, $b = 105$, and $d_1 = 41.25$, and state 2, expressed by

$$f_2(u, v) = -eu, \qquad g_2(u, v) = -cu^2v^2, \qquad (5.99)$$

with $e = 90$, $c = 566.67$, and $d_2 = 20$. Separately, both states are unable to create deterministic patterns (see the constraints reported in Appendix B) and any initial condition evolves to a uniform field.

Patterns emerge if the two states are repeatedly and randomly alternated, with state 1 occurring with probability $P_1 = 0.8$ and state 2 with probability $P_2 = 0.2$. If the switching is sufficiently fast (i.e., $r < 1$), using Eqs. (5.97), we have the average dynamics:

$$\bar{f} = u(P_1avu - P_2e), \qquad \bar{g} = v(P_1b - P_2cu^2v), \qquad \bar{d} = P_1d_1 + P_2d_2, \quad (5.100)$$

which are able to induce pattern formation through Turing instability. In fact, in this case average dynamics (5.100) are the same as those of the deterministic model presented in Appendix B (see Section B.2), and the spatial configuration arising from the dichotomous random switching between states 1 and 2 – e.g., with $k_1 = 1/\tau_1 = 1.25$ and $k_2 = 1/\tau_2 = 5$ – has the same features as those shown in Fig. B.2 in Appendix B.

5.9 Spatiotemporal stochastic resonance

In the third chapter, we described the phenomenon of stochastic resonance in zero-dimensional dynamical systems. Its key feature was associated with the fact that deterministic bistable dynamics can cooperate with a stochastic driver and a weak periodic forcing to induce order in the temporal fluctuations of the system. In the case of spatially extended systems, stochastic resonance also involves spatial coupling, which is generally expressed as a diffusion process. The basic idea of the spatiotemporal

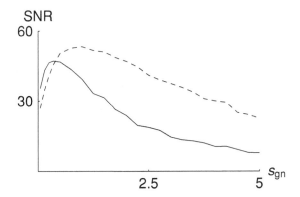

Figure 5.45. Characteristics of the dynamics described by Eq. (5.101) with periodic boundary conditions. The SNR is given for the central oscillator of an array with $N = 101$ elements as a function of the noise intensity s_{gn} and for two different coupling strengths: $D = 0.5$ (solid curve) and $D = 5$ (dashed curve). The parameters are $k = 2.1078$, $k' = 1.4706$, $A = 1.3039$, $f = 0.1162$.

stochastic resonance is that the spatial coupling gives coherence to the local fluctuations in a way that the regular temporal behavior of the single oscillators results in ordered spatial dynamics and in the possible emergence of patterns (see also Sagues et al., 2007, for a review).

The seminal work by Lindner et al. (1995) showed the possible emergence of spatiotemporal stochastic resonance. They considered the 1D model

$$\frac{\partial \phi}{\partial t} = a\phi - a'\phi^3 + A \sin(2\pi ft) + D \frac{\partial^2 \phi}{\partial x^2} + \xi_{gn}(x, t), \qquad (5.101)$$

where a and a' are positive-valued parameters to ensure that the potential is bistable, A and f are the amplitude and the frequency of the periodic forcing, respectively, D is the strength of the diffusive coupling, and $\xi_{gn}(x, t)$ is a Gaussian white (in space and time) noise with strength s_{gn}. In particular, they discretized model (5.101) as an array of N bistable oscillators ϕ_i. Figure 5.45 shows the behavior of a single oscillator in the array and shows the dependence of the SNR on the noise strength for different coupling strengths D. We recall that the SNR is a measure of the intensity of the peak in the power spectrum density with respect to the background noise. A maximum value of the SNR – evaluated at the frequency ω of the periodic forcing – is the typical hallmark of stochastic resonance (see Chapter 3). Figure 5.45 suggests that the effect of the spatial coupling (increasing D) is to shift the occurrence of stochastic resonance to a higher noise intensity and to increase the maximum of the SNR. The presence of higher SNR values in the power spectrum of the spatially coupled system demonstrates that the spatial coupling enhances stochastic resonance, in that the temporal fluctuations of each oscillator become more regular in the presence of the spatial coupling.

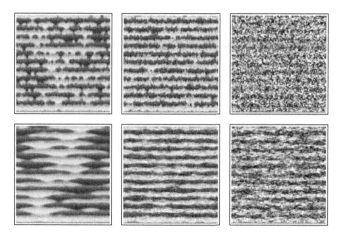

Figure 5.46. Spatiotemporal dynamics of a 101 oscillator array following the dynamics described by Eq. (5.101) with periodic boundary conditions. In each frame the chain is horizontal and time increases upward. A well of the bistable potential is shaded in black and the other in white. The three panels in the upper row correspond to $D = 0.5$ and $s_{gn} = 0.05, 0.35$, and 5, respectively (from left to right). The lower panels refer to $D = 5$ and $s_{gn} = 0.05, 1$, and 5, respectively. In all panels the parameters are $a = 2.1078$, $a' = 1.4706$, $A = 1.3039$, $f = 0.1162$.

Thus Fig. 5.45 shows that the spatial coupling favors the stochastic resonance. However, the relation between the coupling strength D and the SNR is not monotonic. In fact, when the coupling becomes very strong, the oscillators tend to behave as a single oscillator and the dynamics exhibit the same stochastic resonance observable for $N = 1$. In particular, the maximum of the SNR moves toward higher values of D as the number N of oscillators increases. Thus, if the noise intensity remains fixed and the number of oscillators increases, the SNR decreases and the stochastic resonance becomes weaker. Therefore, for a given noise intensity, the response of the system is optimal for intermediate values of N. This relation between noise intensity and system size was elucidated in detail by Pikovsky et al. (2002) for model (5.101).

Figure 5.46 shows the emergence of spatial coherence. For a given strength of the spatial coupling, there is an optimal value of the noise intensity that maximizes the SNR, thereby inducing stochastic resonance. Correspondingly, the chain of oscillators exhibits a collective behavior, i.e., for adequate noise intensities and strength of the spatial coupling, all oscillators move from a potential well to another in a synchronous way. As noted in the case of stochastic resonance in zero-dimensional systems, the spatial coherence weakens and patterns tend to disappear when the noise intensity is reduced or increased with respect to the optimal value: In the former case, transitions become irregular in both time and space, whereas, in the latter case, transitions are continuously activated by the noise and the spatiotemporal pattern becomes random. Marchesoni et al. (1996) explained the enhancement of stochastic resonance in spatially coupled systems as the result of the ability of the coupling to reduce the energy

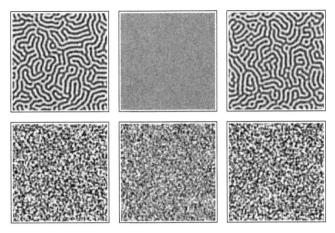

Figure 5.47. Spatiotemporal dynamics of 2D model (5.102). The columns refer to 110, 140, and 195 time units, and the two rows correspond to $s_{gn} = 0.025$ and $s_{gn} = 2.5$, respectively. The other parameters are $\alpha = a' = k_0 = 1$, $a = 0.1$, $\omega = 0.012$, and $D = 1$.

necessary to jump between the two potential wells with respect to the uncoupled case.

Vilar and Rubi (1997) also investigated pattern formation induced by spatiotemporal stochastic resonance. In particular, they focused on the case of the Ginzburg–Landau model, with a Swift–Hohenberg spatial coupling and forced by both an additive noise and a periodic driver:

$$\frac{\partial \phi}{\partial t} = [-a + \alpha \sin(\omega t)]\phi - a'\phi^3 - D(k_0^2 + \nabla^2)^2 \phi + \xi_{gn}(\mathbf{r}, t), \qquad (5.102)$$

where a, α, ω, a', k_0, and D are parameters and ξ_{gn} is a zero-mean Gaussian white noise with strength s_{gn}.

Figure 5.47 shows the resulting spatial configurations of the state variable for two different noise intensities and at different times. When the noise intensity is close to its optimal value, well-defined spatial patterns appear. These patterns have a periodicity in time. Conversely, when the noise is too intense, no patterns occur. In this case, the spatial configuration of the system is random at all times. Notice that patterns emerge periodically as dictated by the periodic forcing, namely with a period equal to $T = \omega^{-1} = 83.3$ time units.

It is interesting to examine also the probability distributions of the field at different times. Figure 5.48 shows two examples of the pdf: The first one refers to a well-patterned case – i.e., an optimal value of noise and a time when the patterns are more evident – and the second pdf corresponds to a case in which noise is too intense and the patterns are absent. We notice that the two fields span nearly the same range of values of ϕ, but the patterned case exhibits a remarkable bimodality, whereas the pdf of the disordered field is unimodal. In spite of bimodality not being a clue of the presence

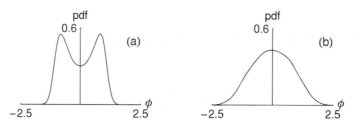

Figure 5.48. Pdf's of the two fields shown in the first column of the Fig. 5.47: (a) refers to $s_{gn} = 0.025$ at $t = 110$ time units, and panel (b) corresponds to the same time but $s_{gn} = 2.5$. Similar pdf's are obtained for $t = 195$ (corresponding to the third column in Fig. 5.47).

of patterns, in model (5.102) the coherent arrangement of the high and low values of ϕ (i.e., the dark and light gray pixels in Fig. 5.47) corresponds to the emergence of two modes in the distribution of ϕ.

In the previous examples, the periodic forcing was always in the time domain. However, spatiotemporal stochastic resonance can be also induced by a purely spatial periodic forcing (Sagues et al., 2007). For example, Vilar and Rubi (2000) studied the case of a Ginzburg–Landau model with an additive diffusive term, a spatially periodic term, and a noise component. They demonstrated the occurrence of a remarkable variety of patterns, both when the random driver varies in space and time and when it varies only in space (so-called *quenched noise*).

5.10 Spatiotemporal stochastic coherence

Similar to the case of stochastic resonance, there is also a spatiotemporal version of stochastic coherence. In the third chapter we showed that a dynamical system able to exhibit deterministic excursions characterized by a typical time scale can be excited by a suitable noise and undergo regular fluctuations. We demonstrated that, when excited from their steady state (or *rest state*), the deterministic dynamics undergo excursions with a characteristic time scale. These excursions may have the same dynamical role as the external periodic forcing in the phenomenon of stochastic resonance.

A classical example of spatiotemporal stochastic coherence can be obtained from the FitzHugh–Nagumo model presented in Chapter 3 with the addition of diffusive spatial coupling (Naiman et al., 1999). The 2D model is

$$\epsilon \frac{\partial \phi_1}{\partial t} = \phi_1 - \frac{\phi_1^3}{3} - \phi_2 + D\nabla^2\phi_1, \tag{5.103}$$

$$\frac{\partial \phi_2}{\partial t} = \phi_1 + a(\mathbf{r}) + \sqrt{\frac{2I}{\tau_{\phi_2}}}\xi_{gn}(\mathbf{r}, t), \tag{5.104}$$

where ϕ_1 and ϕ_2 are the fast and slow variables, respectively, $\epsilon = \tau_{\phi_1}/\tau_{\phi_2} \ll 1$ is the ratio between the typical time scales of the two variables, a are random point-specific numbers greater than one, D is the diffusion coefficient of the activator (i.e., of ϕ_1), $\xi_{gn}(t)$ is a Gaussian white (in space and time) noise with unit intensity, and I is a parameter modulating the noise intensity. When $a > 1$, the deterministic uncoupled version of the FitzHugh–Nagumo model describes an excitable dynamical system, characterized by the fact that, when the steady state is weakly perturbed, the system does not exhibit a monotonic return to the steady state, but a strong transient growth occurs before relaxation to the rest state (see Chapter 3). Thus, when $a > 1$, all points in the domain behave as independent, uncoupled excitable oscillators.

Figure 5.49 shows some examples of numerical solutions of (5.103) and (5.104). The first column shows the case with weak noise: Excursions are seldom activated at random points of the field and, because of the spatial coupling, they propagate as waves across the domain. In this case the ϕ_1 variable in different cells is correlated on only a short time scale dictated by the wave celerity, and no correlation occurs between distant cells. If the noise strength is increased, a pattern instead becomes evident: Firing (i.e., excursions) of distant cells occurs almost in phase, and a spatial coherence emerges. Notice that the pattern is not steady but periodical: The field alternates homogeneous states, when all oscillators are at the peak of the excursion, and patterned states, when islands of firing elements emerge from a basic rest state. Finally, the third column shows what happens when the noise is even stronger. In this case, excursions are continuously activated in the field, no particular patterns emerge, and the field appears to be random. Thus Fig. 5.49 shows that there is an optimal noise level able to give rise to stochastic long-range synchronization in the space–time dynamics of individual cells.

In the third chapter, temporal stochastic coherence was shown to also emerge when a dynamical system close to a Hopf's bifurcation is randomly perturbed. In this case the deterministic system has a stable steady state, but it is not necessarily an excitable system. The role of noise is to induce the system to temporarily cross the bifurcation point, thereby allowing the dynamics to stochastically explore the limit cycle. In this way, the noise acts on the system as a precursor of the limit cycle and unveils the existence of a periodic behavior already in subthreshold conditions before the bifurcation point is reached. For suitable (intermediate) noise intensities, random perturbations and the deterministic periodicity associated with the limit cycle can then cooperate to generate a regular temporal pattern.

A similar behavior can occur also in the spatiotemporal version of the system. Even in this case the deterministic system is in a (now spatially) homogeneous stable state close to a bifurcation point where the homogeneous state becomes unstable. When the control parameter crosses a given threshold, the system evolves deterministically toward a patterned state. Similar to the temporal case, the effect of noise is to disturb the stable state and to temporarily drive the system across a bifurcation point,

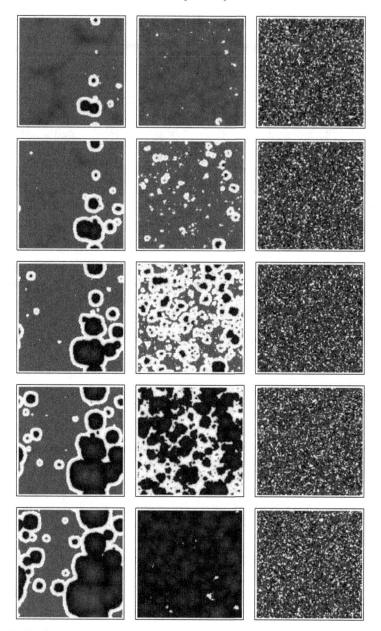

Figure 5.49. Outcomes of model (5.103) and (5.104). The columns correspond to the noise intensity s_{gn} equal to 0.562×10^{-4}, 1.562×10^{-4}, and 2.5×10^{-3}, respectively. The rows refer to 14, 14.5, 15, 15.5, and 16 time units. The parameters are $\epsilon = 0.01$, $D = 0.05$, and $\tau_{\phi_2} = 1$. The parameter a is uniformly distributed in the interval $[1.03, 1.10]$.

Figure 5.50. Examples of field described by the CDIMA reaction modeled by Eqs. (5.105) and (5.106). The three fields refer to the noise intensity s_{gn} equal to $0.001, 0.2$, and 0.6, respectively ($t = 60$ time units in all cases). The parameters are $a = 16, b = 301, c = 0.6, d = 1.07$, and $\bar{\varphi} = 0.35$. In this case the critical φ value is $\varphi_c = 2.3$ (Sanz-Anchelergues et al., 2001)

thereby allowing the system to explore the patterned condition. Thus we can expect an intermediate level of spatiotemporal noise to induce the emergence of patterns even in subthreshold conditions, i.e., when the control parameter has not reached the bifurcation point.

An example is given by the model (Carrillo et al., 2004)

$$\frac{\partial \phi_1}{\partial t} = a - c\phi_1 - 4\frac{\phi_1 \phi_2}{1 + \phi_1^2} - \varphi + \nabla^2 \phi_1, \tag{5.105}$$

$$\frac{\partial \phi_2}{\partial t} = b \left[c\phi_1 - \frac{\phi_1 \phi_2}{1 + \phi_1^2} + \varphi + D\nabla^2 \phi_2 \right], \tag{5.106}$$

where ϕ_1 and ϕ_2 are the state variables and φ is a control parameter. Originally proposed to describe the photosensitivity of the CDIMA reaction (CDIMA stands for chlorine dioxide iodide malonic acid), this model has an homogeneous stable state [$\phi_{1,st} = (a - 5\varphi)/5c$ and $\phi_{2,st} = a(1 + \phi_{1,st}^2)/5\phi_{1,st}$] that becomes unstable to spatial perturbation when φ is lower than a given critical value φ_c, which depends on the system's parameters. When $\varphi < \varphi_c$, a Turing instability occurs and a pattern with a nonzero wave number appears.

We consider the case in which the mean level of the control parameter $\bar{\varphi} = 0.35$ is greater than φ_c – i.e., the homogeneous steady state is deterministically stable – and it is disturbed by a random white Gaussian noise with intensity s_{gn}. Figure 5.50 shows the possible scenarios associated with different noise intensities: If the noise intensity is low, the crossings of the bifurcation point are rare and the field remains fairly homogeneous (see the left-hand panel of Fig. 5.50). In contrast, when the noise is suitably tuned on intermediate intensities, spatiotemporal stochastic coherence occurs and evident spatial patterns emerge (central panel of Fig. 5.50). Finally, when the noise intensity further increases, these patterns are again destroyed by the noise.

6

Noise-induced patterns in environmental systems

6.1 Introduction

A number of environmental processes exhibit the tendency to develop highly organized geometrical features generally referred to as *patterns*. For example, in arid and semiarid landscapes the vegetation cover is often sparse and exhibits spectacular organized spatial features (e.g., Macfadyen, 1950) that can be either spatially periodic or random. These patterns exhibit amazing regular configurations of vegetation stripes or spots separated by bare-ground areas. In some cases patterns may spread over relatively large areas (up to several square kilometers) (White, 1971; Eddy et al., 1999; Valentin et al., 1999; Esteban and Fairen, 2006), and can be found on different soils and with a broad variety of vegetation species and life-forms (i.e., grasses, shrubs, or trees) (Worral, 1959, 1960; White, 1969, 1971; Bernd, 1978; Mabbutt and Fanning, 1987; Montana, 1992; Lefever and Lejeune, 1997; Bergkamp et al., 1999; Dunkerley and Brown, 1999; Eddy et al., 1999; Valentin et al., 1999).

Because vegetation patterns are observed even when topography and soils do not exhibit relevant heterogeneity, their formation represents an intriguing case of self-organized biological systems, which results from completely intrinsic dynamics (Lejeune et al., 1999). Self-organization has been also observed in a number of atmospheric and geomorphic processes. Notable examples include the dynamics underlying the formation of ordered systems of clouds (e.g., Krueger and Fritz, 1961), dunes and ripples (e.g., Elbelrhiti et al., 2005; Colombini and Stocchino, 2008; Seminara, 2010; Fourriere et al., 2010), frost boils (Gleason et al., 1986; Krantz, 1990), river meandering (e.g., Ikeda and Parker, 1989), sinuous coastlines (Ashton et al., 2001), or fringed peatlands (e.g., Eppinga et al., 2008). In all of these cases spatial self-organization has been explained as the result of *symmetry-breaking instability*, which leads to the emergence of stable heterogeneous configurations.

To investigate these processes scientists capitalized on the understanding of pattern-forming mechanisms gained in other fields, such as fluid dynamics

(Dubois-Violette et al., 1978; Chandrasekhar, 1981; DiPrima and Swinney, 1981; Cross and Hohenberg, 1993; Drazin, 2002), chemistry (Turing, 1952), and biology (Murray, 2002). This broad body of literature on patterns in nature inspired a number of studies proposing a variety of models to explain pattern formation in the environmental sciences, particularly in the case of vegetation.

One of the early examples of process-based models of self-organized vegetation can be found in Watt (1947), who invoked mechanisms of reallocation of nutrients and water to explain the emergence of patchy vegetation distributions: This model showed that, as nutrient and water availabilities decrease, plants tend to grow in clumps. The emergence of these aggregated structures is motivated by the need to concentrate scarce resources (e.g., soil moisture, soil nutrients) in smaller areas, thereby increasing the likelihood of vegetation survival within vegetated patches that are richer in resources. In subsequent years, the idea that the mechanisms underlying vegetation pattern formation are intrinsically dynamic and originate from interactions among plant individuals was better articulated and formalized (e.g., Greig-Smith, 1979; Thiery et al., 1995). These studies paved the way for a new generation of models (Lefever and Lejeune, 1997; Klausmeier, 1999; von Hardenberg et al., 2001; Rietkerk et al., 2002), explaining vegetation patterns as the result of self-organization emerging from symmetry-breaking instability, i.e., as a process in which the existence of both cooperative and inhibitory interactions at two slightly different spatial ranges may induce the appearance of heterogeneous distributions of vegetation with wavelengths determined by the interactions between the two spatial scales (Lefever and Lejeune, 1997; Lejeune et al., 1999; Lejeune and Tlidi, 1999; Barbier et al., 2006).

These phenomena are often interpreted as the result of deterministic mechanisms (see Appendix B) of symmetry-breaking instability, whereas stochastic environmental drivers are usually considered as the source of random perturbations in the ordered states of the system. Thus random environmental fluctuations have been frequently associated with disturbances able to destroy patterns formed by deterministic dynamics (e.g., Rohani et al., 1997). However, in Chapter 5 we showed that random fluctuations are also able to play a *constructive* role in the dynamics of nonlinear systems, in that they can induce new dynamical behaviors that did not exist in the deterministic counterpart of the system (e.g., Horsthemke and Lefever, 1984). In particular, stochastic fluctuations have been associated with the emergence of new ordered states in dynamical systems, both in time (e.g., Horsthemke and Lefever, 1984) and in space (Garcia-Ojalvo and Sancho, 1999). Thus random environmental drivers are not necessarily in contraposition to pattern formation. Indeed, patterns have been shown (van den Broeck et al., 1994; Garcia-Ojalvo and Sancho, 1999; Loescher et al., 2003) to emerge as a result of the randomness inherent in environmental fluctuations. These random drivers may induce a symmetry-breaking instability, which leads to morphogenesis. Alternatively, noise may stabilize transient patterned features temporarily emerging in the underlying deterministic dynamics. We refer the reader to Chapter 5 for more details on order formation in stochastic spatiotemporal systems; deterministic theories are reviewed in Appendix B.

We need to stress that most of these deterministic and stochastic theories have not been experimentally validated in the field. A typical example is the case of vegetation: We are not aware of any conclusive experimental evidence that the vegetation patterns observed in many regions of the world are actually induced by mechanisms of symmetry-breaking instability. It has been noticed (Borgogno et al., 2009) that it is really difficult to test these theories just by comparing model simulations with observed patterns, because the same types of patterns can be obtained with different models. This is because the amplitude equations – which determine important properties of pattern geometry – belong to only a few major classes (see Cross and Hohenberg, 1993; Leppanen, 2005). Thus the claim of relating patterns to processes by developing process-based models capable of reproducing the observed patterns is probably too ambitious. Because most models differ in the processes invoked as key mechanisms of pattern formation (see for example Section 6.3), their validation should be based on the assessment of the significance of these processes (Barbier et al., 2008).

Field validations become even more difficult for patterns induced by noise. In this case patterns would emerge as an effect of the temporal variability of environmental drivers. However, long-term observations of patterned landscapes are rare. It is even harder to envision a field experiment in which we can compare a system undergoing environmental fluctuations with a "control system" with similar properties except for being subjected to time-invariant environmental conditions. Thus the view emerging from the study of noise-induced environmental patterns is that this research field is still relatively new. Theoretical frameworks are far from being experimentally testable, and the application of existing theories of noise-induced pattern formation has just started to appear in the environmental sciences. In fact, most of the literature on self-organized pattern formation is based on the deterministic mechanisms discussed in Appendix B, and only a handful of studies have investigated the possible emergence of environmental patterns as a noise-induced effect.

In this chapter we show that new interesting questions exist on the role of random drivers in the self-organization of spatially extended systems. To this end, we provide some examples of applications to the biogeosciences of relatively recent theories of noise-induced pattern formation. The few existing studies invoke stochastic mechanisms of pattern formation based on noise-induced stabilization of short-lived coherent structures emerging in the underlying deterministic dynamics (Section 6.4); the effect of multiplicative noise in systems undergoing nonequilibrium phase transitions (Section 6.5); repeated random switching between two deterministic dynamics (Section 6.6); spatiotemporal stochastic resonance (Section 6.7); spatiotemporal stochastic coherence (Section 6.8). After a brief discussion on the typical mathematical representation of spatial interactions in environmental systems, in the following sections we review some examples of noise-induced patterns in self-organized environmental dynamics.

6.2 Models of spatial interactions

Spatial interactions in environmental systems typically result from the dependence of a state variable ϕ (e.g., plant biomass, population density, etc.) at any point $\mathbf{r} = (x, y)$ on the value of ϕ at points $\mathbf{r}' = (x', y')$, in the neighborhood of $\mathbf{r}$. The nature of this dependence may in general change as a function of the direction and of the distance between $\mathbf{r}$ and $\mathbf{r}'$. In this chapter we concentrate mainly on the case of systems that are mathematically described by only one state variable, ϕ, in a 2D domain (x, y). At any point $\mathbf{r} = (x, y)$, the state variable $\phi(\mathbf{r}, t)$ undergoes local dynamics, expressed by a function $f(\phi)$ with steady state $\phi = \phi_0$ [i.e., $f(\phi_0) = 0$]. For the sake of simplicity we assume that the local dynamics exhibit only one stable state. To express the effect of spatial interactions on the dynamics of $\phi(\mathbf{r}, t)$, we account for the impact that other points $\mathbf{r}' = (x', y')$ of the domain have on $\phi(\mathbf{r}, t)$. It is sensible to assume that this impact depends on the relative position of the two points $\mathbf{r}$ and $\mathbf{r}'$. Because the strength of the interactions with other sites is likely to vary with distance and direction, a weighting function $\omega(\mathbf{r}, \mathbf{r}')$ is introduced to model spatial interactions as a function of $\mathbf{r}'$ and $\mathbf{r}$. We integrate $\mathbf{r}'$ over the whole domain Ω to account for the interactions of $\phi(\mathbf{r}, t)$ with the values of ϕ at any point $\mathbf{r}'$ in the domain Ω and obtain

$$\frac{\partial \phi(\mathbf{r})}{\partial t} = f(\phi(\mathbf{r})) + \int_{\Omega} \omega(\mathbf{r}, \mathbf{r}')[\phi(\mathbf{r}', t) - \phi_0] d\mathbf{r}'. \tag{6.1}$$

Known as the *neural model*, Eq. (6.1) has been used to simulate pattern recognition by the brain and morphogenesis in a variety of natural systems (Murray, 2002). The right-hand side of (6.1) consists of two terms: $f(\phi)$ describes the local dynamics, i.e., the dynamics of ϕ that would take place in the absence of spatial interactions with other points of the domain. The second term expresses the spatial interactions and depends both on the shape of the weighting function (or *kernel*) and on the values of ϕ in the rest of the domain Ω. If $\omega(\mathbf{r}, \mathbf{r}') > 0$, the spatial interactions affect the dynamics of $\phi(\mathbf{r})$ positively or negatively, depending on whether $\phi(\mathbf{r}')$ is greater or smaller than ϕ_0, respectively. The opposite happens when $\omega(\mathbf{r}, \mathbf{r}') < 0$.

When the processes underlying the spatial interactions are homogeneous (i.e., they do not change from point to point) and isotropic (i.e., they are independent of the direction), the kernel function is independent of $\mathbf{r}$ and exhibits axial symmetry. In this case ω is a function only of the distance, $z = |\mathbf{r}' - \mathbf{r}|$, between the two interacting points [i.e., $\omega(\mathbf{r}, \mathbf{r}') = \omega(z)$]. It will be shown that even though the underlying mechanisms are homogeneous, they can lead to pattern formation, i.e., to nonhomogeneous distributions of the state variable.

In some cases, the spatial interactions modulated by the kernel function act only within short distances. Therefore $\omega(z)$ quickly tends to zero for increasing values of z. Thus, depending on the shape of $\omega(z)$, conditions leading to pattern formation in neural models can be formalized through a Taylor expansion of the integral term of Eq. (6.1) for small values of z (Murray, 2002):

$$\frac{\partial \phi}{\partial t} - f(\phi) + \omega_0(\phi - \phi_0) + \omega_2 \nabla^2 \phi + \omega_4 \nabla^4 \phi + \cdots, \tag{6.2}$$

where ∇^4 is the biharmonic operator (see Subsection 5.1.2.2) and w_m are the mth-order moments of the kernel function

$$\omega_m = \frac{1}{m!} \int_\Omega z^m \omega(z) \mathbf{d}z, \qquad m = 0, 2, 4, \ldots. \tag{6.3}$$

In Eq. (6.2) we assume that the dynamics are isotropic, i.e., that the kernel function has axial symmetry. Thus, because in this case the odd-order moments of $\omega(z)$ are zero, we do not include the odd-order terms in the Taylor expansion.

When ω_4 is "sufficiently small" the biharmonic term in Eq. (6.2) can be negligible and the equation may be rewritten in the form of a classical reaction–diffusion equation (or *Fisher's equation*):

$$\frac{\partial \phi}{\partial t} \simeq f_1(\phi) + \omega_2 \nabla^2 \phi, \tag{6.4}$$

with $f_1(\phi) = f(\phi) + \omega_0(\phi - \phi_0)$. Because in Eq. (6.4) only the low-order terms of the Taylor expansion are retained, these dynamics are driven only by short-range interactions in the neighborhood of (x, y).

Conversely, when ω_4 is not "small" the biharmonic term cannot be neglected and the dynamics are expressed by Eq. (6.2). If the moments of order higher than ω_4 are negligible, Eq. (6.2) can be truncated to the fourth order (*long-range diffusion equation*). Thus ω_2 modulates the effect of short-range interactions, and moment ω_4 multiplies the biharmonic term, which accounts for long-range interactions. Notice how the Swift–Hohenberg spatial coupling presented in Chapter 5 and in Section 6.6 can always be written in the same form as truncated equation (6.2) with a suitable kernel function.

6.3 Examples of pattern-forming processes

A common feature of deterministic models of symmetry-breaking instability (see Appendix B) is that the emergence of periodic patterns arises from the balance between positive (activation) and negative (inhibition) interactions (e.g., Shnerb et al., 2003). For example, in the neural model both pattern emergence and pattern geometry are determined by the interplay between short-range facilitation (or cooperation) and long-range competition (or inhibition).

In landscape ecology, cooperative and synergistic short-range effects are usually associated with positive feedbacks resulting from the ability of some species or functional types to create environmental conditions that favor plant establishment, growth, and survival. These feedbacks typically operate within a short range. For example, cooperation among neighboring individuals may lead to the concentration of resources in vegetated areas where plant individuals find more favorable conditions for establishment and survival (Charley and West, 1975; Schlesinger et al., 1990; Greene, 1992; Wilson and Agnew, 1992; Bhark and Small, 2003; D'Odorico et al., 2007a). The aerial parts of plant individuals that have already been established in a certain patch may favor the growth of other plants in the same area by limiting soil-moisture losses associated with evapotranspiration (Vetaas, 1992; Thiery et al., 1995; Lejeune et al., 2004; Zeng et al., 2004; D'Odorico et al., 2007a) either through a mulching effect (i.e., soil evaporation limited by wilted leaves and litter) or shading (i.e., when the foliage shades the ground surface, thereby limiting evaporation) (Zeng and Zeng, 1996; Scholes and Archer, 1997; Zeng et al., 2004; Caylor et al., 2006; Borgogno et al., 2007). Moreover, the formation of physical and biological crusts on bare soil may further reduce the infiltration of surface water (e.g., Fearnehough et al., 1998). Physical crusts, typically 1–3 mm thick, are generated by rain splashing (Esteban and Fairen, 2006), whereas biological crusts are formed by microorganisms such as cyanobacteria, which exude mucilaginous secretions that bind together soil grains and organic fractions (e.g., Meron et al., 2004). These crusts greatly reduce the soil-infiltration capacity, thereby decreasing the soil moisture available in the underlying soil layers, with consequent limitations on the establishment and growth of perennial vegetation (Fearnehough et al., 1998). Because soil crusts, on the other hand, seldom develop beneath vegetation canopies because of the reduced raindrop impact (Boeken and Orenstein, 2001; Meron et al., 2004; Borgogno et al., 2007) and the limited light available to the photosynthetic activity of biological crusts (Walker et al., 1981; Greene, 1992; Joffre and Rambal, 1993; Greene et al., 1994, 2001), a positive feedback exists between presence of vegetation and absence of crusts.

In vegetated areas the protection against evapotranspiration and soil-crust formation enhances surface-water infiltration which, in turn, favors vegetation growth. The associated increase in root density, in turn, enhances the soil infiltration capacity (Walker et al., 1981; Greene, 1992; Joffre and Rambal, 1993; Greene et al., 1994, 2001; HilleRisLambers et al., 2001; Okayasu and Aizawa, 2001; Gilad et al., 2004; Yizhaq et al., 2005; Borgogno et al., 2007). Moreover, a dense canopy of established plants provides protection against herbivores (e.g., birds), thereby favoring plant reproduction and growth (Lejeune et al., 2002) in densely vegetated areas (propagation by reproduction effect) where higher rates of seed production and germination occur (e.g., Lefever and Lejeune, 1997; Lejeune et al., 1999; Lejeune and Tlidi, 1999; Lefever et al., 2000; Couteron and Lejeune, 2001).

Species able to modify the abiotic environment, redistribute resources, and facilitate the growth of other species as well as their own are known as *ecosystems engineers*

(Jones et al., 1994; Gilad et al., 2007). For example, the improvement of conditions existing in the microenvironment underneath the canopy of so-called *nurse plants* (Nigering et al., 1963; Kefi et al., 2007) favors the establishment and growth of other plants (e.g., Garcia-Moya and McKell, 1970; Burke et al., 1998; Aguiar and Sala, 1999). Vegetation cover may decrease the amplitude of temperature fluctuations, reduce the exposure to solar radiation, wind desiccation, and soil erosion, or prevent soil-crust formation (Eldridge and Greene, 1994; Smit and Rethman, 2000; Greene et al., 2001). Moreover, plant individuals located in the middle of vegetated patches are protected against fires and grazing.

As noted, it has been also found that water or nutrient (or both) availability is higher in the areas located under the canopy of existing plant individuals than in the surrounding bare soil (Charley and West, 1975). These nutrient-rich areas are known as *fertility islands* or *resource islands* (Schlesinger et al., 1990). Mechanisms commonly invoked to explain the formation of these heterogeneous distribution of resources include the ability of the canopies to trap nutrient-rich airborne soil particles, the accumulation of sediments transported by wind and water, the sheltering effect of vegetation against the erosive action of wind and water, and the presence of nitrogen-fixing species (Garcia-Moya and McKell, 1970; Charley, 1972; Archer, 1989; Schlesinger et al., 1990; Breman and Kessler, 1995; Li, 2007). Sometimes fertility islands lead to the formation of aperiodic vegetation patterns in the form of stable spatial configuration corresponding to isolated vegetation patches (spots), usually called *localized structures* or *localized patches* (Lejeune et al., 2002).

Other examples of ecosystem engineering relevant to pattern formation include the ability of deep-rooted plants to facilitate shallow-rooted species by increasing surface-soil moisture through "hydraulic-lift" mechanisms (Richards and Caldwell, 1987), the reduction in fire pressure resulting from the encroachment of woody vegetation at the expenses of grass fuel (e.g., Anderies et al., 2002; van Langevelde et al., 2003; D'Odorico et al., 2006b), and the ability of alpine or subalpine vegetation and desert shrubs to maintain warmer microclimate conditions and reduce frost-induced mortality.

Similar facilitative mechanisms exist also in wetland environments, including salt marshes, where vegetation may prevent salt accumulation by limiting soil evaporation (shading effect), riparian corridors, and wetland forests, where vegetation can favor the aeration of anoxic soils through soil drainage by plant uptake and transpiration (Wilde et al., 1953; Chang, 2002; Ridolfi et al., 2006). It has also been found that on cobble beaches, dense stands of *spartina alterniflora* occupying the lower intertidal zone can protect other plant communities from intense wave action (Bruno, 2000; van de Koppel et al., 2006).

On the other hand, competitive or inhibitory effects typically occur within a longer range. Competition for water and nutrients is generally exerted by means of the root system (Aguilera and Lauenroth, 1993; Belsky, 1994; Breman and Kessler, 1995; Breshears et al., 1997; Martens et al., 1997; Couteron and Lejeune, 2001). In fact, the lateral roots extend beyond the edges of the crown (Casper et al., 2003; Barbier et al., 2008) and extract water and nutrients from the intercanopy areas (Martens et al., 1997; Lejeune et al., 2004). Hence the typical range of competitive interactions is larger than that of facilitation. For example, for trees and shrubs, the ratio between the radii of the footprint of canopy and root systems may be as small as as 1/10 (Lejeune et al., 2004). Thus the root system is able to deplete soil resources (i.e., water and nutrients) from the intercanopy areas and to compete for resources with other plant individuals (Lefever and Lejeune, 1997; Yokozawa et al., 1998; Lejeune et al., 1999; Lejeune and Tlidi, 1999; Couteron and Lejeune, 2001; Lejeune et al., 2002; Rietkerk et al., 2004; Yizhaq et al., 2005; Barbier et al., 2006). This competition for resources usually reduces the growth rate of competing individuals, thereby leading to a net effect of inhibition through competition (Shnerb et al., 2003).

Thus facilitative interactions occur within the range of the crown area and are mostly associated with positive feedbacks induced by the canopy (e.g., mulching, shading, protection against fires, grazing, and wind action), whereas negative interactions are exerted mainly as resource competition by roots and typically occur at larger distances. The sign and range of these interactions justify the choice of kernel functions shaped as in Figs. B.3 and 6.9 in Section 6.6: The kernel shows positive interactions within the range of the canopy scale (short-range) and negative interactions within the range of the typical lateral root length (long-range), and the magnitude of these interactions vanishes (i.e., $\omega \to 0$) as the distance between the interacting plants exceeds the typical extent of lateral roots. Nevertheless, spatial patterns can also emerge when the kernel is "upside down" with respect to the case of Fig. B.3 (e.g., Borgogno et al., 2009). In fact, in some cases facilitation can occur at a long range [e.g., buffering from intense wave action in intertidal communities (van de Koppel et al., 2006)] and competition can occur at a short range [e.g., competition for light that is typically due to interactions among canopies (Caylor et al., 2005; van de Koppel et al., 2006)].

Although all these mechanisms of cooperation and competition are generally isotropic (i.e., they operate in the same way in all directions), in some environments the presence of a slope or of a dominant wind direction may lead to anisotropy in the spatial dynamics. In fact, if the wind regime exhibits a prevailing direction, asymmetry may emerge in the cooperative and competitive mechanisms. For example, the persistent existence of a cone-shaped wind shadow downwind of tree–shrub clumps would provide a favorable protected environment for the establishment and growth of other plant individuals (short-range positive feedback), while plant individuals located at larger distances would remain with no protection and would consequently be prone to higher mortality rates (Puigdefabregas et al., 1999; Yokozawa et al., 1999; Ravi et al., 2009). Similarly, Borgogno et al. (2009) discussed the role of advective flow on the dynamics of two diffusive species and its role in the phenomenon of differential-flow instability. Advective flows can originate as an effect of runoff in sloping terrains. During intense rainfall events, water and sediments run off bare areas and are intercepted and trapped by vegetated patches. This supplementary input of limiting resources favors plant growth on the uphill side of vegetated patches, thereby securing more efficient trapping during subsequent rainstorm events. Runoff and erosion are therefore viewed as a fundamental mechanism to maintain striped [Figs. 5.1(a)–5.1(f)] configurations over hillsides (Thiery et al., 1995; Dunkerley, 1997a, 1997b; Okayasu and Aizawa, 2001; Sherrat, 2005; D'Odorico et al., 2007a; Barbier et al., 2008). Thus rainfall onto an unvegetated area generates overland flow, which transports water in the downhill direction until it reaches a vegetated area, where it infiltrates the ground and is taken up by vegetation. The relatively moist soil on the uphill side of a stripe creates opportunities for uphill expansion of the vegetation band at the expenses of the downhill side, which remains deprived of the resources necessary for vegetation survival. The overall dynamics lead to the uphill migration of vegetated bands (Sherrat, 2005). A similar mechanism can explain the banded patterns of trees in Tierra del Fuego [Argentina; see Fig. 5.1(h)], where a sawtooth pattern of tree heights is observed in the wind direction (Puigdefabregas et al., 1999). Taller trees provide more protected favorable conditions for seedling establishment and tree growth in the leeward direction. At the same time, the strong winds uproot and kill the taller upwind trees, leading to an overall downwind migration of the sawtooth pattern.

6.4 Patterns induced by additive noise

The ability of noise to induce ordered states in dynamical systems is commonly explained as an effect of multiplicative noise acting in conjunction with nonlinearities. However, in Chapter 5 we showed how additive noise can also play a crucial role by stabilizing short-lived patterns emerging in deterministic dynamics. In this section we apply a simple stochastic model [based on Eq. (5.29)] of noise-induced pattern formation driven by additive noise. This model (see also Section 5.3) does not invoke any nonlinearity. It involves only three linear components: (i) a deterministic local dynamics term, which linearly damps the system to zero; (ii) an additive noise able to hamper this tendency of the deterministic dynamics to converge to zero; and (iii) a linear (diffusive) spatial coupling, which provides spatial coherence. This model provides a possible noise-induced mechanism for pattern formation in dryland vegetation. In fact, we can model the dynamics of vegetation biomass v as the result of a linear decrease, acting in conjunction with spatial interactions modeled as linear diffusion, and an additive stochastic forcing representing a random environmental driver:

$$\frac{\partial v}{\partial t} = av + D\nabla^2 v + \xi_{\text{gn}}. \tag{6.5}$$

The first term, av, accounts for the local deterministic decay of plant biomass that would occur in the absence of spatial interactions or environmental fluctuations

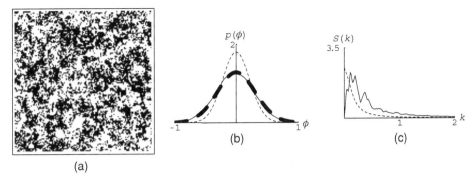

Figure 6.1. Model (6.5), $a = -1$, $s_{gn} = 5$, $D = 50$, simulations at $t = 100$ time units. (a) Numerical simulations on a 2D square 128×128 lattice, with periodic boundary conditions. Black and white tones are used for positive and negative values of the field, respectively. (b) pdf's of v (solid line, numerical simulation; dashed curve, standard mean field; thick dashed curve, corrected mean field). (c) Azimuthal-averaged power spectrum (solid curve, numerical simulations; dashed curve, structure function).

(a is a negative-valued coefficient). The second term expresses the spatial interactions, where D is a parameter indicating the strength of the spatial coupling. The third term is a random environmental driver (e.g., plant growth or death associated with fluctuating levels of soil-water availability), which is here modeled as a white (in time and space) Gaussian noise with zero mean and correlation, $\langle \xi_{gn}(\mathbf{r}, t)\xi_{gn}(\mathbf{r}, t')\rangle = 2s_{gn}\delta(|\mathbf{r} - \mathbf{r}'|)\delta(t - t')$, where s_{gn} is the noise intensity. The emergence of spatial patterns from Eq. (6.5) was discussed in Section 5.3. In the absence of noise the dynamics would be expressed by a linear Fisher's equation and would converge toward the homogeneous deterministic steady state, $v = 0$, for any initial conditions.

To assess pattern formation we use two analytical tools: the mean-field analysis and the structure function (see Chapter 5). The mean-field analysis allows us to obtain the analytical expression of the pdf at steady state. We use both the standard mean-field analysis and the corrected version that provides a better approximation (see Box 5.4). We find that the steady-state probability distribution of v is Gaussian with zero mean and variance equal to $s_{gn}/(2 + 2D)$. The structure function is a prognostic tool able to assess the presence of a selected wavelength in the spatial field. In the case of Eq. (6.5) the steady-state structure function

$$S_{st}(k) = \frac{s_{gn}}{2(Dk^2 - a)} \tag{6.6}$$

shows that the dynamics do not select any spatial periodicity, but there is a range of wave numbers close to zero, which compete to produce multiscale patterns.

Figure 6.1 shows the onset of patterns in the dynamics expressed by Eq. (6.5). Only the results pertaining to the simulations at $t = 100$ time units are given; other results for the same model with different parameters are shown in Fig. 5.16. As expected from the structure-function analysis (i.e., no dominant wavelength different from zero exists), no clear periodicity is visible but many wavelengths are present to produce multiscale patterns, which become steady after a few time units. Moreover, no phase transition occurs because the pdf remains unimodal and with zero mean. In Fig. 6.1 we compare the numerical simulations with the analytical results for the steady-state structure function and pdf of ϕ from mean-field analysis. For the pdf, the corrected mean-field prediction, obtained with a halved diffusion coefficient (see Box 5.4) (thick dashed curve overlaid by the solid curve) is more precise than the results of the standard mean-field analysis (thin dashed curve), which tends to underestimate the variance of the distribution. The numerical simulations show that the multiscale patterns emerging from Eq. (6.5) persist in the steady state of the system and do not disappear. We stress again that in these dynamics pattern formation is only due to the presence of an additive noise acting in conjunction to a spatial coupling. Indeed, the deterministic local dynamics tend to damp the field variable to zero, and the spatial coupling is unable to destabilize this uniform steady state of the system. Thus at steady state the deterministic dynamics underlying Eq. (6.5) do not exhibit pattern formation. The properties of stochastic dynamics (6.5) were discussed in detail in Chapter 5 (Section 5.3). It is interesting to notice that the cooperation between noise and spatial coupling leads to the emergence of temporally stable spatial coherence, as indicated by the power-spectrum structure, which strongly differs from the spatially uncorrelated initial condition. We recall that in these dynamics additive noise is able to keep the system away from the homogeneous state, $\phi = 0$, and the spatial coupling induces spatial coherence. Therefore, in this system, pattern formation is clearly noise induced and arises from a synergism between additive noise and spatial coupling. The mechanism presented in this section differs from the effect of noise-induced stabilization of transient patterns investigated by Butler and Goldenfeld (2009). In fact, Butler and Goldenfeld (2009) focused on a predator–prey system exhibiting Turing instability and showed how additive noise can act on the system as a precursor of a phase transition, which expands the region of the parameter space in which pattern-forming Turing instability may occur. Conversely, in the dynamics presented in this section additive noise does not play the role of a precursor of a phase transition close to a (deterministic) bifurcation point. In fact, no bifurcation exists in the deterministic counterpart of these dynamics (see Chapter 5). Rather, in this case the additive noise unveils the ability of the deterministic system to induce transient periodic patterns, and hampers their disappearance.

To apply Eq. (6.5) to the case of vegetation dynamics we consider a system forced by an additive noise with mean different from zero (to account for random environmental drivers with means different from zero):

$$\frac{\partial v}{\partial t} = -v + D\nabla^2 v + \xi_{\text{gn}} + \mu = -v + D\nabla^2 v + \xi'_{\text{gn}}, \qquad (6.7)$$

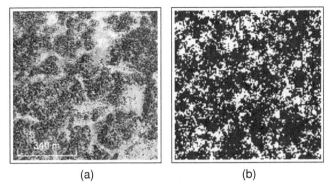

(a)   (b)

Figure 6.2. (a) Example of aerial photograph showing vegetation pattern in New Mexico ($34°$ $46'57.60''$N, $108°20'35.56''$O) and (b) numerical simulation of model (6.7) at $t = 100$, $\mu = 0.1$, $D = 80$, $s_{\mathrm{gn}} = 2$. Google Earth imagery © Google Inc. Used with permission.

where $\xi'_{\mathrm{gn}} = \xi_{\mathrm{gn}} + \mu$ is a white Gaussian noise with mean equal to μ and intensity s_{gn}. In this case the pdf of v is the same as before but with the mean shifted by μ. Equation (6.7) expresses the temporal evolution of the existing vegetation (v) as the result of a local linear decay, vegetation's spatial interactions (the diffusive term $D\nabla^2 v$), and a random growth term (the noise term $\xi_{\mathrm{gn}} + \mu$), which is on average the same at all points in space (μ), though it exhibits fluctuations (ξ_{gn}) in space and time because of local variability in soil properties or microclimate conditions. The distribution of vegetated sites in semiarid environments exhibits spatial configurations resembling those shown in Fig. 6.1. If we assume a constant noise intensity s_{gn}, we can obtain the coefficient of variation CV_v of the field – defined as the ratio between the standard deviation and the mean – by using the mean-field analysis. It is found that $\mathrm{CV}_v = \sqrt{s_{\mathrm{gn}}}/(\mu\sqrt{2 + 2D})$ decreases with increasing values of μ. This means that an arid ecosystem (low μ) would exhibit stronger fluctuations (i.e., more variability) in vegetation cover than a subhumid system (high μ), consistent with the stronger contrast between vegetated and nonvegetated zones typically observed in arid areas. In Fig. 6.2 we compare a real vegetation pattern with a numerical simulation of Eq. (6.7). These patterns exhibit irregular boundaries, and no clear periodicity can be detected because many wavelengths are involved (multiscale patterns).

Models of noise-induced pattern formation and nonequilibrium phase transitions typically invoke the presence of multiplicative noise (van den Broeck et al., 1994; Garcia-Ojalvo et al., 1996; Grinstein et al., 1996; Muller et al., 1997; Sieber et al., 2007) along with a high-order diffusion term (e.g., the Swift–Hohenberg coupling term; see Section 6.2). The influence of additive noise on the transition to ordered states has mainly been investigated in systems affected by the concurrent action of a multiplicative noise (Landa et al., 1998; Zaikin et al., 1999) with only a few exceptions (Zaikin and Schimansky-Geier, 1998; Dutta et al., 2005). Because these models use complicated nonlinear terms to represent both the local deterministic dynamics and the multiplicative-noise terms, their process-based interpretation is often not straightforward. The example presented in this section shows how additive noise may cause morphogenesis even in linear systems and with spatial coupling expressed by (short-range) diffusion.

6.5 Patterns induced by multiplicative white shot noise

In this section we provide an example of patterns induced by multiplicative shot noise in a nonlinear system. We consider the case of mesic and subhumid savannas where the persistence of mixed tree–grass plant communities is maintained by fires (Sankaran et al., 2004; Higgins et al., 2000). In the absence of fires these ecosystems tend to be completely dominated by woody vegetation (e.g., Scholes and Archer, 1997; Higgins et al., 2000; van Wijk and Rodriguez-Iturbe, 2002). In these savanna ecosystems, woody-plant encroachment is limited by disturbances such as fires rather than by resource availability (see also Subsection 4.3.1). Moreover, because of the competitive advantage of trees and shrubs, the dynamics of woody vegetation are not affected by grasses. Thus only the dynamics of woody biomass v are modeled, whereas grass biomass is assumed to be proportional to $v_{max} - v$, where v_{max} is the ecosystem carrying capacity. In this case vegetation dynamics can be modeled with only one equation. D'Odorico et al. (2007b) expressed the temporal variability of v as a growth–death process with tree encroachment modeled as a (deterministic) diffusion process:

$$\frac{\partial v(\mathbf{r}, t)}{\partial t} = \alpha \left[v(\mathbf{r}, t) + \epsilon \right] \left[v_{max} - v(\mathbf{r}, t) \right] \tag{6.8}$$

$$+ D\nabla^2 v(\mathbf{r}, t) - \xi_{sn}[t, v(\mathbf{r}, t)],$$

where $\mathbf{r} = (x, y)$ is the coordinate vector in a 2D domain, D is the diffusion coefficient associated with tree encroachment, and ∇^2 is the Laplace operator. The growth rate is modeled as a deterministic (logistic) function, i.e., with a rate proportional to the existing woody biomass v and the available resources, $v_{max} - v$. The parameter α measures the reproduction rate of the logistic growth, and $\epsilon \ll 1$ prevents the dynamics from remaining locked at $v = 0$ after all the tree biomass at a site is destroyed by intense fires. Diffusion would tend to induce a loss of woody biomass from locations with relatively high vegetation biomass surrounded by areas with lower vegetation densities, whereas the logistic growth compensates for this loss. Thus the model mimics a system in which woody vegetation tends (locally) to carrying capacity and spreads (laterally) into the areas with lower woody biomass. The stochastic component of the process is due to fire-induced tree death associated with random and intermittent fire occurrences. These occurrences are modeled as a Poisson process in time, $\xi_{sn}(t, v)$, with each fire killing an exponentially distributed random amount ω of tree biomass (with mean ω_0) or the whole existing biomass $v(\mathbf{r}, t)$, whichever is less. To account for the positive feedback between fire occurrences at a point and the local vegetation, the rate λ of the Poisson process is expressed as a state-dependent function (van Wilgen et al., 2003; D'Odorico et al., 2006b), $\lambda = \lambda_0 + bv$, with $b \leq 0$ (i.e., higher tree densities are associated with less-frequent fire occurrences).

D'Odorico et al. (2007b) investigated the conditions associated with the emergence of phase transitions and spatial patterns in this system. Although the deterministic counterpart of this process (i.e., the Fisher's equation) is not capable of generating patterns (see Appendix B), self-organized patches of trees may emerge in the presence of the stochastic forcing (i.e., random fire dynamics). To investigate the properties of stochastic process (6.8), we assume that the state variable v is normalized with respect to the ecosystem carrying capacity (i.e., $0 \leq v \leq 1$) and set $v_{max} = 1$ in Eq. (6.8). We calculate the pdf of v by applying the framework presented in Chapter 5 for the case of dynamics driven by WSN. Thus we use the mean-field approximation (Box 5.4) in a finite-difference representation of Eq. (6.8) and obtain the steady-state probability distribution of v (Porporato and D'Odorico, 2004):

$$p(v; \langle v \rangle) = \frac{C}{\rho(v)} \exp\left[\frac{v}{\omega_0} - \int_v \frac{\lambda(u)}{\rho(u)} du\right], \tag{6.9}$$

where $\langle v \rangle$ is the mean of v across the whole field, $\rho(v) = \alpha(v + \epsilon)(1 - v) + D(\langle v \rangle - v)$, and C is the normalization constant. $p(v; \langle v \rangle)$ is defined within the domain $[0, v_{lim}]$, where $v_{lim} \leq 1$ is the positive root of $\rho(v) = 0$. The probability distribution $p(v; \langle v \rangle)$ of v depends on the unknown value of $\langle v \rangle$. Therefore the self-consistency equation is typically used to determine $\langle v \rangle$:

$$\langle v \rangle = \int_0^{v_{lim}} v p(v; \langle v \rangle) \, dv. \tag{6.10}$$

Figure 6.3 shows the steady-state values $\langle v \rangle_{st}$ of an average tree biomass obtained through the mean-field approximation of Eq. (6.8). For low values of D, the system has only one steady state, which is either at $v = v_{lim}$ or $v = 0$, depending on the rates of tree growth and fire occurrence. Multiple steady states exist when D exceeds a critical value (e.g., the bifurcation point at $D = 0.14$ in Fig. 6.3), indicating the occurrence of a (noise-induced) phase transition when the spatial coupling is relatively strong. Thus, because of the stronger spatial coupling, for relatively high values of D, the system converges toward one out of two mutually exclusive steady states. The dependence on the initial condition can be investigated through numerical simulations (Fig. 6.4). Obtained from Eq. (6.8) without invoking the mean-field approximation, the numerical simulations support the approximated analytical results presented in

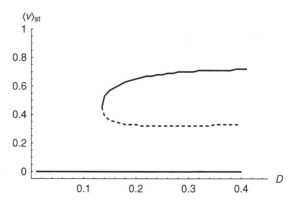

Figure 6.3. Steady-state average woody biomass as a function of the diffusion coefficient for $\alpha = 0.45$, $\lambda_0 = 0.65$, $b/\lambda_0 = -0.9$, $\omega_0 = 0.4$, $\epsilon = 0.0001$. These results are obtained from the numerical solution of Eq. (6.10). Figure taken from D'Odorico et al. (2007b).

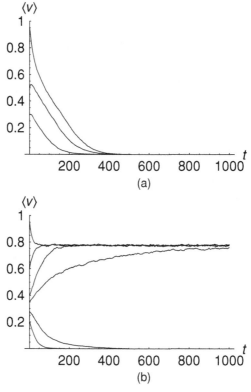

Figure 6.4. Numerical simulation (in a 128×128 grid) of the temporal evolution of mean tree biomass as a function of the initial condition for (a) $D = 0.1$ and (b) $D = 0.3$; the other parameters are the same as in Figure 6.3. Figure taken from D'Odorico et al. (2007b).

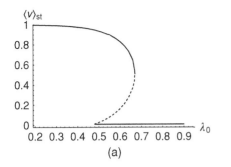
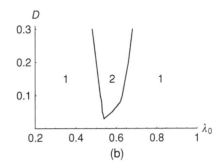

(a) (b)

Figure 6.5. (a) Stable (solid curve) and unstable (dashed curve) steady states of $\langle v \rangle_{st}$ as functions of the noise parameter λ_0, for $D = 0.30$, $\alpha = 0.45$, $b/\lambda_0 = -0.9$, $\omega_0 = 0.4$, $\epsilon = 0.0001$; (b) number of (stable) steady states of the system. Figure taken from D'Odorico et al. (2007b).

Fig. 6.3. Moreover, the numerical simulations show that the intermediate steady state in Fig. 6.3 (dashed curve) is unstable.

To investigate the dependence of multiple steady states on noise intensity, $\langle v \rangle_{st}$ was plotted as a function of the noise parameter λ_0, maintaining a constant ratio b/λ_0. The results (Fig. 6.5) show that in the absence of fires (i.e., $\lambda_0 = 0$) the system has only one steady configuration corresponding to landscapes completely dominated by trees. For relatively large values of λ_0 fires prevent the establishment of woody vegetation and the system has only one steady state at $\langle v \rangle_{st} = 0$. In intermediate conditions the system undergoes phase transitions. The parameter b represents the "strength" of the feedback between vegetation and fires. In the absence of such feedback (i.e., $b = 0$, not shown) the system exhibits neither phase transitions nor pattern formation. Figure 6.5(b) shows the number of stable states of the system (i.e., of stable solutions of the self-consistency equation) as a function both of λ_0 and D. The disappearance of multiple (statistically) steady states for low and high values of noise intensity suggests that the phase transition is *reentrant* (Porporato and D'Odorico, 2004). In the region of the parameter space located between these two phase transitions, ordered states emerge. Once the possibility of noise-induced bimodality and phase transitions has been detected, the existence of patterned states can be assessed numerically. Examples of vegetation patterns generated by the model (starting from random initial conditions) are shown in Fig. 6.6.

Patterns produced by this model are not stationary and emerge only in the transient from the initial condition to the asymptotic state of uniform vegetation. Although transient conditions may last for very long time – especially close to the bifurcation point – the mechanism proposed in this study explains only the initiation of vegetation patterns, and other processes need to be invoked to stabilize and maintain spatial organization.

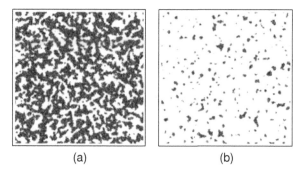

(a)   (b)

Figure 6.6. Examples of model-generated patterns (trees or shrubs in black; grass or bare soil in white). (a) "Pseudo-labyrinthine" pattern (in an 800×800 lattice, with the same parameters as in Fig. 6.5, $D = 0.3$, uncorrelated random initial condition with mean 0.33, with simulation interrupted at time $t = 350$); (b) spot patterns [same parameters as in Fig. 6.6(a) but with mean initial condition $= 0.28$]. Figure taken from D'Odorico et al. (2007b).

6.6 Random switching between two deterministic dynamics

6.6.1 Swift–Hohenberg process driven by dichotomous noise

A recurrent process of pattern formation in environmental fluid dynamics is associated with Rayleigh–Bénard convection (Chandrasekhar, 1981). This phenomenon leads to the emergence of organization in atmospheric convection, which is often evidenced by well-defined cloud patterns (Fig. 6.7). The organization results from (symmetry-breaking) thermoconvective instability typically observed when a fluid overlies a hot surface (Chandrasekhar, 1981). To study the effect of hydrodynamic fluctuations in systems exhibiting Rayleigh–Bénard convective cells, Swift and Hohenberg (1977) developed a stochastic model based on a biharmonic equation forced by additive white noise. The deterministic counterpart of this model is frequently used to investigate deterministic patterns emerging from symmetry-breaking instability of biharmonic dynamics (Cross and Hohenberg, 1993).

In Chapter 5 we introduced the Swift–Hohenberg model and discussed the ability of additive and multiplicative Gaussian noise to generate patterns in Swift–Hohenberg dynamics. Here we consider the case of a Swift–Hohenberg system driven by DMN and the emergence of patterns induced by the random switching between two deterministic Swift–Hohenberg dynamics. The two dynamics have the same (Swift–Hohenberg) spatial coupling term and two different local kinetics, $f_{1,2}(\phi)$:

$$\frac{\partial \phi(\mathbf{r}, t)}{\partial t} = f_{1,2}[\phi(\mathbf{r}, t)] - D(\nabla^2 + k_0^2)^2 \phi(\mathbf{r}, t). \qquad (6.11)$$

Buceta and Lindenberg (2002a) considered the case with

$$f_{1,2}(\phi) = -A_{1,2}\left(\phi^3 \pm \phi^2 - \phi \mp 1\right) \qquad (6.12)$$

Figure 6.7. Aerial view of an organized cloud system.

and investigated pattern formation induced by the switching between states 1 and 2 [Eqs. (6.11) and (6.12)]. As discussed in Section 5.8, each of these dynamics (6.11) is unable to generate patterns. However, Buceta and Lindenberg (2002a) noticed that the random switching between these two "disordered" states may lead to order and pattern formation, similar to the case of paradoxical games (or *Parrondo paradox*), whereby the alternation of two losing games may lead to the emergence of a winning one (Harmer and Abbott, 1996). If the system is in states 1 and 2 with probabilities P and $1 - P$, respectively, the dynamics resulting from the random switching between these states can be expressed as

$$\frac{\partial \phi(\mathbf{r}, t)}{\partial t} = f_1[\phi(\mathbf{r}, t)]\xi_{\text{dn}}(t) + f_2[\phi(\mathbf{r}, t)](1 - \xi_{\text{dn}}(t)) - D(\nabla^2 + k_0^2)^2 \phi(\mathbf{r}, t),$$
(6.13)

where $\xi_{\text{dn}}(t)$ is a DMN assuming values 0 and 1 with probability $1 - P$ and P, respectively [hence $\langle \xi_{\text{dn}}(t) \rangle = P$]. As noted in Section 5.8 in the case of random switching between two Turing's models, if the rate of random switching is relatively fast with

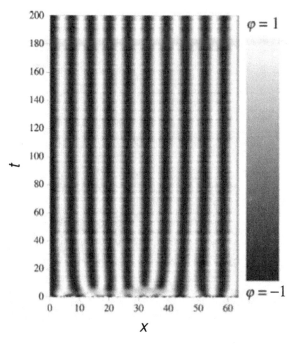

Figure 6.8. Spatiotemporal patterns of ϕ obtained as solutions of the 1D version of (6.11) calculated for $A_1 = A_2 = 1$. Figure taken from Buceta et al. (2002a).

respect to the rate of convergence to equilibrium in each of the two states, we can replace the noise term in (6.13) with its average value. We obtain the (deterministic) equation

$$\frac{\partial \phi(\mathbf{r}, t)}{\partial t} = f_1[\phi(\mathbf{r}, t)]P + f_2[\phi(\mathbf{r}, t)](1 - P) - D(\nabla^2 + k_0^2)^2 \phi(\mathbf{r}, t). \qquad (6.14)$$

Thus, in this case, we can investigate pattern formation by referring to the average dynamics, using the methods presented for the deterministic case [Eqs. (5.18) and (5.19)]. Although for $P = 0$ and $P = 1$ the potential function defined as $V'(\phi) = -\bar{f}(\phi) = -[Pf_1(\phi) + (1 - P)f_2(\phi)]$ has only one stable state, for intermediate values of P bistability may emerge. When $A_1 = A_2 = 1$ and $P = 0.5$, the local dynamics are expressed by the function $\bar{f} = \phi - \phi^3$. In this system, stable patterns emerge, as shown in Fig. 6.8 for the case of a 1D domain.

We need to stress again that, even though in this example patterns emerge as a result of the random switching between two deterministic dynamics, these patterns are not necessarily noise induced. In fact, Buceta and Lindenberg (2002a) showed that the same patterns would emerge even when the random forcing ξ_{dn} in (6.13) is replaced with a deterministic periodic function driving the switching between the two states, 1 and 2. Thus, in this system, dichotomous noise is used as a random mechanism to drive the repeated alternation between two deterministic states. However, deterministic mechanisms would lead to comparable results. A similar discussion on the role of random and periodic drivers was presented in Chapter 3 (Subsection 3.2.1.3) in the context of noise-induced transitions in zero-dimensional systems driven by DMN.

The main difference with the stochastic case is that the spatial patterns emerging from periodic switching are not always time independent but exhibit a pulsating behavior (i.e., periodic oscillations in time) if the period of the external forcing is of the same order as the relaxation time to equilibrium in deterministic states 1 and 2 (Buceta and Lindenberg, 2002a; Buceta et al., 2002a).

6.6.2 *Random switching between stressed and unstressed conditions in vegetation*

In this subsection we follow D'Odorico et al. (2006c) and show how patterns may emerge as a result of the random switching between two deterministic dynamics, similar to those discussed in the previous example. The main difference is that in this case the spatial coupling is expressed by an integral term as in the neural models discussed in Section 6.2 and Appendix B. We consider in particular the case of dryland vegetation and show how vegetation patterns could emerge as an effect of random interannual rainfall fluctuations, which are typically strong in dryland ecosystems (e.g., Noy-Meir, 1973; Nicholson, 1980; D'Odorico et al., 2000).

The spatial and temporal variabilities of vegetation biomass v (normalized between 0 and 1) are modeled as a random sequence of two deterministic dynamics corresponding to (i) drought-induced vegetation decay, and (ii) unstressed vegetation growth. In both cases, vegetation dynamics at any point $\mathbf{r}(x, y)$ are expressed as the sum of two terms accounting for the local dynamics, $f_{1,2}[v(\mathbf{r}, t)]$, and spatial interactions, $\mathcal{L}[v(\mathbf{r}, t), v(\mathbf{r}', t)]$, with the surrounding vegetation existing at all points $\mathbf{r}'$ in the neighborhood of $\mathbf{r}$:

$$\frac{\partial v(\mathbf{r}, t)}{\partial t} = f_{1,2}[v(\mathbf{r}, t)] + \zeta \mathcal{L}[v(\mathbf{r}, t), v(\mathbf{r}', t)], \tag{6.15}$$

where ζ is a dimensionless coefficient determining the relative importance of spatial versus local dynamics. Two different functions, $f_1(v)$ and $f_2(v)$, are used to describe the local dynamics: The loss of vegetation occurring in water-stressed conditions is assumed to be proportional to the existing biomass, and the unstressed growth of v is proportional to the existing vegetation biomass v and to the available resources $1 - v$:

$$f_1[v(\mathbf{r}, t)] = -\alpha_1 v(\mathbf{r}, t), \tag{6.16}$$

$$f_2[v(\mathbf{r}, t)] = \alpha_2 v(\mathbf{r}, t)[1 - v(\mathbf{r}, t),], \tag{6.17}$$

where α_1 is the mortality rate per unit v and α_2 is the reproduction rate (e.g., Murray, 2002) of the logistic equation.

Following Lefever and Lejeune (1997), D'Odorico et al. (2006c) modeled the spatial interactions as the combined effect of facilitation and competition mechanisms:

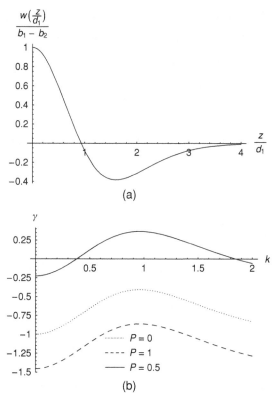

Figure 6.9. (a) Example of kernel function [Eq. (6.19)] used to model the spatial dynamics of facilitation–competition, (b) relation between growth factor γ and wave number k of linear perturbations of v_0 in solutions of (6.21) calculated for different values of P, i.e., of the probability of not being in water-stressed conditions ($\zeta = 0.4$; $\chi = 2.0$; $\epsilon = 0.25$; $\eta = 1.454$). Figure taken from D'Odorico et al. (2006c).

Near-neighbor interactions facilitate vegetation survival and growth because of the favorable environment existing in the subcanopy soils, e.g., higher moisture contents (Walker et al., 1981; Greene et al., 1994; Scholes and Archer, 1997; Zeng et al., 2004). Long-range interactions are dominated by competitions between lateral root systems, which expand beyond the vertical projection of the canopy (e.g., Casper et al., 2003; Caylor et al., 2006). At any point $\mathbf{r}(x, y)$ in a 2D domain Ω, the effect of interactions with vegetation at another point $\mathbf{r}'(x, y)$ is assumed to be proportional to the biomass of the neighboring vegetation and to a weight function $w(z)$ of the distance, $z = |\mathbf{z}| = |\mathbf{r} - \mathbf{r}'|$, between $\mathbf{r}'$ and $\mathbf{r}$. This weight function [Fig. 6.9(a)] is positive for relatively small values of $\mathbf{r}$ (facilitation) and negative at larger distances (competition). In the absence of relief, the spatial interactions [and hence the function $w(z)$] depend on the length but not on the direction of the displacement vector $\mathbf{z}$, i.e., the process is isotropic. An integral formulation (see Section 6.2) is used to account for the effect of the interactions with vegetation existing at all points $\mathbf{r}'(x', y')$ in Ω

(e.g., Murray, 2002). For each of the two climate-controlled states (i.e., stressed and unstressed conditions) vegetation dynamics can be expressed as

$$\frac{\partial v(\mathbf{r}, t)}{\partial t} = f_{1,2}[v(\mathbf{r}, t)] + \int_{\Omega} w(|\mathbf{r}' - \mathbf{r}|)v(\mathbf{r}', t)\, d\mathbf{r}', \tag{6.18}$$

where the subscripts 1 and 2 denote the local dynamics of stressed and unstressed vegetation, respectively, and v represents vegetation biomass, which is normalized with respect to the ecosystem carrying capacity in state 2 (i.e., $0 \le v \le 1$). The local dynamics are expressed by $f_{1,2}(v)$ as in Eqs. (6.16) and (6.17). The weighting (or *kernel*) function $w(z)$ is modeled here as the difference between two Gaussian functions (e.g., Murray, 2002):

$$w(z) = b_1 \exp\left[-\left(\frac{z}{d_1}\right)^2\right] - b_2 \exp\left[-\left(\frac{z}{d_2}\right)^2\right], \tag{6.19}$$

where b_1 and b_2 express the relative importance of facilitation and competition processes and d_1 and d_2 are related to the radii of canopy and root footprints, respectively. The kernel function $w(z)$ qualitatively has the shape shown in Fig. 6.9 when $d_1 < d_2$ and $b_1 > b_2$.

We use DMN as a mechanism to drive the switching between the two local dynamics [Eqs. (6.16) and (6.17)]: With probability $(1 - P)$ vegetation is water stressed and its dynamics are modeled by Eqs. (6.16) and (6.18). With probability P plants are unstressed and vegetation growth is modeled by Eqs. (6.17) and (6.18). Thus the effect of (large-scale) random, interannual rainfall fluctuations cause the switching between these dynamics. This alternation simultaneously occurs at all points in the 2D domain.

When DMN is used to model the random switching, vegetation dynamics can be expressed by the stochastic integral–differential equation

$$\frac{\partial v(\mathbf{r}, t)}{\partial t} = f_+[v(\mathbf{r}, t)] + f_-[v(\mathbf{r}, t)]\xi_{dn} \tag{6.20}$$

$$+ \int_{\Omega}\left[b_1\exp\left(-\frac{|\mathbf{r}' - \mathbf{r}|^2}{d_1^2}\right) - b_2\exp\left(-\frac{|\mathbf{r}' - \mathbf{r}|^2}{d_2^2}\right)\right]v(\mathbf{r}', t)d\mathbf{r}',$$

where ξ_{dn} is a DMN assuming values of -1 and 1 with probability P and $(1 - P)$, respectively, and $f_{\pm}[v(\mathbf{r}, t)] = \frac{1}{2}\{f_1[v(\mathbf{r}, t)] \pm f_2\,[v(\mathbf{r}, t)]\}$.

We focus on the case in which neither one of the dynamics of stressed and unstressed vegetation [Eqs. (6.15)–(6.17)] is – separately – able to generate patterns, whereas the random switching induces pattern formation (D'Odorico et al., 2006c). The response of (woody) vegetation to water-stress conditions is relatively slow (a few decades) compared with the year-to-year climate variability considered in this study (e.g., Archer et al., 1988; Barbier et al., 2006). Thus we can investigate pattern emergence by replacing the (stochastic) local term in (6.20) with its average. In dimensionless form, Eq. (6.20) becomes

$$\frac{\partial v(\tilde{\mathbf{r}}, t)}{\partial \tau} = -(1 - P)v(\tilde{\mathbf{r}}, t) + P\eta v(\tilde{\mathbf{r}}, t)[1 - v(\tilde{\mathbf{r}}, t)]$$

$$+ \zeta \int_{\Omega} \left[\exp\left(-|\tilde{\mathbf{r}}' - \tilde{\mathbf{r}}|^2\right) - \epsilon \exp\left(-\frac{|\tilde{\mathbf{r}}' - \tilde{\mathbf{r}}|^2}{\chi^2}\right) \right] v(\tilde{\mathbf{r}}', t)\, d\tilde{\mathbf{r}}', \quad (6.21)$$

with

$$\tau = \alpha_1 t; \quad \tilde{\mathbf{r}} = \frac{\mathbf{r}}{d_1}; \quad \zeta = \frac{b_1 d_1^2}{\alpha_1}; \quad \epsilon = \frac{b_2}{b_1}; \quad \chi = \frac{d_2}{d_1}; \quad \eta = \frac{\alpha_2}{\alpha_1}, \quad (6.22)$$

where ζ and $\epsilon < 1$ represent the relative importance of local–spatial dynamics [see Eq. (6.15)] and of competition–facilitation processes, respectively; $\chi > 1$ depends on the ratio between the radii of root system and canopy footprints, and η expresses the relative importance of logistic growth and stress-induced mortality. To simplify the notation, in what follows we drop the tilde "$\sim$" and indicate by the dimensionless coordinate vector $\mathbf{r}(x, y)$. The homogeneous steady states are obtained as solutions of (6.21) for $v = v_0 = \text{const}$:

$$- (1 - P)v_0 + P\eta v_0(1 - v_0)$$

$$+ \zeta v_0 \int_{\Omega} \left\{ \exp\left[-|\mathbf{r}' - \mathbf{r}|^2\right] - \epsilon \exp\left[-|\mathbf{r}' - \mathbf{r}|^2/\chi^2\right] \right\} d\mathbf{r}' = 0. \quad (6.23)$$

In particular, we study the linear stability of the homogeneous stable state

$$v_0 = 0 \quad \left(\text{if } P \leq \frac{1}{\eta + 1} \right), \quad (6.24)$$

$$v_0 = 1 - \frac{1 - P}{\eta P} + \frac{\zeta \pi}{\eta P}(1 - \epsilon \chi^2) \quad \text{(otherwise)},$$

by seeking for solutions of (6.21) in the form of a sum of v_0 with a perturbation term δ_v,

$$v = v_0 + \delta_v = v_0 + \hat{\delta}_v e^{\gamma \tau + i \mathbf{k} \cdot \mathbf{r}}, \quad (6.25)$$

where $\hat{\delta}_v$ is the amplitude of the perturbation, γ is its growth factor, $\mathbf{k}$ is the wave-number vector, $i = \sqrt{-1}$ is the imaginary unit, and "·" is the scalar-product operator. $v = v_0$ is linearly unstable when $\gamma > 0$, because any disturbance δ_v of v_0 would indefinitely grow with time. To determine the relation between growth factor and wave number in (6.25), we insert Eq. (6.25) into (6.21), and obtain (after a Taylor expansion for small values of $\hat{\delta}_v$)

$$\gamma(k) = -(1 - P) + \eta P(1 - 2v_0) + \zeta W(k)$$

$$= -(1 - P) + \eta P(1 - 2v_0) + \zeta \pi \left(e^{-\frac{k^2}{2}} - \epsilon \chi^2 e^{-\frac{k^2 \chi^2}{2}} \right), \quad (6.26)$$

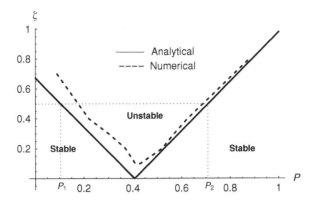

Figure 6.10. Analytical and numerical marginal stability curve $\zeta = \zeta^*$ as a function of the probability P of being in unstressed conditions (with $\chi = 2.0$; $\epsilon = 0.25$; and $\eta = 1.454$). Patterns emerge for $\zeta > \zeta^*$. Figure taken from D'Odorico et al. (2006c).

with $k = |\mathbf{k}|$ and $W(k) = \int_{\Omega} w(|\mathbf{r}|)e^{i\mathbf{k}\cdot\mathbf{r}}d\mathbf{r}$. Figure 6.9(b) shows that, for given values of ζ, ϵ, and η, $\gamma(k)$ is negative both for $P = 0$ and for $P = 1$, indicating that the corresponding homogeneous solutions are stable. Thus no patterns emerge in the absence of climate fluctuations (i.e., with no switching) because the system evolves toward a homogeneous state. $\gamma(k)$ may be positive for intermediate values of P, indicating how climate-driven random switching between deterministic dynamics may indeed trigger instability.

The marginal stability conditions are shown in Fig. 6.10: The solid line [Eq. (6.26)] separates stable from unstable states in the parameter space. The V-shape is due to the discontinuous dependence of v_0 on P [Eq. (6.25)]. For a given value of ζ (Fig. 6.10, dotted curve) there are two values, P_1 and P_2, of P (Fig. 6.10) marking the transition between stable and unstable states. For $P < P_1$ the unvegetated state is stable; for $P_1 < P < P_2$ the system tends to a spatially heterogeneous stable state with organized vegetated patches bordered by bare ground. For $P > P_2$ the homogeneous state v_0 is stable. The linear-stability analysis of the state $v = v_0$ does not account for the existence of a bound for v at $v_0 = 0$. The numerical simulations carried out to investigate the effect of this bound show that conditions of instability and pattern formation are reached for values of P_1 (Fig. 6.10, dashed curve) that are slightly different from those predicted by the analytical methods. However, the existence of a bound at $v = 0$ does not qualitatively change the stability of the unvegetated state. Interestingly, Fig. 6.10 (solid line) also shows the emergence of a completely nondeterministic limit behavior in the local dynamics for $\zeta = 0$ (i.e., with no deterministic spatial interactions) and $P \approx 0.4$, though this behavior disappears when the bound of the dynamics at $v = 0$ is accounted for (Fig. 6.10).

The parameter P (i.e., the probability of not being in water-stressed conditions) increases along a rainfall gradient and can be considered as a surrogate variable

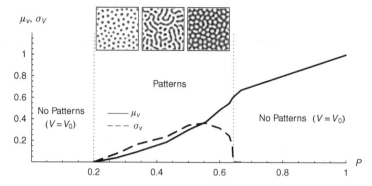

Figure 6.11. Dependence of mean and standard deviation of v on the probability P of not being in water-stressed conditions ($\zeta = 0.4$ and same parameters as Fig. 6.10). For $0.2 < P < 0.64$ the spatial standard deviation is larger than zero, suggesting the emergence of spatially heterogeneous vegetation, i.e., pattern formation. The patterns generated by the model are shown in the insets and include spotted vegetation (inset on the left, $P = 0.375$), labyrinthine patterns (central inset, $P = 0.50$), and spotted bare-ground gaps (inset on the right, $P = 0.6$). Figure taken from D'Odorico et al. (2006c).

for mean annual rainfall. Figure 6.11 shows the mean and standard deviation of v as functions of P. For relatively low values of P the system is in water-stressed conditions for most of the time and vegetation is unable to establish and grow. Thus the system tends to a uniformly unvegetated state (i.e., $v = 0$; $\sigma_v = 0$). For relatively high values of P the system is unstressed for most of the time and vegetation is able to reach a uniformly vegetated state. In these conditions $\sigma_v = 0$ and no patterns emerge. In intermediate conditions neither one of the two uniform states (vegetated or unvegetated conditions) can be attained because the repeated switching between the two dynamics does not allow the system to reach the steady states of the underlying deterministic dynamics. In this case – depending on the spatial interactions – vegetation may attain a spatially heterogeneous stable state with a sparse canopy separated by barren areas. The typical sequence of (noise-induced) patterns along the gradient in mean annual precipitation (i.e., in the P parameter) is shown in Fig. 6.11 (insets). This sequence ranges from spotted vegetation, to labyrinthine patterns, to spotted gaps. The widespread occurrence of this type of pattern in dryland ecosystems has been well documented (e.g., Ludwig and Tongway, 1995; Couteron and Lejeune, 2001; Barbier et al., 2006; Borgogno et al., 2009).

The instability of the state v_0 is associated with the emergence of spatial patterns when the most unstable mode, k_{max}, is different from zero (see Murray, 2002, p. 488). k_{max} can be determined from Eq. (6.26), $k_{max} = \sqrt{2\log(\epsilon\chi^4)/(\chi^2 - 1)}$. The condition of marginal stability [$\gamma_{max} = \gamma(k_{max}) = 0$] becomes

$$\zeta = \zeta^* = \frac{[(1 - p) - \eta P + 2\eta P v_0](\epsilon\chi^4)^{\frac{\chi^2}{\chi^2 - 1}}}{\pi\epsilon\chi^2(\chi^2 - 1)}, \tag{6.27}$$

where v_0 is a function of P [see Eq. (6.25)]. Because in Fig. 6.9(b) (bottom) the most unstable mode k_{max} is greater than zero (i.e., $\chi > \epsilon^{-\frac{1}{4}}$), the instability (i.e., $\zeta > \zeta^*$) of the homogeneous stable state v_0 is associated with the emergence of spatial patterns.

6.7 Spatiotemporal stochastic resonance in predator–prey systems

Some of the early stochastic resonance models of noise-induced pattern formation in ecosystems were developed within the context of spatially explicit predator–prey systems (Spagnolo et al., 2004). We present here the case of three interacting species: two preys, $A_{1,2}(\mathbf{r}, t)$, and one predator, $A_3(\mathbf{r}, t)$. Following Spagnolo et al. (2004), we consider the discrete representation of the dynamics of these interacting species; for each point $\mathbf{r_i}$ in a 2D square lattice, the biomass of the three species at time $t + 1$ can be related to their values at time t through the difference equations (Spagnolo et al., 2004):

$$A_1(\mathbf{r_i}, t + 1) = \mu A_1(\mathbf{r_i}, t) [1 - \nu A_1(\mathbf{r_i}, t) - \beta(t) A_2(\mathbf{r_i}, t) - \gamma A_3(\mathbf{r_i}, t)]$$
$$+ A_1(\mathbf{r_i}, t) \xi_1(\mathbf{r_i}, t) + D \sum_\delta [A_1(\mathbf{r_\delta}, t) - A_1(\mathbf{r_i}, t)],$$

$$A_2(\mathbf{r_i}, t + 1) = \mu A_2(\mathbf{r_i}, t) [1 - \nu A_2(\mathbf{r_i}, t) - \beta(t) A_1(\mathbf{r_i}, t) - \gamma A_3(\mathbf{r_i}, t)]$$
$$+ A_2(\mathbf{r_i}, t) \xi_2(\mathbf{r_i}, t) + D \sum_\delta [A_2(\mathbf{r_\delta}, t) - A_2(\mathbf{r_i}, t)],$$

$$A_3(\mathbf{r_i}, t + 1) = \mu' A_3(\mathbf{r_i}, t) \{-\beta' + \gamma'[A_1(\mathbf{r_i}, t) + A_2(\mathbf{r_i}, t)]\}$$
$$+ A_3(\mathbf{r_i}, t) \xi_3(\mathbf{r_i}, t) + D \sum_\delta [A_3(\mathbf{r_\delta}, t) - A_3(\mathbf{r_i}, t)] \qquad (6.28)$$

where the parameters γ and γ' express the predator–prey interactions, and μ' are the reproduction rate of the logistic growth, ν is a parameter determining the resources used by the same species, D is a diffusion coefficient, $\xi_{1,2,3}(\mathbf{r_i}, t)$ are independent zero-mean white-Gaussian-noise terms with intensities $s_{1,2,3}$, respectively. The sum $\sum_\delta$ is calculated over the four nearest neighbors of $\mathbf{r_i}$. The interaction between the two preys is expressed by the time-dependent coefficient $\beta(t)$, which undergoes periodical fluctuations with amplitude α and angular frequency ω_0:

$$\beta(t) = 1 + \epsilon + \alpha \cos(\omega_0, t),$$

where ϵ is a parameter expressing the displacement of the mean from the unit value. Thus $\beta(t)$ fluctuates around the value $\beta = 1 + \epsilon$, thereby inducing the periodic alternation between conditions of coexistence of the two preys ($\beta < 1$) and exclusion of one of them ($\beta > 1$). Spagnolo et al. (2004) showed that in this system spatial patterns emerge as an effect of the random forcing, as shown in Fig. 6.12. If homogeneous initial conditions are used the two preys exhibit anticorrelated spatial patterns (i.e., mutual exclusion of A_1 and A_2), whereas the predator has a spatial distribution

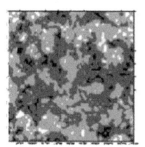 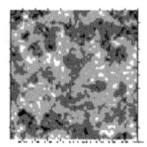 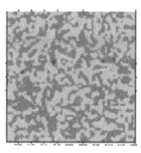

Figure 6.12. Emergence of stationary patterns of prey [A_1 (left), A_2 (center)], and predator (A_3, right) biomass in the case with $\epsilon = -0.01$, $\mu = 2$, $\mu' = 1$, $\nu = 1$, $\beta' = 0.01$, $\nu_0 = (\omega_0/2\pi) = 10^{-3}$, $\alpha = 0.1$, $s_1 = s_2 = s_3 = 0.5 \times 10^{-16}$, $D = 0.01$, $\gamma = 3 \times 10^{-2}$, $\gamma' = 2.05 \times 10^2$. Darker gray corresponds to higher biomass densities. Figure taken from Spagnolo et al. (2004).

that is correlated with both preys (Fig. 6.12). Conversely, if the initial distribution of the two preys has a peak and the predator has a homogeneous initial condition, the system converges to a configuration with strong cross correlation between the two preys (i.e., coexistence of A_1 and A_2). These patterns emerge when the noise intensity exceeds a certain critical level. However, patterns disappear for relatively large values of the noise intensity, consistent with the theory of stochastic resonance (Chapters 3 and 5).

6.8 Spatiotemporal coherence resonance in excitable plankton systems

In this section we consider a spatially extended version of the excitable phytoplankton–zooplankton system presented in Chapter 4 (Subsection 4.8.4). We recall that it is a system with three state variables: phytoplankton susceptible to infection Z_s, infected phytoplankton Z_i, and zooplankton Z_z. Phytoplankton biomass $Z_{s,i}$ undergoes logistic growth and is harvested by the zooplankton. Moreover, Z_s is turned into Z_i (infection process), while Z_i undergoes disease-induced mortality. The zooplanton biomass grows proportionally to the rate of phytoplankton harvesting and decays proportionally to Z_z. Sieber et al. (2007) investigated the effect of multiplicative noise on the temporal dynamics of this system and found that coherence resonance may emerge from the cooperation between noise and the underlying nonlinear deterministic dynamics close to a Hopf bifurcation. Thus noise is able to unveil a characteristic time scale of the deterministic dynamics and induce coherent periodical fluctuations in the time domain.

We now consider a 2D extension (Sieber et al., 2007) of this system with spatial coupling expressed by a diffusion term:

$$\frac{\partial Z_k(\mathbf{r}, t)}{\partial t} = f_k[Z_s(\mathbf{r}, t), Z_i(\mathbf{r}, t), Z_z(\mathbf{r}, t)] + D\nabla^2 Z_k(\mathbf{r}, t) + Z_k(\mathbf{r}, t)\xi_k(\mathbf{r}, t), \quad (6.29)$$

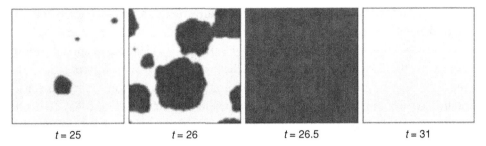

$t = 25$ $t = 26$ $t = 26.5$ $t = 31$

Figure 6.13. Spatiotemporal evolution of the excitation of susceptible biomass $Z_s(\mathbf{r}, t)$ calculated with $D = 5 \times 10^{-2}$, $s_{gn} = 2 \times 10^{-4}$, and the same parameters as in Subsection 4.8.4 in Chapter 4. Darker gray corresponds to higher biomass densities. Figure taken from Sieber et al. (2007).

with $k = s, i, z$. The functions $f_k[Z_s(\mathbf{r}, t), Z_i(\mathbf{r}, t), Z_z(\mathbf{r}, t)]$ are defined as the right-hand sides of Eqs. (4.80) in Chapter 4; D is a diffusion coefficient, and ξ_k is a zero-mean Gaussian noise with intensity s_{gn} and no autocorrelation either in space or time. Numerical simulations of (6.29) (Sieber et al., 2007) show that with adequate levels of noise the temporal dynamics of the spatial mean exhibit periodic fluctuations similar to those detected in the zero-dimensional model (Subsection 4.8.4, Chapter 4). The random forcing displaces the system from its homogeneous *rest* state by inducing local excitations, which then spread over the whole domain as an effect of diffusion. Thus, depending on the diffusion coefficient and the size of the domain, a state of global excitation may be reached before the system returns back to its homogeneous rest state, as shown in Fig. 6.13. Therefore both diffusion and local excitations induced

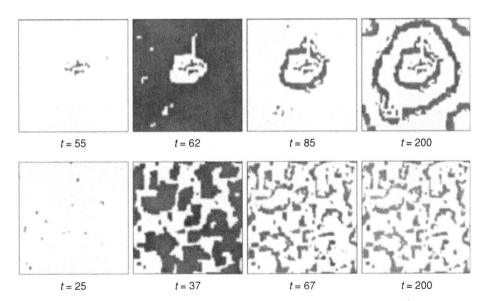

$t = 55$ $t = 62$ $t = 85$ $t = 200$

$t = 25$ $t = 37$ $t = 67$ $t = 200$

Figure 6.14. Emergence of stationary patterns of Z_s with $s_{gn} = 3 \times 10^{-4}$ (top row) and $s_{gn} = 5 \times 10^{-4}$ (bottom row) and the same parameters as in Fig. 6.13. Darker gray corresponds to higher biomass densities. Figure taken from Sieber et al. (2007).

by spatially (and temporally) uncorrelated noise are crucial to the emergence of global excitation events.

With stronger noise intensities (i.e., higher values of s_{gn}) stationary spatial patterns emerge with the formation of excited patches (i.e., high plankton population) that increase in density until a stationary configuration is reached, in which no further excitation occurs in the interspaces between excited structures (Fig. 6.14). The configuration resulting from these dynamics exhibits a remarkable spatial organization (Fig. 6.14, bottom). It has also been observed that, with lower levels of noise intensity, pattern formation starts with the excitation of single patches leading to the emergence of concentric-ring-shaped structures (Fig. 6.14, top).

Appendix A

Power spectrum and correlation

The steady-state pdf $p(\phi)$ of a stochastic process $\phi(t)$ is a key piece of information in the study of noise-induced phenomena; however, it does not give indications about the temporal structure of the process. In fact, processes with different temporal evolutions can share the same pdf. Because some noise-induced phenomena underlie changes in the temporal behavior of dynamical systems (e.g., the stochastic resonance), it is useful to introduce two mathematical tools that are commonly used to quantitatively investigate the temporal structure of a signal, namely the power spectrum and the autocorrelation function. In this appendix we recall the basic concepts and some analytical results, referring to specialized textbooks (e.g., Papoulis, 1984) for a more comprehensive description. Moreover, in the following discussion we consider signals in the time domain, though the same results are valid also if the process is sampled in space, e.g., when transects of spatial fields are studied (see Chapter 5). In this case, the power spectrum (also known as *structure function*) and the autocorrelation function are useful tools for investigating the existence of regular patterns in the field.

Let us start from a quite specific case and consider a piecewise continuously differentiable periodic function $\phi(t)$, with period 2π (if the signal has a different period, it may be mapped to a 2π period through a suitable scaling of time). Fourier demonstrated that such periodic functions can be written as the superposition of infinite harmonics (i.e., sinusoidal functions), namely

$$\phi(t) = \frac{a_0}{2} + \sum_{k=1}^{\infty} (a_k \cos kt + b_k \sin kt), \qquad (A.1)$$

where the k coefficients are obtainable by minimizing the mean square deviation between $\phi(t)$ and the summation truncated to the k order, and taking advantage of the fact that the sines and cosines functions form an orthogonal set. It follows that

$$a_0 = \frac{1}{\pi} \int_{-\pi}^{\pi} \phi(t)dt, \quad a_k = \frac{1}{\pi} \int_{-\pi}^{\pi} \phi(t) \cos kt dt, \quad b_k = \frac{1}{\pi} \int_{-\pi}^{\pi} \phi(t) \sin kt dt,$$
$$(A.2)$$

where the values of the k coefficients give the weight of the k harmonics in the original signal $\phi(t)$.

If the signal is not periodic and it is defined in only a finite interval, for example $[-\pi, +\pi]$, the harmonic analysis can be applied after the function $\phi(t)$ is extended periodically beyond the initial interval by means of the relation $\phi(t + 2\pi) = \phi(t)$, by assuming the mean of the two limiting values at the discontinuity points (i.e., at the odd multiplies of π).

The Fourier analysis takes a more compact aspect if the complex notation is introduced by the Euler formula, $e^{i\alpha} = \cos\alpha + i\sin\alpha$. In this way, the sines and cosines functions can be written as

$$\cos kt = \frac{1}{2}(e^{ikt} + e^{-ikt}), \qquad \sin kt = \frac{1}{2}(e^{ikt} - e^{-ikt}), \qquad (A.3)$$

and the Fourier series takes the form

$$\phi(t) = \sum_{k=-\infty}^{+\infty} \alpha_k\, e^{ikt}, \qquad (A.4)$$

where

$$\alpha_k = \frac{1}{2\pi} \int_{-\pi}^{\pi} \phi(t)\, e^{-ikt}\, dt. \qquad (A.5)$$

If the function $\phi(t)$ has a period $T \neq 2\pi$, the angular frequency $\omega = 2\pi/T$ is introduced and t is replaced with $\omega t'$. In this way the Fourier series becomes

$$\phi(t) = \sum_{k=-\infty}^{+\infty} \alpha_k e^{ik\omega t'}, \qquad \alpha_k = \frac{1}{T} \int_{-T/2}^{T/2} \phi(t')e^{-ik\omega t'}\, dt'. \qquad (A.6)$$

Finally, the constraint of periodicity can also be removed by allowing the length of the interval to tend to infinity (i.e., $T \to \infty$). In this case the Fourier integral is replaced with the Fourier transform:

$$\phi(t) = \frac{1}{2\pi} \int_{-\infty}^{\infty} F(\omega)e^{i\omega t}\, d\omega, \qquad (A.7)$$

where $F(\omega)$ is the Fourier transform of $\phi(t)$,

$$F(\omega) = \int_{-\infty}^{\infty} \phi(t)e^{-i\omega t}\, dt. \qquad (A.8)$$

If the rescaled frequency $f = \omega/(2\pi) = 1/T$ is used in place of ω, the previous relations become

$$\phi(t) = \int_{-\infty}^{\infty} F(f)e^{2\pi i f t}\, df, \qquad F(f) = \int_{-\infty}^{\infty} \phi(t)e^{-2\pi i f t}\, dt. \qquad (A.9)$$

Equations (A.7)–(A.9) establish a one-to-one link between $\phi(t)$ and $F(\omega)$ [or $F(f)$], generally indicated as $\phi(t) \Leftrightarrow F(\omega)$ and called a *transform pair*. Consequently we can analyze the same physical process from two different viewpoints: in the time domain as the signal $\phi(t)$ or in the frequency domain through the function $F(\omega)$. Equations (A.7) and (A.8) or (A.9) relate the process in these two domains.

The Fourier transform is a linear operator, and the result is in general a complex number, i.e.,

$$F(\omega) = F_R(\omega) + i\, F_I(\omega) = A(\omega)e^{i\theta(\omega)}, \tag{A.10}$$

where the amplitude $A(\omega)$ is the so-called Fourier spectrum [its square $A^2(\omega) = |F(\omega)|^2$ is the energy spectrum] and $\theta(\omega)$ is the phase angle. A number of important equations relate the time to the frequency domains (Papoulis, 1984). We recall Parseval's theorem,

$$\int_{-\infty}^{\infty} |\phi(t)|^2 \mathrm{d}t = \frac{1}{2\pi} \int_{-\infty}^{\infty} A^2(\omega)\mathrm{d}\omega = \int_{-\infty}^{\infty} |F(f)|^2 \mathrm{d}f, \tag{A.11}$$

expressing the total energy in a signal in time and frequency domains. When the domain of $\phi(t)$ extends over the whole real axis (i.e., $-\infty < t < \infty$) the total energy can be infinite (this is the case, for example, for all periodic signals). In these cases, the mean power of the signal can be used in place of the total energy,

$$\lim_{T\to\infty} \frac{1}{2T} \int_{-T}^{T} |\phi(t)|^2 \mathrm{d}t = \overline{\phi^2}. \tag{A.12}$$

$\overline{\phi^2}$ usually assumes finite values because of the presence of T in the denominator of Eq. (A.12). In this case $A^2(\omega)$ is substituted with the so-called power spectrum (or power spectral density)

$$P(\omega) = \lim_{T\to\infty} \frac{1}{2T} \left| \int_{-T}^{T} \phi(t)e^{i\omega t}\, \mathrm{d}t \right|^2. \tag{A.13}$$

Therefore $P(\omega)\mathrm{d}\omega$ indicates how much power of the signal is contained in the angular frequency interval $[\omega, \omega + \mathrm{d}\omega]$. Moreover, in many applications the one-sided power spectrum, $P_h(\omega) = |P(\omega)| + |P(-\omega)|$ (with $0 \le \omega < \infty$), is used. The one-sided power spectrum, often indicated dropping the subscript h, is the most adopted tool for analyzing the structure of a signal in the frequency domain.

The autocovariance function $\rho(\tau)$ is another powerful tool for investigating the temporal structure of a signal. $\rho(\tau)$ is defined as

$$\rho(\tau) = \int_{-\infty}^{\infty} \phi(t + \tau)\phi(t)\mathrm{d}t. \tag{A.14}$$

It gives a proxy of the interrelations of the signal $\phi(t)$ at two distinct times, t and $t + \tau$. It depends on the time delay (or lag) τ and reflects the memory of the signal: High values of $\rho(\tau)$ are a symptom of a strong link, whereas low values indicate a weak link. However, the autocovariance function shows only the linear links in the signal, whereas the nonlinear ones have to be detected by more complex tools, such as mutual information (e.g., see Kantz and Schreiber, 1997). Similar to Eq. (A.12),

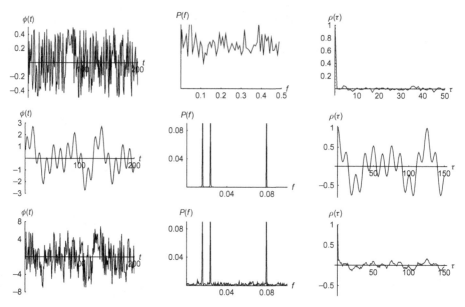

Figure A.1. Examples of power spectrum and autocorrelation function. The rows refer to the white noise (upper row), deterministic signal (A.18) (central row), and the sum of both (bottom row). From left to right, the columns show a portion of the signals, the power spectrum, and the autocorrelation function, respectively. The time is expressed in time units, 4096 samples.

when the signals have infinite energy, the covariance is defined as the mean value of the covariance function:

$$\overline{\rho}(\tau) = \lim_{T\to\infty} \frac{1}{2T} \int_{-T}^{T} \phi(t+\tau)\phi(t)\mathrm{d}t. \tag{A.15}$$

This expression, divided by $\overline{\phi^2}$ [in order to have $\overline{\rho}(\tau) \le 1$] is known as the autocorrelation function, which is commonly used in signal analysis.

The link between autocovariance and spectrum is given by the Wiener–Khinchin theorem;

$$\int_{-\infty}^{\infty} \rho(\tau)e^{i\omega\tau}\mathrm{d}\tau = A^2(\omega), \qquad \int_{-\infty}^{\infty} A^2(\omega)e^{-i\omega\tau}\mathrm{d}\omega = \rho(\tau), \tag{A.16}$$

stating that the energy spectrum is the Fourier transform of the autocovariance, namely the energy spectrum and autocovariance function form a Fourier pair [i.e., $\rho(\tau) \Leftrightarrow A^2(\omega)$]. Similarly, we can relate the autocorrelation function to the power spectrum as

$$\int_{-\infty}^{\infty} \overline{\rho}(\tau)e^{i\omega\tau}\mathrm{d}\tau = P(\omega), \qquad \int_{-\infty}^{\infty} P(\omega)e^{-i\omega\tau}\mathrm{d}\omega = \overline{\rho}(\tau). \tag{A.17}$$

Figure A.1 shows some examples of signals along with the corresponding power spectra and autocorrelation functions. The first row refers to a white noise with zero mean and uniform distribution in the interval $[-0.5, 0.5]$, the second row corresponds to a deterministic periodic signal given by the sum of three sinusoids,

$$\phi(t) = \sin(0.1t) + \sin(0.15t) + \sin(0.5t); \qquad (A.18)$$

finally, in the third case the signal is obtained as the sum of (A.18) and white noise.

The spectrum of the white noise is flat, showing that no particular frequency dominates the signal. This is confirmed also by the autocorrelation function that falls to negligible values for any time lag different from zero. In contrast, in the case of the periodic deterministic signal, the power spectrum has three spikes corresponding to the frequency of the sinusoids (i.e., $0.1/2\pi$, $0.15/2\pi$, and $0.5/2\pi$). Notice that also the autocorrelation shows the existence of regular waves in the signal – in particular, it has the same period as the original signal – but its interpretation is more difficult.

Finally, in the third case we can appreciate the importance of analyzing signals in both the time and frequency domains. In fact, the time series appears quite irregular and the periodic carrier wave is not easily detectable by eye. Similarly, the autocorrelation function does not give clear indications of the periodicity embedded in the signal. In contrast, the power spectrum allows us to recognize the existence of periodic components and to detect their frequency.

Appendix B

Deterministic mechanisms of pattern formation

B.1 Introduction

In this appendix we provide a mathematical description of the three major deterministic models of self-organized pattern formation that are commonly invoked to explain mechanisms of spatial self-organization in the biogeosciences. In these models patterns emerge from a mechanism of *symmetry-breaking instability*, whereby the uniform state of the system becomes unstable, thereby leading to the emergence of spatial patterns. Spatial interactions induce this instability, whereas the resulting patterns are stabilized by suitable nonlinear terms. In Turing and kernel-based models, symmetry breaking is the result of the interactions between short-range activation and long-range inhibition, i.e., of positive and negative feedbacks acting at different spatial scales. In the third class of models (i.e., differential-flow models) symmetry breaking emerges as a result of the differential-flow rate between two (or more) species.

In Turing and differential-flow models, the nonlinearities are local (i.e., they do not appear in the terms expressing spatial interactions), whereas in kernel-based models the nonlinearities can be in general nonlocal, i.e., they can appear as multiplicative functions of the term accounting for spatial interactions (e.g., Lefever and Lejeune, 1997). In a particular class of kernel-based models – known as *neural models* (e.g., Murray and Maini, 1989) – the nonlinearities are only local and do not affect the spatial interactions. In these models the nonlinear terms appear as additive functions of the spatial interaction term. Here we describe the Turing and kernel-based models separately because they use different mathematical representations of the spatial dynamics. However, Borgogno et al. (2009) showed that Turing and neural models are based on mathematical frameworks that are closely related. Both models invoke similar mechanisms of morphogenesis, namely symmetry-breaking instability induced by spatial interactions in activation-inhibition systems and stabilization by local nonlinearities.

B.2 Turing-like instability

In the study of nonlinear chemical systems, Turing (1952) found that the diffusion of two species (reagents) may lead to pattern formation when they have different diffusivities. In the absence of diffusion both species reach a stable and spatially uniform steady state, whereas diffusion may be able to destabilize this state (*diffusion-driven instability*), leading to the formation of spatial patterns. Known as *Turing's instability*, this mechanism seems to be counterintuitive. In fact, diffusion is usually believed to act as a homogenizing process, leading to the dissipation of concentration gradients of the diffusing species. Conversely, *Turing*'s model (1952) shows that diffusion may lead to the emergence of spatial heterogeneity in the coupled nonlinear dynamics of two diffusing species. In the literature on symmetry-breaking instability in chemistry and biology, the two diffusive species are often called *activator* and *inhibitor* and pattern emergence requires (i) nonlinear local dynamics and (ii) a faster diffusion for the inhibitor than for the activator (e.g., Murray and Maini, 1989).

Patterns emerging from Turing's instability are self-organized, in that they originate from the internal dynamics of the system and are not imposed by heterogeneities in the external drivers. Thus this mechanism is often invoked to explain the emergence of self-organized patterns also in fields other than chemistry, such as physics and biology, in systems with two or more diffusing species. Notable examples include convection in fluid mixtures (Platten and Legros, 1984), the formation of shell patterns from pigment diffusion (Murray, 2002), and vegetation pattern formation from diffusion-induced instability in arid landscapes (e.g., HilleRisLambers et al., 2001). The emergence of natural patterns from Turing's instability was experimentally demonstrated in a chemical system (Castets et al., 1990) and in nonlinear optics (e.g., Staliunas and Sanchez-Morchillo, 2000). We are not aware of any similar experiment for the case of environmental patterns. Thus, although models based on Turing's instability are capable of generating patterns resembling those observed in nature, there is no conclusive experimental evidence suggesting that these patterns do emerge from Turing's dynamics. One of the major challenges in the application of Turing's activator-inhibition model to environmental systems arises from the need to recognize two or more leading state variables and to assess whether they do diffuse in space. The diffusive character of the spatial dynamics of both activator and inhibitor is fundamental to the development of a sound Turing-like model of pattern formation, in that diffusion is crucially important to the emergence of symmetry-breaking instability in a Turing system.

We present the mathematical framework of Turing's models for the case of two species, u and v, diffusing across a 2D infinite domain $\{x, y\}$. The dynamics of u and v are modeled by two differential equations involving both diffusive terms and functions of the local values of the state variables (e.g., Murray and Maini, 1989;

Murray, 2002; Henderson et al., 2004):

$$\frac{\partial u}{\partial t} = f(u, v) + \nabla^2 u,$$

$$\frac{\partial v}{\partial t} = g(u, v) + d\nabla^2 v, \tag{B.1}$$

where t is time, f and g are the local reaction kinetics, d is the ratio $d = d_2/d_1$ between the two diffusivities d_1 and d_2 of u and v, respectively, and ∇^2 is the Laplace operator. All variables are in dimensionless units.

Turing (1952) demonstrated that this diffusive system exhibits diffusion-driven instability if (i) in the absence of diffusion the homogeneous steady state is linearly stable (i.e., stable with respect to small perturbations) and (ii) when diffusion is present the homogeneous steady state is linearly unstable. Thus we first need to determine the homogeneous steady state (u_0, v_0) as the solution of Eqs. (B.1) with $\nabla^2 u = \nabla^2 v = 0$ (homogeneous state) and $\partial u/\partial t = \partial v/\partial t = 0$ (steady state): $f(u_0, v_0) = 0$ and $g(u_0, v_0) = 0$. Then we need to impose the condition that this solution be stable in the absence of diffusion. To this end, we can study the stability of (u_0, v_0) with respect to small perturbations,

$$\mathbf{w} = \begin{pmatrix} u - u_0 \\ v - v_0 \end{pmatrix}, \tag{B.2}$$

around the steady state. For small perturbations of the steady homogeneous state (i.e., for $|\mathbf{w}| \to 0$) system (B.1) can be linearized around (u_0, v_0). Using a linear Taylor expansion we have

$$\frac{d\mathbf{w}}{dt} = J\mathbf{w}, \qquad J = \begin{pmatrix} \frac{\partial f}{\partial u} & \frac{\partial f}{\partial v} \\ \\ \frac{\partial g}{\partial u} & \frac{\partial g}{\partial v} \end{pmatrix}\Bigg|_{u_0, v_0}, \tag{B.3}$$

where J is the Jacobian of dynamical system (B.1).

The solutions of this set of equations are in exponential form and express the temporal evolution of the perturbation of the homogeneous steady state, where γ is an eigenvalue of system (B.1), i.e., a solution of the secular polynomial

$$|J - \gamma I| = 0, \tag{B.4}$$

where I is the identity matrix.

When the real part of γ, Re$[\gamma]$, is negative, $|\mathbf{w}|$ tends to zero for $t \to \infty$ and the steady homogeneous state (u_0, v_0) is linearly stable with respect to small perturbations. From the analysis of Eq. (B.4) we obtain that this condition is met when

$$\frac{\partial f}{\partial u} + \frac{\partial g}{\partial v} < 0, \qquad \frac{\partial f}{\partial u}\frac{\partial g}{\partial v} - \frac{\partial f}{\partial v}\frac{\partial g}{\partial u} > 0, \tag{B.5}$$

with all derivatives being calculated in (u_0, v_0) (Murray, 2002).

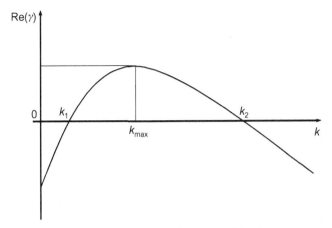

Figure B.1. Example of dispersion relation for a two-diffusive-species monodimensional system. k_1 and k_2 are the extremes of the range of unstable Fourier modes, and k_{max} represents the most unstable Fourier mode.

To study the effect of diffusion on the stability of (u_0, v_0), we consider full system (B.1) and use a Taylor expansion to linearize this set of equations around the homogeneous steady state, (u_0, v_0);

$$\frac{\partial \mathbf{w}}{\partial t} = J\mathbf{w} + D\nabla^2 \mathbf{w}, \qquad D = \begin{pmatrix} 1 & 0 \\ 0 & d \end{pmatrix}. \tag{B.6}$$

The solution of Eqs. (B.6) can be written in the form of a sum of Fourier modes:

$$\mathbf{w}(\mathbf{r}, t) = \mathbf{W}_k e^{\gamma t + i\mathbf{k}\cdot\mathbf{r}}, \tag{B.7}$$

where $\mathbf{k} = (k_x, k_y)$ is the wave-number vector, $\mathbf{r} = (x, y)$ is the coordinate vector, and $\mathbf{W}_k$ are the Fourier coefficients (Murray, 2002).

We can obtain the relation between eigenvalues and wave numbers (known as the dispersion relation, see Fig. B.1) by inserting Eq. (B.7) into (B.6) and searching for nontrivial solutions. Setting $k = \sqrt{k_x^2 + k_y^2}$, we obtain

$$|\gamma I - J + Dk^2| = 0. \tag{B.8}$$

For the state (u_0, v_0) to be unstable with respect to small perturbations, the solution of dispersion relation (B.8) should exhibit positive values of $\mathrm{Re}[\gamma(k)]$ for some wave number $k \neq 0$. Using Eq. (B.8), we find that this condition is met when

$$d\frac{\partial f}{\partial u} + \frac{\partial g}{\partial v} > 0, \qquad \left(d\frac{\partial f}{\partial u} + \frac{\partial g}{\partial v} \right)^2 - 4d \left(\frac{\partial f}{\partial u}\frac{\partial g}{\partial v} - \frac{\partial f}{\partial v}\frac{\partial g}{\partial u} \right) > 0. \tag{B.9}$$

The first of conditions (B.5) combined with the first of conditions (B.9) implies that $d \neq 1$, indicating that the system cannot be unstable with respect to small perturbations if both species have the same diffusivity. If all four conditions (B.5) and (B.9) hold, at least one eigenfunction is unstable with respect to small perturbations and grows exponentially with time as a consequence of the destabilizing effect of diffusion. Dispersion relation (B.8) imposes a specific link between eigenvalues γ and wave numbers k. The wave number k_{max}, corresponding to a maximum positive value of $\text{Re}[\gamma]$, represents the most unstable mode of the system. This implies that if $\text{Re}[\gamma(k_{max})] > 0$, this mode grows faster than the others, and the state of the system for $t \to \infty$ is dominated by k_{max} in the sense that, as $t \to \infty$, only k_{max} dictates the length scale of the spatial pattern.

Because this linear-stability analysis is developed in the limit $|\mathbf{w}| \to 0$ (i.e., under the assumption of small perturbations of the homogeneous, steady state), it cannot provide any information on the state of the system when the perturbation grows in amplitude. In the absence of nonlinearities in $f(u, v)$ and $g(u, v)$, solutions of (B.7) of the linearized model would coincide with the exact solutions of (B.1) even away from the state (u_0, v_0). In this case Eq. (B.7) clearly shows that if the state (u_0, v_0) is unstable the perturbations $\mathbf{w}$ grow indefinitely. Thus suitable nonlinear terms are needed to stabilize the pattern through higher-order terms in the Taylor expansion, which become important when the amplitude of the perturbation is finite. In other words, the system reaches a steady configuration when the exponential growth of the eigenfunction is limited by second-order (or higher) terms that come into play once the perturbation has finite amplitude. In these conditions the (nonlinear) stability of the system can be partly studied through a more complex mathematical framework based on the so-called *amplitude equations*, which investigate the dynamics of the system in the neighborhood of the most unstable mode. This appendix does not review these nonlinear methods, and we refer the interested reader to specific literature on this topic for further details (Cross and Hohenberg, 1993; Leppanen, 2005).

B.2.1 An example of a Turing model

We consider a simple example of a Turing model able to generate spatial patterns in a system with two species, u (activator) and v (inhibitor). To this end, we use Eqs. (B.1) with local kinetic functions:

$$f(u, v) = u(avu - e),$$

$$g(u, v) = v(b - cu^2v), \tag{B.10}$$

where a, b, c, and e are dimensionless positive constants.

The first equation describes the growth or the decay of the activator and accounts for a positive interaction between u and v. In fact, as v increases, the growth rate of species u increases. Moreover, the growth rate of u increases with increasing values of u. The second equation is a generalized logistic growth (e.g., Murray, 2002) with

Figure B.2. Spatial pattern emerging for the variable u in the Turing system in Eqs. (B.10). The parameters are $a = 22$, $b = 84$, $c = 113.33$, $e = 18$, and $d = 27.2$. The parameters a and g do not influence the emergence of spatial patterns (see the end of Section B.2); they influence only the shape of spatial patterns. The simulation is carried out over a domain of 256×256 cells, each cell representing a spatial step $\Delta x = \Delta y = 0.2$.

carrying capacity b and a strong negative influence (inhibition) of species u on the growth rate of v: In fact, as u increases, the second term of the function g decreases nonlinearly.

The homogeneous steady state of this system is $u_0 = ab/ce$, $v_0 = ce^2/ba^2$. The derivatives of the two functions of (B.10) calculated in (u_0, v_0) are $\partial f/\partial u = e$, $\partial f/\partial v = a^3b^2/c^2e^2$, $\partial g/\partial u = -2g^2e^3/ba^3$, and $\partial g/\partial v = -b$, and the four conditions (B.5) and (B.9) leading to diffusion-driven instability become $e - b < 0$, $eb > 0$, $de - b > 0$, and $d^2e^2 - 6bde + b^2 > 0$, respectively.

Figure B.2 shows an example in which these conditions are met and patterns emerge from diffusion-driven instability as a hexagonal arrangement of spots with wavelength

$$\lambda \simeq 2\pi / \sqrt{\frac{1}{1-d}\left(e + b - \frac{1+d}{d}\sqrt{2bde}\right)}, \tag{B.11}$$

in agreement with the wavelength of the most unstable mode obtained through dispersion relation (B.8).

B.3 Kernel-based models of spatial interactions

We classify as *kernel-based models* those modeling frameworks in which spatial interactions are expressed through a kernel function, accounting both for short-range and long-range coupling (see Section 6.2). In most of these models self-organized patterns arise as a result of short-range cooperation (or *activation*) and long-range inhibition. These spatial interactions cause symmetry-breaking instability and the system converges to an asymmetric state, which exhibits patterns. As in the case of Turing models, the convergence to this state is due to suitable nonlinear terms, which prevent the initial (linear) instability from growing indefinitely.

We first consider a particular type of kernel-based models, whereby the nonlinearity is not in the spatial coupling but in an additive term. These models are often known as *neural models* because of their applications to neural systems. Some of the most fascinating and complex pattern-forming processes existing in nature are associated with neural systems. Typical examples include the process of pattern recognition, the transmission of visual information to the brain, and stripe formation in the visual cortex (Murray, 2002). The framework of a neural model is often used to represent other systems, including the case of vegetation dynamics in spatially extended systems (D'Odorico et al., 2006c).

Neural models can in general be developed for systems with more than one state variable. However, unlike Turing models, pattern-forming symmetry-breaking instability can emerge even when the dynamics have only one state variable. Thus we concentrate on the case of neural models that are mathematically described by only one state variable, ϕ, in a 2D domain (x, y). At any point $\mathbf{r} = (x, y)$ of the domain, the variable $\phi(\mathbf{r})$ undergoes local dynamics expressed by a function $f(\phi)$ that is independent of spatial interactions. The local dynamics exhibit a steady state at $\phi = \phi_0$ [i.e., $f(\phi_0) = 0$]. We express the effect of spatial interactions by using kernel functions, as explained in detail in Section 6.2. The impact that other points $\mathbf{r}'(x, y)$ have on the dynamics of $\phi(\mathbf{r}, t)$ depends on the relative position of the two points $\mathbf{r}$ and $\mathbf{r}'$ and is expressed through a weighting (or *kernel*) function, $\omega(\mathbf{r}, \mathbf{r}')$. We integrate $\mathbf{r}'$ over the whole domain Ω to account for the interactions of $\phi(\mathbf{r}, t)$ with all points $\mathbf{r}'$ in Ω:

$$\frac{\partial \phi}{\partial t} = f(\phi) + \int_{\Omega} \omega(\mathbf{r}, \mathbf{r}')[\phi(\mathbf{r}', t) - \phi_0]d\mathbf{r}'. \tag{B.12}$$

The terms in Eq. (B.12) are explained in detail in Section 6.2. In neural models of pattern formation the interactions between cells are typically represented by short-range activation and long-range inhibition (Oster and Murray, 1989). In this case the kernel is positive at small distances, $z = |\mathbf{r} - \mathbf{r}'|$, and becomes negative at greater distances (Fig. B.3). This type of framework has been proposed as a model for spatial interactions within plant communities (e.g., Lefever and Lejeune, 1997; Yokozawa et al., 1999; Couteron and Lejeune, 2001). A kernel with the shape illustrated in

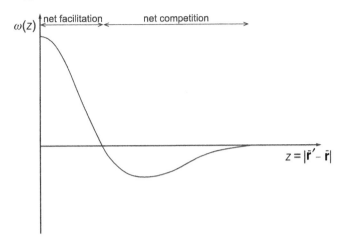

Figure B.3. Typical kernel that exhibits local activation–long-range inhibition.

Fig. B.3 can be obtained for example as the difference between two exponential functions of the form

$$\omega(z) = b_1 \exp\left[-\left(\frac{z}{q_1}\right)^2\right] - b_2 \exp\left[-\left(\frac{z}{q_2}\right)^2\right],$$ (B.13)

where $0 < q_1 < q_2$ and b_1 and b_2 are two coefficients expressing the relative importance of the facilitation and competition components of the kernel (see also Chapter 6).

The dynamics expressed by Eq. (B.12) may lead to pattern formation through mechanisms that resemble those of Turing's instability. In fact, patterns emerge as a result of the spatial interactions, which destabilize the uniform stable state ϕ_0 of the local dynamics. To study the stability of the state $\phi = \phi_0$ with respect to infinitesimal perturbations, we linearize Eq. (B.12) around the steady state $\phi = \phi_0$. Indicating by $\hat{\phi} = \phi - \phi_0$ the amplitude of the ("small") perturbation, we obtain

$$\frac{\partial \hat{\phi}}{\partial t} = \hat{\phi} f'(\phi_0) + \int_\Omega \omega(|\mathbf{r}' - \mathbf{r}|)\hat{\phi}(\mathbf{r}', t)d\mathbf{r}',$$ (B.14)

where $f'(\phi_0)$ is the derivative of the function $f(\phi)$, calculated for $\phi = \phi_0$.

Solutions of Eq. (B.14) can be expressed in the form of integral sums of the harmonics $\hat{\phi}(\mathbf{r}, t) \propto \exp[\gamma t + i\mathbf{k} \cdot \mathbf{r}]$, where each harmonic is a solution of (B.14), $\mathbf{k} = (k_x, k_y)$ is the wave-number vector, and the growth factor γ is an eigenvalue of Eq. (B.14). Substituting this solution into Eq. (B.14), setting $z = |\mathbf{r}' - \mathbf{r}|$, and canceling out the exponential function, we obtain the dispersion relation, that is, the relation between $\mathbf{k}$ and γ in solutions of (B.12) obtained as small perturbations of the state $\phi = \phi_0$:

$$\gamma(k) = f'(\phi_0) + \int_\Omega \omega(z)\exp[i\mathbf{k} \cdot \mathbf{z}]dz = f'(\phi_0) + W(k),$$ (B.15)

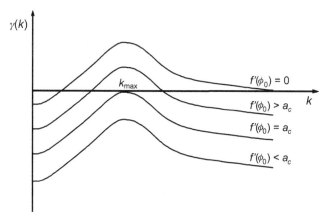

Figure B.4. Dispersion relation $\gamma(k)$ as a function of the wave number k for various values of the bifurcation parameter $f'(\phi_0)$. a_c is the critical value for $f'(\phi_0)$ that discriminates the situations of stability and instability.

with $k = |\mathbf{k}|$. If Ω is infinitely extended in both the x and y directions, $W(k)$ is the Fourier transform of the kernel function. The dispersion relation obtained with kernel (B.13) is shown in Fig. B.4. Notice how the shape of the dispersion relation is entirely determined by $W(k)$, i.e., by the effect of the kernel function on the spatial dynamics, whereas the local dynamics affect Eq. (B.15) only through the constant $f'(\phi_0)$. In fact, changes in this constant determine a vertical shift of the curves in Fig. B.4 without modifying their shape. This vertical shift affects the sign of $\gamma(k)$, thereby determining the stability or instability of the system and the range of unstable modes. All modes with $\gamma < 0$ are linearly stable because they vanish as time t passes. Conversely, all modes with $\gamma > 0$ are linearly unstable and tend to grow with time. However, even in this case, when the amplitude of the unstable modes becomes finite, the assumptions underlying this linear-stability analysis (i.e., that perturbations are "small" or infinitesimal) are no longer valid. Thus the linear-stability analysis does not shed light on the state approached by the system as an effect of the unstable modes. However, as noted for Turing's instability, the dominant wavelength of patterns emerging from this instability is dictated by the most unstable mode, k_{max}, (which grows faster than the other unstable modes, thereby determining some key aspects of pattern geometry). This wavelength depends on only the shape of the kernel function and is not affected by the term $f'(\phi_0)$ [see Eq. (B.15)], even though $f'(\phi_0)$ determines the stability of the system and the emergence of spatial patterns: For relatively low values of $f'(\phi_0)$, $\gamma < 0$ for all wave numbers k (see Fig. B.4), whereas as $f'(\phi_0)$ increases above a critical value, $\gamma(k_{max})$ becomes positive and the mode k_{max} is unstable. Larger values of $f'(\phi_0)$ correspond to broader ranges of unstable wave numbers.

Spatial interactions lead to pattern formation in Eq. (B.12) when the following conditions are met:

1. In the absence of spatial interactions the uniform steady state $\phi = \phi_0$ of the local dynamics is stable. The linear-stability analysis demonstrates that the stability of $\phi = \phi_0$ requires $f'(\phi_0)$ to be negative, as shown by Eq. (B.14), when the integral term is set equal to zero.
2. In the presence of spatial interactions there should be at least one wave number (k_{max}) associated (through the dispersion relation) with a positive value of γ.
3. Because the mode $k = 0$ corresponds to a spatially uniform perturbation of $\phi = \phi_0$ in the deterministic models, instability does not lead to the emergence of any spatial pattern if the most unstable mode k_{max} is zero. Therefore k_{max} needs to be different from zero.

Thus, in a neural model, patterns emerge from spatial interactions when

$$f'(\phi_0) < 0, \qquad W(k_{max}) + f'(\phi_0) > 0, \qquad k_{max} > 0, \qquad \text{(B.16)}$$

where k_{max} is the solution of $W'(k_{max}) = 0$ with $W''(k_{max}) < 0$. We also notice that, if $f(\phi)$ is a linear function of ϕ, Eq. (B.12) is also linear. Thus solutions of Eq. (B.14) are exact expressions (rather than approximations) of the perturbed state of the system $(\hat{\phi})$. In this case, because of the linearity of (B.14), the perturbed state remains an exponential function of γ even when the amplitude of the perturbation is no longer infinitesimal. In other words, if $f(\phi)$ is linear and conditions (B.16) are met, the steady homogeneous state is unstable and any perturbation of $\phi = \phi_0$ grows indefinitely without ever reaching a steady configuration. Thus a suitable nonlinear function $f(\phi)$ is needed for the neural model to have a steady state in which patterns emerge from symmetry-breaking instability. In this case, as soon as the initial perturbation of the steady homogeneous state grows in amplitude, suitable nonlinear terms may come into play and prevent the indefinite growth of the perturbation.

In the particular case of the kernel function expressed by (B.13), the dispersion relation becomes

$$\gamma(k) = f'(\phi_0) + \pi b_1 q_1^2 \exp\left(-\frac{q_1^2 k^2}{2}\right) - \pi b_2 q_2^2 \exp\left(-\frac{q_2^2 k^2}{2}\right); \qquad \text{(B.17)}$$

the most unstable mode is

$$k_{max} = q_1 \sqrt{\frac{2\ln(\epsilon \chi^4)}{\chi^2 - 1}}, \qquad \text{(B.18)}$$

with $\epsilon = b_2/b_1$ and $\chi = q_2/q_1$.

As noted, one of the first models of vegetation self-organization (Lefever and Lejeune, 1997) used a kernel-based framework that resembles that of Eq. (B.12), with spatial interactions involving both short-range activation and long-range inhibition. The model by Lefever and Lejeune (1997) differs from a neural model in that the nonlinearities are not strictly local but modulate the spatial interactions.

B.3.1 Biharmonic approximation of neural models

In Chapter 6, Section 6.2, we presented an approximated representation of Eq. (B.12), based on a Taylor expansion of the integral term in the neighborhood of $\phi = \phi_0$ (i.e., for small values of z). If the kernel function has axial symmetry and only the terms up to the fourth order are retained, Eq. (B.12) can be approximated as (Murray, 2002)

$$\frac{\partial \phi}{\partial t} \approx f(\phi) + \omega_0(\phi - \phi_0) + \omega_2 \nabla^2 \phi + \omega_4 \nabla^4 \phi, \tag{B.19}$$

where the terms on the right-hand-side are the same as in Eq. (6.2) and w_m is the mth-order moment of the kernel function defined in Eq. (6.3). In the case of axial symmetric kernel functions, the odd-order moments of $\omega(z)$ are zero.

It can be shown that a second-order expansion (i.e., a Fisher equation) is unable to lead to persistent patterns (e.g., Murray, 2002) and that the biharmonic term $\omega_4 \nabla^4 \phi$ is needed in the series expansion to obtain [with approximation (B.19)] deterministic patterns that do not vanish with time. In fact, linear-stability analysis of the state $\phi = \phi_0$ with respect to a perturbation $\hat{\phi}(\mathbf{r}, t) \propto e^{\gamma t + i \mathbf{k} \cdot \mathbf{r}}$ leads to the dispersion relation:

$$\gamma(k) = f'(\phi_0) + \omega_0 - 2\omega_2 k^2 + 4\omega_4 k^4. \tag{B.20}$$

In the absence of the biharmonic term (i.e., when $\omega_4 = 0$), the most unstable mode $k_{\max}$ is zero and no patterns emerge. In the case of biharmonic approximation (B.19) (i.e., when $\omega_4 \neq 0$), the most unstable mode can be easily obtained from Eq. (B.20) as $k_{\max} = 0.5\sqrt{\omega_2/\omega_4}$. Patterns emerge when $k_{\max}$ is real and different from zero (i.e., ω_2 and ω_4 need to have the same sign), and $\gamma(k_{\max}) > 0$:

$$\gamma(k_{\max}) = f'(\phi_0) + \omega_0 - \frac{\omega_2^2}{4\omega_4} > 0. \tag{B.21}$$

In addition, the stability of $\phi = \phi_0$ in the absence of spatial dynamics requires $f'(\phi_0)$ to be negative, as in the first of conditions (B.16). Moreover, in many applications ϕ is always nonnegative. This condition is met when $\omega_0 < 0$. Because in this case ω_0 and $f'(\phi_0)$ are both negative, Eq. (B.21), combined with the requirement that ω_2 and ω_4 have the same sign, implies that pattern formation occurs only if ω_2 and ω_4 are also negative. However, the condition that ω_0, ω_2, and ω_4 are negative is only necessary and not sufficient for pattern formation as condition (B.21) would still need to be met for the instability to emerge.

B.4 Patterns emerging from differential-flow instability

The third major deterministic model of self-organized pattern formation associated with symmetry-breaking instability is due to differential flow. This mechanism resembles Turing's dynamics, in that it involves two diffusing species, u and v (activator and inhibitor, respectively). However, unlike Turing's model, diffusion is not important to the destabilization of the homogeneous state. In this case, one or both species are subjected to advective flow (or "drift"), and instability emerges as a result of the differential-flow rate of the two species (Rovinsky and Menzinger, 1992). Although diffusion is not fundamental to the emergence of differential-flow instability, it plays a crucial role in imposing an upper bound to the range of unstable modes k and determines the wavelength of the most unstable mode (Rovinsky and Menzinger, 1992). As a result of the drift, patterns generated by this process are not time independent, as are those associated with Turing's instability. Rather, they exhibit traveling waves in the flow direction. Self-organized patterns of this type have been observed in nature, mainly in chemical systems ("the Belousov–Zhabotinsky reaction," Rovinsky and Zhabotinsky, 1984). The same mechanism was also invoked to explain ecological patterns subject to drift, including banded vegetation (Klausmeier, 1999; Okayasu and Aizawa, 2001; von Hardenberg et al., 2001; Shnerb et al., 2003; Sherrat, 2005). We note that this mechanism of pattern formation induced by differential flow is often classified as a Turing model in that in both models the dynamics can be expressed by the same set of reaction–advection–diffusion equations. In the case of Turing models, instability is induced by the Laplacian term, whereas in the case of differential-flow, instability it is the gradient term that causes instability. For the sake of clarity here we discuss the case of differential-flow instability separately.

We introduce the mathematical model of differential-flow instability (e.g., Rovinsky and Menzinger, 1992) assuming that only one of the two species undergoes a drift, and we orient the x axis in the direction of the advective flow. The activator-inhibitor dynamics can be expressed as

$$\frac{\partial u}{\partial t} = f(u, v) + p\frac{\partial u}{\partial x} + d_1 \nabla^2 u,$$

$$\frac{\partial v}{\partial t} = g(u, v) + d_2 \nabla^2 v, \tag{B.22}$$

where p is the drift velocity, and d_1 and d_2 are the diffusivities of u and v, respectively. Notice that when $p = 0$ Eqs. (B.22) can be written in the same form as Eqs. (B.1).

When $p \neq 0$ the conditions on d_1 and d_2 for the emergence of patterns from Eqs. (B.22) are less restrictive than those for Turing's instability. To stress the fact that patterns emerge from the differential-flow rates of u and v, we first consider the conditions leading to instability in the absence of diffusion and set $d_1 = d_2 = 0$. The homogeneous steady state (u_0, v_0), obtained as solution of the equation set $f(u_0, v_0) = g(u_0, v_0) = 0$, is stable when conditions (B.5) are met. To determine the conditions in which the differential flow destabilizes the state (u_0, v_0), we linearize $f(u, v)$

and $g(u, v)$ around (u_0, v_0), and seek solutions of the linearized equations in the form of

$$u = \hat{u} + u_0,$$
$$v = \hat{v} + v_0. \tag{B.23}$$

We obtain

$$\frac{\partial \hat{u}}{\partial t} = f_u \hat{u} + f_v \hat{v} + p\frac{\partial \hat{u}}{\partial x},$$

$$\frac{\partial \hat{v}}{\partial t} = g_u \hat{u} + g_v \hat{v}, \tag{B.24}$$

where $f_u = \partial f(u, v)/\partial u$, $f_v = \partial f(u, v)/\partial v$, $g_u = \partial g(u, v)/\partial u$, and $g_u = \partial g(u, v)/\partial v$. The solution of equation set (B.24) can be expressed as a sum (or integral sum in spatially infinite domains) of Fourier modes $\hat{u}_k = U_k \exp(\gamma t + i\mathbf{k} \cdot \mathbf{r})$ and $\hat{v}_k = V_k \exp(\gamma t + i\mathbf{k} \cdot \mathbf{r})$, where U_k and V_k are the Fourier coefficients of the kth mode. Because Eqs. (B.24) need to be satisfied for each mode k, we have

$$\gamma U_k = f_u U_k + f_v V_k + ipU_k k_x,$$
$$\gamma V_k = g_u U_k + g_v V_k. \tag{B.25}$$

Nontrivial solutions of equation set (B.25) exist when its determinant is zero:

$$\gamma^2 - (f_u + g_v + ipk_x)\gamma + f_u g_v - f_v g_u + ipk_x g_v = 0. \tag{B.26}$$

Notice how in this case γ is a complex number. The emergence of instability requires the real part of γ to be positive. Traveling-wave patterns require the imaginary part of γ to be different from zero. It has been noticed (Rovinsky and Menzinger, 1992) that Eq. (B.26) does not lead to the selection of any finite value for the most unstable wave number in that γ is a monotonically increasing function of k. Thus the wave-number interval of the unstable modes has no upper bound. However, the addition to Eqs. (B.25) of a diffusion term to either the first or the second equation [or to both, as in Eqs. (B.22)] imposes an upper bound to the range of unstable modes. In this case the most unstable mode corresponds to a finite value of the wave number.

B.4.1 Case study: A differential-flow ecological model of pattern formation

We present, as an example of differential-flow instability, a model developed to study the formation of patterns in young mussel beds (van de Koppel et al., 2005). This model was used by Borgogno et al. (2009) to describe a system involving trees or grasses. Two (dimensionless) state variables, representing nutrient concentration

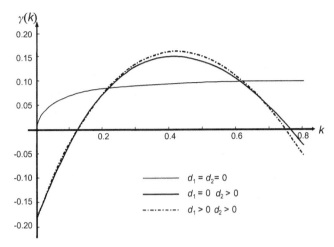

Figure B.5. Different dispersion relations for the case of differential-flow instability, Eqs. (B.27). The parameters for the bold solid curve are $\varphi = 0.72, \eta = 6.10, \delta = 5.14$, $p = -1.315, d_1 = 0$, and $d_2 = 1$.

u and vegetation density v, are used. The dynamics of the two variables are expressed as

$$\frac{\partial u}{\partial t} = \varphi(1 - u) - uv + p\nabla u + d_1 \nabla^2 u,$$

$$\frac{\partial v}{\partial t} = \eta uv - \delta \frac{v}{1 + v} + d_2 \nabla^2 v. \tag{B.27}$$

The first term on the right-hand side of the first equation represents the rate of increase in nutrient concentration, the second term accounts for the consumption of nutrient by biomass, and the third term is the loss of nutrients by advection; the last term models the spreading of u by diffusion. The first term on the right-hand side of the second equation represents the nutrient-dependent rate of biomass growth; the second term represents the state-dependent mortality rate, and the third term accounts for the diffusion-like spatial spreading of biomass. The steady homogeneous state $[u_0 = \eta\varphi - \delta/(\eta(\varphi - 1)), v_0 = \varphi(\delta - \eta)/(\eta\varphi - \delta)]$ is stable in the absence of drift and diffusion when conditions (B.5) are met. Drift-induced instability occurs if the drift term is able to destabilize the homogeneous state (u_0, v_0) even when the Laplacian terms are set equal to zero (see Section B.4). In this case dispersion relation (B.26) provides the range of Fourier modes that are destabilized by drift (see Fig. B.5). As noted by Rovinsky and Menzinger (1992), in the absence of a diffusion term the interval of the unstable modes has no upper bound [Eq. (B.26)]. When a diffusion term is added to the first equation (i.e., $d_1 \neq 0$), the dispersion relation becomes

$$\gamma^2 + (d_1 k^2 - f_u - g_v - ipk_x)\gamma + (f_u g_v - f_v g_u + ipk_x g_v - d_1 k^2 g_v) = 0. \tag{B.28}$$

Figure B.6. Spatial pattern emerging for the variable u using model (B.22). The parameters are $\eta = 6.10$, $\delta = 5.14$, $\varphi = 0.72$, $p = -1.315$, $d_1 = 0$, and $d_2 = 1$. The simulation is carried out over a domain of 256×256 cells, each cell representing a spatial step $\Delta x = \Delta y = 0.8$.

The plot of this relation (see Fig. B.5) shows that in this case the interval of the unstable wave numbers has an upper bound and the most unstable mode has a finite wave number. When a diffusive term is also added to the second equation (i.e., $d_1 \neq 0$, $d_2 \neq 0$) as in Eqs. (B.27), the dispersion relation becomes

$$\gamma^2 + (d_1 k^2 + d_2 k^2 - f_u - g_v - i p k_x)\gamma \qquad (B.29)$$
$$+ (f_u g_v - f_v g_u + i p k_x g_v - i d_2 p k_x k^2 - d_1 k^2 g_v - d_2 k^2 f_u + d_1 d_2 k^4) = 0,$$

with no substantial differences in the amplitude of the interval of unstable modes (see Fig. B.5).

An example of spatial patterns emerging with this model is shown in Fig. B.6.

Appendix C

List of symbols and acronyms

C.1 Greek symbols

α	average of the random heights of jumps in the compound Poisson process
γ	growth factor
$\Gamma[\cdot]$	gamma function
$\delta[\cdot]$	Dirac delta function
Δ_1, Δ_2	states of dichotomic noise
θ	threshold value
$\Theta[\cdot]$	unit step function
$\kappa_{1dn}, \kappa_{2dn}$	first and second cumulants of dichotomous noise
$\kappa_{1gn}, \kappa_{2gn}$	first and second cumulants of Gaussian white noise
$\kappa_{1sn}, \kappa_{2sn}$	first and second cumulants of shot noise
λ	rate of the shot-noise process
ξ	noise term
ξ_{cn}	Gaussian colored-noise process
ξ_{dn}	dichotomous-noise process
ξ_{gn}	Gaussian-white-noise process
ξ_{ou}	Ornstein–Uhlenbeck process
$\xi_{per}(t)$	periodical zero-mean forcing
ξ_{sn}	shot-noise process
ξ'_{sn}	zero-average shot-noise process
$\rho(\cdot)$	drift term in the additive-shot-noise process
τ_1, τ_2	mean permanence time in a state of the dichotomic noise
τ_c	relaxation time
τ_{ou}	correlation time of the Ornstein–Uhlenbeck process
ϕ	state variable
ϕ_m	mode or antimode of the steady-state pdf of ϕ

ϕ_{st}	stationary point of the ϕ dynamics
ϕ_-	lower boundary of the ϕ dynamics
ϕ_+	upper boundary of the ϕ dynamics
$\omega(\mathbf{r})$	kernel function in the integral operator

C.2 Latin symbols

$B_1[\cdot]$	first-order modified Bessel function
C	integration constant
D	coefficient expressing the strength of the spatial coupling
$f(\phi)$	deterministic function (drift term)
$f_1(\phi)$, $f_2(\phi)$	drift term in states 1 and 2
$F(t)$	time-dependent forcing
$g(\phi)$	deterministic function (multiplicative term)
$g_S(\phi)$	deterministic function arising from the relation $\langle g(\phi)\xi(t)\rangle = s\,\langle g_S(\phi)\rangle$
$G(\mathbf{r})$	correlation function
h	random height of a jump in the compound Poisson process
$h_{\mathcal{L}}$	Fourier transform of the spatial operator $\mathcal{L}(\phi)$
$\mathcal{I}$	integral scale
i_p	intensity of the peak in the structure function $S(\mathbf{k})$
$\mathbf{k}$	wave-number vector, $\mathbf{k} = (k_x, k_y)$
k_1, k_2	transition rates of dichotomous noise
K_{dn}	cumulant generating function of dichotomous noise
$\mathcal{L}[\phi]$	operator expressing the spatial coupling of the ϕ dynamics
$l(\phi_i, \phi_j)$	finite-difference equivalent of $\mathcal{L}[\phi]$
M_{dn}	moment-generating function of dichotomous noise
n_1, n_2	residence time in the above- (below-) threshold state
$P_1(t)$, $P_2(t)$	state probability of dichotomous noise
$p_\Phi(\phi)$ or $p(\phi)$	probability density function (pdf) of the variable ϕ
$P_\Phi(\phi)$ or $P(\phi)$	cumulative distribution function of the variable ϕ
$q(t)$	external random driver
$\mathbf{r}$	coordinate vector, $\mathbf{r} = (x, y)$
s_{dn}	intensity of dichotomous noise
s_{gn}	intensity of Gaussian white noise
s_{sn}	intensity of shot noise
$S(\mathbf{k})$	structure function
t	time
$V(\phi)$	deterministic potential of the state variable
$\mathcal{V}(\phi)$	probabilistic potential of the state variable
$V_1(\phi)$, $V_2(\phi)$	potential of the state variable in states 1 and 2

x, y	spatial coordinates
$W(t)$	Wiener process
z	distance function, $z = \|\mathbf{r} - \mathbf{r}'\|$
$Z(t)$	compound Poisson process

C.3 Mathematical symbols

∂	partial derivative
d	ordinary derivative
$'$	derivative with respect to the argument, e.g.,
	$F'(\phi_m) = \mathrm{d}F(\phi)/\mathrm{d}\phi\|_{\phi=\phi_m}$
ϱ	functional derivative
$\langle \cdot \rangle$	expectation
$\|\cdot\|$	absolute value
$\sum$	summation
$\circ(\cdot)$	term of a lower order of magnitude of the argument
∇^2	Laplacian operator, $\nabla^2\phi = \partial^2\phi/\partial x^2 + \partial^2\phi/\partial y^2$
∇^4	biharmonic operator, $\nabla^4\phi = \partial^4\phi/\partial x^4 + 2\partial^4\phi/\partial x^2\partial y^2 + \partial^4\phi/\partial y^4$

C.4 Acronyms

1D	one-dimensional
2D	two-dimensional
3D	three-dimensional
CDIMA	chlorine dioxide iodide malonic acid
CV	coefficient of variation
DMN	dichotomous Markov noise
DO	Dansgaard–Oeschger
O-U	Ornstein–Uhlenbeck
pdf	probability density function
SNR	signal-to-noise ratio
SST	sea-surface temperature
UCNA	unified colored-noise approximation
VPT	van den Broeck–Parrondo–Toral
WSN	white shot noise

Bibliography

Aguiar, M. R. and Sala, O. E. (1999). Patch structure, dynamics and implications for the functioning of arid ecosystems. *Tree* **14**: 273–277.

Aguilera, M. O. and Lauenroth, W. K. (1993). Seedling establishment in adult neighborhoods intraspecific constraints in the regeneration of the bunchgrass Bouteloua-gracilis. *J. Ecol.* **81**: 253–261.

Ahnert, F. (1988). Modelling landform change. In Anderson, M., editor, *Modelling Geomorphological Systems*. Wiley, New York, pp. 375–400.

Ajdai, A. and Prost, J. (1992). Mouvement induit par un potentiel periodique de basse symmetrie: dielectrophorese pulsee. *C. R. Acad. Sci. Paris Sér. II*, **315**: 1635.

Alfieri, L., Claps, P., D'Odorico, P., Laio, F., and Over, T. M. (2008). An analysis of the soil moisture feedback on convective and stratiform precipitation. *J. Hydrometeorol.* **9**: 280–291.

Allen, J. (1984). *Sedimentary Structures: Their Character and Physical Basis*. Elsevier, Amsterdam.

Alley, R. B., Anandakrishnan, S., and Jung, P. (2001). Stochastic resonance in the North Atlantic. *Paleoceanography* **16**: 190–198.

Anderies, J. M., Janssen, M. A., and Walker, B. H. (2002). Grazing, management, resilience and the dynamics of fire-driven rangeland system. *Ecosystems* **5**: 23–44.

Archer, S. (1989). Have Southern Texas savannas been converted to woodlands in recent history? *Am. Naturalist* **134**: 545–561.

Archer, S., Scifres, C., Bassham, C. R., and Maggio, R. (1988). Autogenic succession in a subtropical savanna: Conversion of grassland to thorn woodland. *Ecol. Monogr.* **58**: 111–127.

Argyris, J., Faust, G., and Haase, M. (1994). *An Exploration of Chaos*. North-Holland, Amsterdam.

Ashton, A., Murray, A. B., and Arnault, O. (2001). Formation of coastline features by large-scale instabilities induced by high-angle waves. *Nature (London)* **414**: 296–300.

Astunian, R. D. and Hanggi, P. (2002). Brownian motors. *Phys. Today* **55**: 33–39.

Bak, P. (1996). *How Nature Works: The Science of Self-Organized Criticality*. Copernicus/Springer-Verlag, New York.

Barbier, N., Couteron, P., Lefever, R., Dublauwe, V., and Lejeune, O. (2008). Spatial decoupling of facilitation and competetion at the origin of gapped vegetation patterns. *Ecology* **89**: 1521–1531.

Barbier, N., Couteron, P., Lejoly, J., Deblauwe, V., and Lejeune, O. (2006). Self-organized vegetation patterning as a fingerprint of climate and human impact on semi-arid ecosystems. *J. Ecol.* **94**: 537–547.

Barzykin, A., Seki, K., and Shibata, F. (1998). Periodically driven linear system with multiplicative colored noise. *Phys. Rev. E* **57**: 6555–6563.

Beck, C. (2004). Superstatistics: Theory and applications. *Continuum Mech. Thermodyn.* **16**: 293–304.

Bellwood, D. R., Houghes, T. P., Folke, C., and Nystrom, M. (2004). Confronting the coral reef crisis. *Nature (London)* **429**: 827–833.

Belsky, A. J. (1994). Influences of trees on savanna productivity – Tests of shade, nutrients, and tree-grass competition. *Ecology* **75**: 922–932.

Bena, I. (2006). Dichotomous Markov noise: Exact results for out-of-equilibrium systems. *Int. J. Mod. Phys.* **20**: 2825–2888.

Bena, I., Van den Broeck, C., Kawai, R., and Lindenberg, K. (2003). Drift by dichotomous Markov noise. *Phys. Rev. E* **68**: 041111.

Benda, L. and Dunne, T. (1997). Stochastic forcing of sediment supply to channel networks from landsliding and debris flow. *Water Resources Res.* **33**: 2849–2863.

Benedetti-Cecchi, L., Bertocci, I., Vaselli, S., and Maggi, E. (2006). Temporal variance reverses the impact of high mean intensity of stress in climate change experiments. *Ecology* **87**: 2489–2499.

Bengtsson, J., Angelstam, P., Elmqvist, T., Emanuelsson, U., Folke, C., Ihse, M., Moberg, F., and Nystrom, M. (2003). Reserves, resilience and dynamic landscapes. *Ambio* **32**: 389–396.

Benzi, R., Parisi, G., Sutera, A., and Vulpiani, A. (1982a). Stochastic resonance in climatic changes. *Tellus* **34**: 10–16.

Benzi, R., Parisi, G., Sutera, A., and Vulpiani, A. (1982b). A theory of stochastic resonance in climate change. *SIAM J. Appl. Math.* **43**: 565–578.

Berdichevsky, V. and Gitterman, M. (1999). Stochastic resonance in linear systems subject to multiplicative and additive noise. *Phys. Rev. E* **60**: 1494–1499.

Bergkamp, G., Cerdá, A., and Imeson, A. C. (1999). Magnitude-frequency analysis of water redistribution along a climate gradient in Spain. *Catena* **37**: 129–146.

Bernd, J. (1978). The problem of vegetation stripes in semi-arid Africa. *Plant. Res. Develop.* **8**: 37–50.

Bertocci, I., Vaselli, S., Maggi, E., and Benedetti-Cecchi, L. (2007). Changes in temporal variance of rocky shore organism abundances in response to manipulation of mean intensity and temporal variability of aerial exposure. *Marine Ecol. Progr. Ser.* **338**: 11–20.

Bhark, E. W. and Small, E. E. (2003). Association between plant canopies and the spatial patterns of infiltration in shrubland and grassland of the Chihuahuan Desert, New Mexico. *Ecosystems* **6**: 185–196.

Bharucha-Reid, A. T. (1960). *Elements of the Theory of Markov Processes and Their Applications*. McGraw-Hill, New York.

Bird, E. (2000). *Coastal Geomorphology: An Introduction*. Wiley, Chichester, UK.

Boeken, B. and Orenstein, D. (2001). The effect of plant litter on ecosystem properties in a Mediterranean semi-arid shrubland. *J. Veg. Sci.* **12**: 825–832.

Borgogno, F., D'Odorico, P., Laio, F., and Ridolfi, L. (2007). The effect of rainfall interannual variability on the stability and resilience of dryland plant ecosystems. *Water Resource Res.* **43**: W06411.

Borgogno, F., D'Odorico, P., Laio, F., and Ridolfi, L. (2009). Mathematical models of vegetation pattern formation in ecohydrology. *Rev. Geophys.* **47**: RG1005.

Bras, R. L. and Rodriguez-Iturbe, I. (1994). *Random Functions and Hydrology*. Dover, New York.

Braumann, C. A. (1983). Populations growth in random environments. *Bull. Mater. Biol.* **45**: 635–641.

Braumann, C. A. (2007). Harvesting in a random environment: Ito or Stratonovich calculus. *J. Theor. Biol.* **244**: 424–432.

Braun, H., Christi, M., Rahmstorf, S., Ganopolski, A., Mangini, A., Kubatzki, C., Roth, K., and Kromer, B. (2005a). Possible solar origin of the 1,470-year glacial climate cycle demonstrated in a coupled model. *Nature (London)* **438**: 208–211.

Braun, H., Ganopolski, A., Christi, M., and Chialvo, D. R. (2005b). A simple conceptual model of abrupt glacial climate events. *Nonlin. Process. Geophys.* **14**: 709–721.

Breman, H. and Kessler, J. (1995). *Woody Plants in Agro-Ecosystems of Semi-Arid Regions*. Springer-Verlag, New York.

Breshears, D. D., Myers, O. B., Johnson, S. R., Meyer, C. W., and Martens, S. N. (1997). Differential use of spatially heterogeneous soil moisture by two semiarid woody species: Pinus edulis and Juniperus monosperma. *J. Ecol.* **85**: 289–299.

Brovkin, V., Claussen, M., Petoukhov, V., and Ganopolski, A. (1998). On the stability of the atmosphere-vegetation system in the sahara/sahel region. *J. Geophys. Res.* **103**: 31613–31624.

Brubaker, K. L. and Entekhabi, D. (1995). An analytic approach to modeling land atmosphere interaction. 1. Construct and equilibrium behavior. *Water Resource Res.* **31**: 619–632.

Bruno, J. F. (2000). Facilitation of cobble beach plant communities through habitat modification by spartina alterniflora. *Ecology* **81**: 1179–1192.

Buceta, J., Ibanes, M., Sancho, J., and Lindenberg, K. (2003). Noise-driven mechanism for pattern formation. *Phys. Rev. E* **67**: 021113.

Buceta, J. and Lindenberg, K. (2002a). Stationary and oscillatory spatial patterns induced by global periodic switching. *Phys. Rev. Lett.* **88**: 024103-1–024103-4.

Buceta, J. and Lindenberg, K. (2002b). Switching-induced Turing instability. *Phys. Rev. E* **66**: 0462022-1–046202-6.

Buceta, J. and Lindenberg, K. (2003). Spatial patterns induced by purely dichotomous disorder. *Phys. Rev. E* **68**: 011103.

Buceta, J. and Lindenberg, K. (2004). Comprehensive study of phase transitions in relaxational systems with field-dependent coefficients. *Phys. Rev. E* **69**: 011102.

Buceta, J., Lindenberg, K., and Parrondo, J. M. R. (2002a). Pattern formation induced by nonequilibrium global alternation of dynamics. *Phys. Rev. E* **66**: 036216-1–036216-11.

Buceta, J., Lindenberg, K., and Parrondo, J. M. R. (2002b). Spatial patterns induced by random switching. *Fluctuation and Noise Letters* **2**: L21–L29.

Budyko, M. I. (1969). The effect of solar radiation variations on the climate of the earth. *Tellus* **21**: 611–619.

Burke, I. C., Lauenroth, W. K., Vinton, M. A., Hook, P. B., Kelly, R. H., Epstein, H. E., Aguiar, M. R., Robles, M. D., Aguilera, M. O., Murphy, K. L., and Gill, R. A. (1998). Plant–soil interactions in temperate grasslands. *Biogeochemistry* **42**(1-2): 121–143.

Butler, T. and Goldenfeld, N. (2009). Robust ecological pattern formation induced by demographic noise. *Phys. Rev. E.* **80**: 030902(R).

Caddemi, S. and Di Paola, M. (1996). Ideal and physical white noise in stochastic analysis. *Int. J. Non-Linear Mech.* **31**: 581–590.

Campbell, N. (1909a). Discontinuities in light emission. *Proc. Cambr. Philos. Soc.* **15**: 310–338.

Campbell, N. (1909b). The study of discontinuous phenomena. *Proc. Cambr. Philos. Soc.* **15**: 117–136.

Camporeale, C. and Ridolfi, L. (2006). Riparian vegetation distribution induced by river flow variability: A stochastic approach. *Water Resource Res.* **42**: W10415.

Carrillo, O., Ibanes, M., Garcia-Ojalvo, J., Casademunt, J., and Sancho, J. M. (2003). Intrinsic noise-induced phase transitions: Beyond the noise interpretation. *Phys. Rev. E* **67**: Article 046110.

Carrillo, O., Santos, M. A., Garcia-Ojalvo, J., and Sancho, J. M. (2004). Spatial coherence resonance near pattern-forming instabilities. *Europhys. Lett.* **65**: 452–458.

Carson, M. A. and Kirkby, M. J. (1972). *Hillslope Form and Process.* Cambridge University Press, Cambridge.

Casper, B. B., Schenk, H. J., and Jackson, R. B. (2003). Defining a plant's belowground zone of influence. *Ecology* **84**: 2313–2321.

Castets, V., Dulos, E., Boissonade, J., and Dekepper, P. (1990). Experimental evidence of a sustained standing Turing-type nonequilibrium chemical-pattern. *Phys. Rev. Lett.* **64**: 2953–2956.

Caylor, K. K., D'Odorico, P., and Rodriguez-Iturbe, I. (2006). On the ecohydrology of structurally heterogeneous semiarid landscapes. *Water Resource Res.* **42**(W07424): doi:10.1029/2005WR04683.

Caylor, K. K., Shugart, H. H., and Rodriguez-Iturbe, I. (2005). Tree canopy effects on simulated water stress in Southern African savannas. *Ecosystems* **8**: 17–32.

Chandrasekhar, S., editor (1981). *Hydrodynamic and Hydromagnetic Stability.* Clarendon, Oxford.

Chang, M. (2002). *Forest Hydrology: An Introduction to Water and Forests.* CRC Press, Boca Raton, FL.

Chapin F. S., III, Walker, B. H., Hobbs, R. J., Hooper, D. U., Lawton, J. H., Sala, O. E., and Tilman, D. (1997). Biotic controls of the functioning of ecosystems. *Science* **277**: 500–504.

Charley, J. (1972). The role of shrubs in nutrient cycling. In *Wildland Shrubs: Their Biology and Utilization*, pp. 182–203. U. S. Department of Agriculture, Forest Service, General Technical Report INT-1.

Charley, J. and West, N. E. (1975). Plant-induced soil chemical patterns in some shrub-dominated semi-desert ecosystem of Utah. *J. Ecol.* **63**: 945–963.

Charney, J. C. (1975). Dynamics of deserts and droughts in the Sahel. *Q. J. R. Meteorol. Soc.* **101**: 193–202.

Chesson, P. L. (1982). The stabilizing effect of a random environment. *J. Math. Biol.* **15**: 1–36.

Chesson, P. L. (1994). Multispecies competition in variable environments. *Theor. Popul. Biol.* **45**: 227–276.

Chesson, P. L. (2000). Mechanisms of maintenance of species diversity. *Annu. Rev. Ecol. Syst.* **31**: 343–366.

Christophorov, L. N. (1996). Dichotomous noise with feedback and charge-conformational interactions. *J. Biol. Phys.* **22**: 197–208.

Colombini, M. and Stocchino, A. (2008). Finite-amplitude river dunes. *J. Fluid Mech.* **611**: 283–306.

Connell, J. H. (1978). Diversity in tropical rain forests and coral reefs: High diversity of trees and corals is maintained only in a non-equilibrium state. *Science* **199**: 1302–1310.

Couteron, P. and Lejeune, O. (2001). Periodic spotted patterns in semi-arid vegetation explained by a propagation-inhibition model. *Ecology* **89**: 616–628.

Cox, D. R. and Miller, H. D. (1965). *The Theory of Stochastic Processes.* Methuen, London.

Cross, M. C. and Hohenberg, P. C. (1993). Pattern formation outside of equilibrium. *Rev. Mod. Phys.* **65**: 851–1112.

Dale, M. (1999). *Spatial Pattern Analysis in Plant Ecology.* Cambridge University Press, Cambridge.

Daly, E. and Porporato, A. (2006). State-dependent fire models and related renewal processes. *Phys. Rev. E* **74**: 041112.

Daly, E. and Porporato, A. (2007). Intertime jump statistics of state-dependent Poisson processes. *Phys. Rev. E* **75**: 011119.

De Angelis, D. L. (1975). Stability and connectance in food web models. *Ecology* **56**: 238–243.

DeLonge, M., D'Odorico, P., and Lawrence, D. (2008). Feedbacks between phosphorous deposition and canopy cover: The emergence of multiple states in dry tropical forests. *Global Change Biol.* **14**: 154–160.

Demaree, G. R. and Nicolis, C. (1990). Onset of sahelian drought viewed as a fluctuation-induced transition. *Q. J. R. Meteorol. Soc.* **116**: 221–238.

Denisov, S., Horsthemke, W., and Hanggi, P. (2009). Generalized Fokker–Planck equation: Derivation and exact solutions. *Eur. Phys. J. B* **68**: 567–575.

Dietrich, W. E., Reiss, R., Hsu, M. L., and Montgomery, D. (1995). A process-based model for colluvial soil depth and shallow lanslides using digital elevation data. *Hydrol. Process* **9**: 383–400.

DiPrima, R. C. and Swinney, H. L. (1981). Instabilities and trasition in flow between concentric rotating cylinders. In Swinney, H. L. and Gollub, J. P., editors, *Hydrodynamic Instabilities and the Transition to Turbulence*, Springer-Verlag, Berlin, p. 139.

Ditlevsen, P. D., Kristensen, M. S., and Andersen, K. (2005). The recurrence time of Dansgaard-Oeschger events and limits on the possible periodic component. *J. Clim.* **18**: 2594–2603.

D'Odorico, P. (2000). A possible bistable evolution of soil thickness. *J. Geophys. Res. (Solid Earth)* **105**(B11): 25927–25935.

D'Odorico, P., Caylor, K., Okin, G. S., and Scanlon, T. M. (2007a). On soil moisture-vegetation feedbacks and their possible effects on the dynamics of dryland ecosystems. *J. Geophys. Res.* **112**: G04010.

D'Odorico, P. and Fagherazzi, S. (2003). A probabilistic model of rainfall-triggered shallow landslides in zero-order basins. *Water Resource Res.* **39**: 1262.

D'Odorico, P., Laio, F., Porporato, A., Ridolfi, L., and Barbier, N. (2007b). Noise-induced vegetation patterns in fire-prone savannas. *Biogeosciences* **112**: G02021.

D'Odorico, P., Laio, F., and Ridolfi, L. (2005). Noise-induced stability in dryland plant ecosystems. *Proc. Natl. Acad. Sci. USA* **102**: 10819.

D'Odorico, P., Laio, F., and Ridolfi, L. (2006a). Patterns as indicators of productivity enhancement by facilitation and competition in dryland vegetation. *J. Geophys. Res. Biogeosi.* **111**: Article G03010.

D'Odorico, P., Laio, F., and Ridolfi, L. (2006b). A probabilistic analysis of fire-induced tree-grass coexistence in savannas. *Am. Naturalist* **167**: E79–E87.

D'Odorico, P., Laio, F., and Ridolfi, L. (2006c). Vegetation patterns induced by random climate fluctuations. *Geophys. Res. Lett.* **33**: L19404.

D'Odorico, P., Laio, F., Ridolfi, L., and Lerdau, M. T. (2008). Biodiversity enhancement induced by environmental noise. *J. Theor. Biol.* **255**: 332–337.

D'Odorico, P. and Porporato, A. (2004). Preferential states in soil moisture and climate dynamics. *Proc. Natl. Acad. Sci. USA* **101**: 8848–8851, 10.1073/pnas.0401428101.

D'Odorico, P. and Porporato, A., editors (2006). *Dryland Ecohydrology*. Springer-Verlag, New York.

D'Odorico, P., Porporato, A., and Ridolfi, L. (2001). Transitions between stable states in the dynamics of soil development. *Geophys. Res. Lett.* **28**: 595–598.

D'Odorico, P., Ridolfi, L., Porporato, A., and Rodriguez-Iturbe, I. (2000). Preferential states of seasonal soil moisture: the impact of climate fluctuations. *Water Resource Res.* **36**: 2209–2219.

Doering, C. and Horsthemke, W. (1985). A comparison between transitions induced by random and periodic fluctuations. *J. Stat. Phys.* **38**: 763–783.

Doob, J. (1942). The Brownian movement and stochastic equations. *Ann. Math.* **43**: 351–369.

Drazin, P. (2002). *Introduction to Hydrodynamic Stability*. Cambridge University Press, New York.

Dubois-Violette, E., Durand, G., Guyon, E., Manneville, P., and Pieranski, P. (1978). Instabilities in nematic liquid crystals. In Liebert, L., editor, *Solid State Physics*. Academic, New York, Vol. Suppl. 14, p. 147.

Dunkerley, D. L. (1997a). Banded vegetation: Development under uniform rainfall from a simple cellular automaton model. *Plant Ecol.* **129**: 103–111.

Dunkerley, D. L. (1997b). Banded vegetation: Survival under drought and grazing pressure based on a simple cellular automaton model. *J. Arid Environ.* **35**: 419–428.

Dunkerley, D. L. and Brown, K. J. (1999). Banded vegetation near Broken Hill, Australia: Significance of surface roughness and soil physical properties. *Catena* **37**(1-2): 75–88.

Dutta, S., Riaz, S., and Ray, D. (2005). Noise-induced instability: An approach based on higher-order moments. *Phys. Rev. E* **71**: 036216.

Easterling, D. R., Evans, J. L., Groisman, P., Karl, T. R., Kunkel, K. E., and P. Ambenje, K. E. (2000a). Observed variability and trends in extreme climate events: A brief review. *Bull. Am. Meteorol. Soc.* **81**: 417–425.

Easterling, D. R., Meehl, G. A., Parmesan, C., Changnon, S. A., Karl, T. R., and Mearns, L. O. (2000b). Climate extremes: Observations, modeling, and impacts. *Science* **289**: 2068–2074.

Eddy, J., Humphreys, G. S., Hart, D. M., Mitchell, P. B., and Fanning, P. C. (1999). Vegetation arcs and litter dams: Similarities and differences. *Catena* **37**(1-2): 57–73.

Edwards, S. F. and Wilkinson, D. R. (1982). The surface statistics of a granular aggregate. *Proc. R. Soc. London Ser. A* **381**: 17–31.

Elbelrhiti, H., Claudin, P., and B., A. (2005). Field evidence of surface-wave-induced instability of sand dunes. *Nature (London)* **437**: 720–723.

Eldridge, D. J. and Greene, R. S. B. (1994). Microbiotic soil crusts – A review of their roles in soil ecological processes in the rangelands of Australia. *Austr. J. Soil Res.* **32**: 389–415.

Elmqvist, T., Folke, C., Nystrom, M., Peterson, G., Bengtsson, L., Walker, B., and Norberg, J. (2003). Response diversity and ecosystem resilience. *Fronteers Ecol. Environ.* **1**: 488–494.

Eltahir, E. A. B. and Bras, R. L. (1996). Precipitation recycling. *Rev. Geophys.* **34**: 367–378.

Eppinga, M., Rietkerk, M., Borren, W., Lapshina, E., Bleuten, W., and Wassen, M. (2008). Regular surface patterning of peatlands: Confronting theory with field data. *Ecosystems* **11**: 520–538.

Esteban, J. and Fairen, V. (2006). Self-organized formation of banded vegetation patterns in semi-arid regions: A model. *Ecol. Complex.* **3**: 109–118.

Fearnehough, W., Fullen, M. A., Mitchell, D. J., Trueman, I. C., and Zhang, J. (1998). Aeolian deposition and its effect on soil and vegetation changes on stabilised desert dunes in Northern China. *Geomorphology* **23**: 171–182.

Fernandez, I., Lopez, J. M., Pacheco, J. M., and Rodriguez, C. (2002). Hopf bifurcations and slow-fast cycles in a model of phytoplankton dynamics. *Rev. Acad. Canaria Ciencias* **14**: 121–137.

Feynman, R. P., Leighton, R. B., and Sands, M. (1963). *The Feynman Lectures on Physics*. Addison-Wesley, Reading, MA, Vol. 1.

Findell, K. L. and Eltahir, E. A. B. (1997). An analysis of the soil moisture-rainfall feedback, based on direct observations from Illinois. *Water Resource Res.* **33**: 725–735.

Findell, K. L. and Eltahir, E. A. B. (2003). Atmospheric controls on soil moisture-boundary layer interactions. Part i. Framework development. *J. Hydrometeorol.* **4**: 552–569.

Folke, C., Carpenter, S., Walker, B., Scheffer, M., Elmqvist, T., Gunderson, L., and Holling, C. S. (2004). Regime shifts, resilience and biodiversity in ecosystem management. *Annu. Rev. Ecol. Evol. Syst.* **35**: 557–581.

Fourriere, A., Claudin, P., and Andreotti, B. (2010). Bedforms in a turbulent stream: Formation of ripples by primary linear instability and of dunes by nonlinear pattern coarsening. *J. Fluid Mech.* **649**: 287–328.

Fox, R. (1986). Uniform convergence to an effective Fokker–Planck equation for weakly colored noise. *Phys. Rev. A* **34**: 4525.

Gammaitoni, L., Hänggi, P., Jung, P., and F., M. (1998). Stochastic resonance. *Rev. Mod. Phys.* **70**: 223–287.

Ganopolski, A. and Rahmstorf, S. (2001). Rapid changes of glacial climate simulated in a coupled climate model. *Nature (London)* **409**: 153–158.

Ganopolski, A. and Rahmstorf, S. (2002). Abrupt glacial climate changes due to stochastic resonance. *Phys. Rev. Lett.* **88**: 038501–1,4.

Ganopolski, A., Rahmstorf, S., Petoukhov, V., and Claussen, M. (1998). Simulation of modern and glacial climates with a coupled global model of intermediate complexity. *Nature (London)* **391**: 351–356.

Garcia-Moya, E. and McKell, C. M. (1970). Contribution of shrubs to the nitrogen economy of a desert-wash plant community. *Ecology* **51**: 81–88.

Garcia-Ojalvo, J., Parrondo, J. M. R., Sancho, J. M., and Van den Broeck, C. (1996). Reentrant transition induced by multiplicative noise in the time-dependent Ginzburg–Landau model. *Phys. Rev. E* **54**: 6918–6921.

Garcia-Ojalvo, J. and Sancho, J. M. (1999). *Noise in Spatially Extended Systems*. Springer-Verlag, New York.

Gardiner, C. W. (1983). *Handbook of Stochastic Methods*. Springer, Berlin.

Gardner, M. R. and Ashby, W. R. (1970). Connectance of large dynamics (cybernetic) systems: Critical values for stability. *Nature (London)* **228**: 784.

Gilad, E., von Hardenberg, J., Provenzale, A., Shachak, M., and Meron, E. (2004). Ecosystems engineers: From pattern formation to habitat creation. *Phys. Rev. Lett.* **93**: 098105–1–098105–4.

Gilad, E., von Hardenberg, J., Provenzale, A., Shachak, M., and Meron, E. (2007). A mathematical model of plants as ecosystem engineers. *J. Theor. Biol.* **244**: 680–691.

Gilligan, C. and van den Bosch, F. (2008). Epidemiological models for invasion and persistence of pathogens. *Annu. Rev. Phytopathol.* **46**: 385–418.

Gleason, K., Krantz, W. B., Caine, N., George, J. H., and Gunn, R. D. (1986). Geometrical aspects of sorted patterned ground in recurrently frozen soil. *Science* **232**: 216–220.

Goel, N., Maitra, S., and Montroll, E. (1971). On the Lotka–Volterra and other non-linear models of interacting populations. *Rev. Mod. Phys.* **43**: 231–276.

Goldammer, J. G. and de Ronde, C., editors (2004). *Wildland Fire Management Handbook for Sub-Sahara Africa*. Global Fire Monitoring Center Cape Town, South Africa.

Goldwasser, L. and Roughgarden, J. (1993). Construction and analysis of a large Caribbean food web. *Ecology* **74**: 1216–1233.

Greene, R. S. B. (1992). Soil physical properties of three geomorphic zones in a semiarid mulga woodland. *Austr. J. Soil Res.* **30**: 55–69.

Greene, R. S. B., Kinnell, P. I. A., and Wood, J. T. (1994). Role of plant cover and stock trampling on runoff and soil erosion from semiarid wooded rangelands. *Austr. J. Soil Res.* **32**: 953–973.

Greene, R. S. B., Valentin, C., and Esteves, M. (2001). Runoff and erosion processes. In Valentin, C., Tongway, D., Seghieri, J., and d'Herbes, J. M., editors, *Banded Vegetation Patterning in Arid and Semi-Arid Environment-Ecological Processes and Consequences for Management*, Vol. 149 of Ecological Studies Series. Springer-Verlag, New York, pp. 52–76.

Greenman, J. V. and Benton, T. G. (2003). The amplification of environmental noise in population models: Causes and consequences. *Am. Naturalist* **161**: 225–239.

Greig-Smith, P. (1979). Pattern in vegetation. *J. Ecol.* **67**: 775–779.

Grinstein, G., Munoz, M., and Tu, Y. (1996). Phase structure of systems with multiplicative noise. *Phys. Rev. Lett.* **76**: 4376–4379.

Gunderson, L. H. (2000). Ecological resilience – In theory and application. *Annu. Rev. Ecol. Syst.* **31**: 425–439.

Hahn, H., Nitzan, A., Ortoleva, P., and Ross, J. (1974). Threshold excitations, relaxation oscillations, and effect of noise in an enzyme reaction. *Proc. Natl. Acad. Sci. USA* **71**: 4067–4071.

Han, Y., Li, J., and Chen, S. (2005). Effect of asymmetric potential and Gaussian colored noise on stochastic resonance. *Commun. Theor. Phys.* **44**: 226–230.

Hanggi, P. (1978). Correlation functions and masterequations of generalized (non-Markovian) Langevin equations. *Z. Phys. B* **31**: 407–416.

Hanggi, P. (1989). Colored noise in continuous dynamical systems: A functional calculus approach. In Moss, F. and P.V.E., M., editors, *Noise in Nonlinear Dynamical Systems*. Cambridge University Press, New York, pp. 307–328.

Hanggi, P. and Jung, P. (1995). Colored noise in dynamical systems. In Prigogine, I. and S.A., R., editors, *Advances in Chemical Physics*. Wiley, New York, pp. 239–326.

Hanggi, P., Mroczkowski, T., Moss, F., and McClintock, P. (1985). Bistability driven by colored noise: Theory and experiment. *Phys. Rev. A* **32**: 695–698.

Harmer, J. and Abbott, D. (1996). Losing strategies can win by Parrondo's paradox. *Phys. Rev. Lett.* **85**: 5226.

Harrison, M. A., Lai, Y. C., and Holt, R. (2001). Dynamical mechanism for coexistence of dispersing species. *J. Theor. Biol.* **213**: 53–72.

Heimsath, A. M., Dietrich, W. E., Nishiizumi, K., and Finkel, R. C. (1997). The soil production function and landscape equilibrium. *Nature (London)* **388**: 358–361.

Heino, M. (1998). Noise colour, synchrony, and extinctions in spatially structured populations. *Oikos* **83**: 368–375.

Heino, M., Ripa, J., and Kaitala, V. (2000). Extinction risk under coloured environmental noise. *Ecography* **23**: 177–184.

Henderson, T. C., Venkataraman, R., and Choikim, G. (2004). Reaction-diffusion patterns in smart sensor networks. In *Proceedings of the 2004 IEEE International Conference on Robotics and Automation*. IEEE, New York, pp. 654–656.

Higgins, S. I., Bond, W. J., and Trollope, W. S. W. (2000). Fire, resprouting and variability: A recipe for tree–grass coexistence in savanna. *J. Ecol.* **88**: 213–229.

HilleRisLambers, R., Rietkerk, M., van den Bosch, F., Prins, H. H. T., and de Kroon, H. (2001). Vegetation pattern formation in semi-arid grazing systems. *Ecology* **82**: 50–61.

Hipel, K. W. and McLeod, A. I. (1994). *Time Series Modeling of Water Resources and Environmental Systems*. Elsevier, Amsterdam.

Holling, C. S. (1973). Resilience and stability of ecological systems. *Annu. Rev. Ecol. Evol. Syst.* **4**: 1–23.

Holt, R. D. (1984). Spatial heterogeneity, indirect interactions and the coexistence of prey species. *Am. Naturalist* **124**: 377–406.

Holt, R. D. and McPeek, M. A. (1996). Chaotic population dynamics favors the evolution of dispersal. *Am. Naturalist* **144**: 741–771.

Hongler, M.-O. (1979). Exact time dependent probability density for a non-linear non-Markovian stochastic process. *Helv. Phys. Acta* **52**: 280–287.

Horsthemke, W. and Lefever, R. (1984). *Noise-Induced Transitions: Theory and Applications in Physics, Chemistry and Biology*. Springer, Berlin.

Howard, J. (1997). Molecular motors: Structural adptations to cellular functions. *Nature (London)* **389**: 561–567.

Hughes, A. R., Byrnes, J. E., Kimbro, D. L., and Stachowicz, J. J. (2007). Reciprocal relationships and potential feedbacks between biodiversity and disturbance. *Ecol. Lett.* **10**: 849–864.

Huston, M. A. (1979). A general hypothesis of species diversity. *Am. Naturalist* **113**: 81–101.

Hutt, A., Longtin, A., and Schimansky-Geier, L. (2008). Additive noise-induced turing transitions in spatial systems with application to neural fields and the Swift–Hohenberg equation. *Physica D* **237**: 755–773.

Ikeda, S. and Parker, G. (1989). *River Meandering*. American Geophysical Union, Water Resources Monograph, Washington, D.C.

Jeltsch, F., Milton, S. J., Dean, W. R. J., and Van Rooyen, N. (1996). Tree spacing and coexistence in semiarid savannas. *J. Ecol.* **84**: 583–595.

Joffre, R. and Rambal, S. (1993). How tree cover influences the water balance of Mediterranean rangelands. *Ecology* **74**: 570–582.

Johnson, N. L., Kotz, S., and Balakrishnan, N. (1994). *Continuous Univariate Distributions*. Wiley, New York.

Jones, C. G., Lawton, J. H., and Shachak, M. (1994). Organisms as ecosystem engineers. *Oikos* **69**: 373–386.

Julicher, F., Ajdari, A., and Prost, J. (1997). Modeling molecular motors. *Rev. Mod. Phys.* **69**: 1269–1281.

Jung, P. and Hanggi, P. (1987). Dynamic systems – A unified colored-noise approximation. *Phys. Rev. A* **35**: 4464–4466.

Kantz, H. and Schreiber, T. (1997). *Nonlinear Time Series Analysis*. Cambridge University Press, Cambridge.

Katul, G., Porporato, A., and Oren, R. (2007). Stochastic dynamics of plant–water interactions. *Annu. Rev. Ecol. Evol. Syst.* **38**: 767–791.

Katz, R. W. and Brown, B. G. (1992). Extreme events in a changing climate: Variability is more important than averages. *Clim. Change* **21**: 289–302.

Kefi, S., Rietkerk, M., van Baalen, M., and Loreau, M. (2007). Local facilitation, bistability and transitions in arid ecosystems. *Theor. Popul. Biol.* **71**: 367–379.

Kendall, M. and Stuart, A. (1977). *The Advanced Theory of Statistics*. Charles Griffin, London.

Kent, M., Moyeed, R., C.L., R., Pakeman, R., and Weaver, R. (2006). Geostatistics, spatial rate of change analysis and boundary detection in plant ecology and biogeography. *Progr. Phys. Geogr.* **30**: 201–231.

Kim, Y. and Eltahir, E. A. B. (2004). Role of topography in facilitating coexistence of trees and grasses within savannas. *Water Resource Res.* **40**: W07505.

Kirkby, M. J. (1971). Hillslope process-response models based on the continuity equation. In *Slopes and Form Process*. Institute of British Geography, Spec. Publ. 3, London, pp. 15–30.

Kitahara, K., Horsthemke, W., Lefever, R., and Inaba, Y. (1980). Phase diagrams of noise induced transitions. *Prog. Theor. Phys.* **64**: 1233–1247.

Klausmeier, C. A. (1999). Regular and irregular patterns in semiarid vegetation. *Science* **284**: 1826–1828.

Kot, M. (2001). *Elements of Mathematical Ecology*. Cambridge University Press, Cambridge.

Krantz, W. B. (1990). Self-organization manifest as patterned ground in recurrently frozen soils. *Earth Sci. Rev.* **29**: 117–130.

Krueger, A. and Fritz, S. (1961). Cellular cloud patterns revealed by Tiros I. *Tellus* **13**: 1–12.

Krug, J. (1997). Origins of scale invariance in growth processes. *Adv. Phys.* **46**: 139–282.

Krumbein, W. C. and Graybill, F. A. (1965). *An Introduction to Statistical Models in Geology.* McGraw-Hill, New York.

Kuznetsov, P., Stratonovic, R., and Thikhonov, V. (1965). *Nonlinear Transformations of Stochastic Processes.* Pergamon, Oxford.

Laakso, J., Kaitala, V., and Ranta, E. (2001). How does environmental variation translate into biological processes? *Oikos* **92**: 119–122.

Lachenbruch, A. H. (1961). Depth and spacing of tension cracks. *J. Geophys. Res.* **66**: 4273–4292.

Lai, Y. C. and Liu, Y. R. (2005). Noise promotes species diversity in nature. *Phys. Rev. Lett.* **94**: 038102.

Laio, F., Porporato, A., Ridolfi, L., and Rodriguez-Iturbe, I. (2001). Plants in water-controlled ecosystems: Active role in hydrologic processes and response to water stress. ii. Probabilistic soil moisture dynamics. *Adv. Water Res.* **24**: 707–723.

Laio, F., Ridolfi, L., and D'Odorico, P. (2008). Noise-induced transitions in state-dependent dichotomous noise. *Phys. Rev. E* **78**: 031137.

Lancaster, N. (1995). *Geomorphology of Desert Dunes.* Routledge, London.

Landa, P., Zaikin, A., and Schimansky-Geier, L. (1998). Influence of additive noise on noise-induced phase transitions in nonlinear chains. *Chaos, Solitons, Fractals* **9**: 1367–1372.

Lawton, J. H. and Brown, V. K. (1993). Noise promotes species diversity in nature. In E. D. Schulze and H. A. Mooney, editors, *Biodiversity and Ecosystem Function.* Springer, Berlin, pp. 255–270.

Lefever, R. (1990). Stochastically perturbed chemical systems. In Gray, P., Nicolis, G., Baras, F., Borkmans, P., and Scott, S., editors, *Spatial Inhomogeneities and Transient Behaviour.* Manchester University Press, Proceedings in Nonlinear Science, Manchester, UK, pp. 479–505.

Lefever, R. and Lejeune, O. (1997). On the origin of tiger bush. *Bull. Math. Biol.* **59**: 263–294.

Lefever, R., Lejeune, O., and Couteron, P. (2000). Generic modelling of vegetation patterns. A case study of tiger bush in sub-saharian sahel. In Maini, P. K. and Othmer, H. G., editors, *Mathematical Models for Biological Pattern Formation.* Springer, New York, pp. 83–112.

Lejeune, O., Couteron, P., and Lefever, R. (1999). Short range co-operativity competing with long range inhibition explains vegetation patterns. *Acta Oecol.* **20**: 171–183.

Lejeune, O. and Tlidi, M. (1999). A model for the explanation of vegetation stripes (tiger bush). *J. Veg. Sci.* **10**: 201–208.

Lejeune, O., Tlidi, M., and Couteron, P. (2002). Localized vegetation patches: A self-organized response to resource scarcity. *Phys. Rev. E* **66**: 010901–4.

Lejeune, O., Tlidi, M., and Lefever, R. (2004). Vegetation spots and stripes: Dissipative structures in arid landscapes. *Int. J. Quantum Chem.* **98**: 261–271.

Leppanen, T. (2005). The theory of Turing pattern formation. In Kaski, K. and Barrio, R. A., editors, *Current Topics in Physics in Honor of Sir Roger Elliot.* Imperial College Press, London, pp. 190–227.

Levins, R. (1969). The effect of random variations of different types on population growth. *Proc. Natl. Acad. Sci. USA* **62**: 1061–1065.

Lewontin, R. C. and Cohen, D. (1969). On population growth in a randomly varying environment. *Proc. Natl. Acad. Sci. USA* **62**: 1056–1060.

Li, J. M. (2007). Escape over fluctuating potential barrier with complicated dichotomous noise. *J. Phys. A* **40**: 621–635.

Lindner, B., Garcia-Ojalvo, J., and Neiman, A. (2004). Effects of noise in excitable systems. *Phys. Rep.* **392**: 321–424.

Lindner, J., Meadows, B. K., Ditto, W., Inchiosa, M., and Bulsara, A. (1995). Array enhanced stochastic resonance and spatiotemporal synchronization. *Phys. Rev. Lett.* **75**: 3–6.

Loescher, H. W., Oberbauer, S. F., Gholz, H. L., and Clark, D. B. (2003). Environmental controls on net ecosystem-level carbon exchange and productivity in a Central American tropical wet forest. *Global Change Biol.* **9**: 396–412.

Loreau, M. (1992). Time scale of resource dynamics and coexistence through time partitioning. *Theor. Popul. Biol.* **41**: 401–412.

Lovejoy, S. (1982). Area-perimeter relation for rain and cloud areas. *Science* **216**: 185–187.

Ludwig, D., Jones, D. D., and Holling, C. S. (1978). Quantitative analysis of insect outbreak systems: The spruce budworm and forest. *J. Anim. Ecol.* **47**: 315–332.

Ludwig, J. A. and Tongway, D. J. (1995). Spatial-organization of landscapes and its function in semiarid woodlands, Australia. *Landscape Ecol.* **10**: 51–63.

Mabbutt, J. A. and Fanning, P. C. (1987). Vegetation banding in arid western australia. *J. Arid Environ.* **12**: 41–59.

Macfadyen, W. (1950). Vegetation patterns in the semi-desert plains of British Somaliland. *Geogr. J.* **166**: 199–211.

Mackey, R. L. and Currie, D. J. (2001). The diversity-disturbance relationship: Is it generally strong and peaked? *Ecology* **82**: 3479–3492.

Magnasco, M. O. (1993). Forced thermal ratchets. *Phys. Rev. Lett.* **71**: 1477–1481.

Mangioni, S., Deza, R., S., W. H., and Toral, R. (1997). Disordering effects of color in nonequilibrium phase transitions induced by multiplicative noise. *Phys. Rev. Lett.* **79**: 2389–2393.

Mankin, R., Ainsaar, A., Haljas, A., and Reiter, E. (2002). Trichotomous noise-induced catastrophic shifts in symbiotic ecosystems. *Phys. Rev. E* **65**: 051108.

Mankin, R., Ainsaar, A., and Reiter, E. (1999). Trichotomous noise-induced transitions. *Phys. Rev. E* **60**: 1374–1380.

Mankin, R., Sauga, A., Ainsaar, A., Haljas, A., and Paunel, K. (2004). Colored-noise-induced transitions in symbiotic ecosystems. *Phys. Rev. E* **69**: 061106.

Manneville, P., editor (1990). *Dissipative Structures and Weak Turbulence*. Academic, London.

Manor, A. and Shnerb, N. (2008). Facilitation, competition, and vegetation patchiness: From scale free distribution to patterns. *J. Theor. Biol.* **253**: 838–842.

Marchesoni, F., Gammaitoni, L., and Bulsara, A. (1996). Spatiotemporal stochastic resonance in a ϕ^4 model of kink–antikink nucleation. *Phys. Rev. Lett.* **75**: 2609–2612.

Martens, S. N., Breshears, D. D., Meyer, C. W., and Barnes, F. J. (1997). Scales of above-ground and below-ground competition in a semi-arid woodland detected from spatial pattern. *J. Veg. Sci.* **8**: 655–664.

Masoliver, J. and Weiss, G. E. (1994). Telegrapher's equations with variable propagation speeds. *Phys. Rev. E* **49**: 3852–3854.

Mason, B. J. (1976). Towards the understanding and prediction of climatic variation. *Q. J. R. Meteorol. Soc.* **102**: 473–499.

May, R. M. (1972). Will a large complex system be stable? *Nature (London)* **238**: 413–414.

May, R. M. (1973). *Stability and Complexity in Model Ecosystems*. Princeton University Press, Princeton, NJ.

McFadden, J. A. (1959). The probability density of the output of a filter when the input is a random telegraphic signal: differential equation method. *IRE Trans. Inf. Theory* **5**: 228–233.

McNamara, B., Wiesenfeld, K., and Roy, R. (1988). Observation of stochastic resonance in a ring laser. *Phys. Rev. Lett.* **60**: 2626–2629.

McNaughton, S. J. (1977). Diversity and stability of ecological communities: A comment on the role of empiricism in ecology. *Am. Naturalist* **111**: 515–525.

Meron, E., Gilad, E., von Hardenberg, J., Shachak, M., and Zarmi, Y. (2004). Vegetation patterns along a rainfall gradient. *Chaos, Solitons Fractals* **19**: 367–376.

Montana, C. (1992). The colonization of bare areas in 2-phase mosaics of an arid ecosystem. *J. Ecol* **80**: 315–327.

Mulder, C. P. H., Uliassi, D. D., and Doak, D. F. (2001). Physical stress and diversity-productivity relationships: The role of positive interactions. *Proc. Natl. Acad. Sci USA* **98**: 6704–6708.

Muller, R., Lippert, K., Kuhnel, A., and Behn, U. (1997). First-order nonequilibrium phase transition in a spatially extended system. *Phys. Rev. E* **56**: 2658–2662.

Murray, J. D. (2002). *Mathematical Biology*. Springer, Berlin.

Murray, J. D. and Maini, P. K. (1989). Pattern formation mechanisms – A comparison of reaction diffusion and mechanochemical models. In Goldbeter, A., editor, *Cell to Cell Signaling: From Experiments to Theoretical Models*. Academic, New York, pp. 159–170.

Naeem, S. and Li, S. (1997). Biodiversity enhances ecosystem reliability. *Nature (London)* **390**: 507–509.

Naiman, A., Schimansky-Geier, L., Cornell-Bell, A., and Moss, F. (1999). Noise-induced phase synchronization in excitable media. *Phys. Rev. Lett.* **87**: 4896–4899.

Nicholson, S. E. (1980). The nature of rainfall fluctuations in subtropical West Africa. *Mon. Weather Rev.* **108**: 473–487.

Nicholson, S. E. (2000). Land surface processes and sahel climate. *Rev. Geophys.* **38**: 117–139.

Nigering, W. A., Whittaker, R. H., and Lowe, C. H. (1963). The Saguaro: A population in relation to environment. *Science* **142**: 25–51.

Noy-Meir, I. (1973). Desert ecosystems: Environments and producers. *Annu. Rev. Ecol. Syst.* **4**: 25–51.

Noy-Meir, I. (1975). Stability of grazing systems: An application of predator–prey graphs. *J. Ecol.* **63**: 459–481.

Okayasu, T. and Aizawa, Y. (2001). Systematic analysis of periodic vegetation patterns. *Progr. Theor. Phys.* **106**: 705–720.

O'Loughlin, C. L. and Pearce, A. J. (1976). Influence of cenozoic geology on mass movement and sediment yield response to forest removal, North Westland, New Zealand. *Bull. Int. Assoc. Eng. Geol.* **14**: 41–46.

Oster, G. F. and Murray, J. D. (1989). Pattern-formation models and developmental constraints. *J. Exper. Zool.* **251**: 186–202.

Papoulis, A. (1984). *Signal Analysis*. McGraw-Hill, Auckland, Australia.

Parrondo, J. M. R., van den Broeck, C., Buceta, J., and DeLaRubia, F. J. (1996). Noise-induced spatial patterns. *Physica A* **224**: 153–161.

Parzen, E. (1967). *Stochastic Processes*. Holden-Day, San Francisco.

Pawula, R. F. (1967). Generalizations and extensions of the Fokker–Planck–Kolmogorov equations. *IEEE Trans. Inf. Theory* **13**: 33–41.

Pawula, R. F. (1977). Probability density and level-crossing of first order nonlinear-systems driven by random telegraph signal. *Int. J. Control* **25**: 283–292.

Pawula, R. F., Porra, J. M., and Masoliver, J. (1993). Mean first passage times for systems driven by gamma and McFadden dichotomous noise. *Phys. Rev. E* **47**: 189–201.

Pelletier, J. D. (2003). Coherence resonance in ice ages. *J. Geophys Res.* **108**(D20): 4645.

Petchey, O. L., Gonzalez, A., and Wilson, H. B. (1997). Effects on population persistence: The interaction between environmental noise colour, intraspecific competition and space. *Proc. R. Soc. London Ser. B* **264**: 1841–1847.

Pikovsky, A., Zaikin, A., and de la Casa, M. A. (2002). System size resonance in coupled noisy systems and in the Ising model. *Phys. Rev. E* **88**: 050601.

Pikovsky, A. S. and Kurths, J. (1997). Coherence resonance in a noise-driven excitable system. *Phys. Rev. Lett.* **78**: 775–778.

Pimm, S. L. (1984). The complexity and stability of ecosystems. *Nature (London)* **307**: 1321–326.

Pirrotta, A. (2005). Non-linear systems under parametric white noise input: Digital simulation and response. *Int. J. Non-Linear Mech.* **40**: 1088–1101.

Platten, J. K. and Legros, J. C. (1984). *Convection in Liquids.* Springer-Verlag, Berlin.

Polis, G. A. (1991). Complex trophic interactions in deserts: an empirical critique of food-web theory. *Am. Naturalist* **138**: 123–155.

Polyanin, A. D., Zaitsev, V. F., and Moussiaux, A. (2002). *Handbook of First Order Partial Differential Equations.* Taylor & Francis, London.

Porporato, A. and D'Odorico, P. (2004). Phase transitions driven by state-dependent Poisson noise. *Phys. Rev. Lett.* **92**: 110601.

Porporato, A., Laio, F., Ridolfi, L., and Rodriguez-Iturbe, I. (2001). Plants in water-controlled ecosystems: Active role in hydrologic processes and response to water stress. iii. Vegetation water stress. *Adv. Water Resources* **24**: 725–744.

Puigdefabregas, J., Gallart, F., Biaciotto, O., Allogia, M., and del Barrio, G. (1999). Banded vegetation patterning in a subantarctic forest of Tierra del Fuego, as an outcome of the interaction between wind and tree growth. *Acta Oecol.* **20**: 135–146.

Ravi, S., D'Odorico, P., Wang, L., and Collins, S. (2009). Form and function of grass ring patterns in arid grasslands: The role of abiotic controls. *Oecologia* **158**:–555, doi:10.1007/s00442–008–1164–1.

Reimann, P. (2002). Brownian motors: Noisy transport far from equilibrium. *Phys. Rep.* **361**: 57–265.

Reimann, P. and Hanggi, P. (2002). Introduction to the physics of Brownian motors. *Appl. Phys.* **75**: 169–178.

Rennermalm, A., Słgaard, H., and C., N. (2005). Interannual variability in carbon dioxide flux from a high arctic fen estimated by measurements and modelling. *Arctic, Antarctic Alpine Res.* **37**: 545–556.

Rice, S. (1944). Mathematical analysis of random noise. *Bell Syst. Tech. J.* **23**: 282332.

Rice, S. (1945). Mathematical analysis of random noise. *Bell Syst. Tech. J.* **24**: 146–156.

Richards, J. H. and Caldwell, M. M. (1987). Hydraulic lift – Substantial nocturnal water transport between soil layers by Artemisia-Tridentata roots. *Oecologia* **73**: 486–489.

Ridolfi, L., D'Odorico, P., and Laio, F. (2006). Effect of vegetation-water table feedbacks on the stability and resilience of plant ecosystems. *Water Resource Res.* **42**: W01201.

Ridolfi, L., D'Odorico, P., and Laio, F. (2007). Vegetation dynamics induced by phreatophyte-water table interactions. *J. Theor. Biol.* **248**: 301–310.

Ridolfi, L., D'Odorico, P., Porporato, A., and Rodriguez-Iturbe, I. (2000). Impact of climate variability on the vegetation water stress. *J. Geophys. Res. (Atmospheres)*, **105**(D14): 18013–18025.

Rietkerk, M., Boerlijst, M. C., van Langevelde, F., HilleRisLambers, R., van de Koppel, J., Kumar, L., Klausmeier, C. A., Prins, H. H. T., and de Roos, A. M. (2002). Self-organisation of vegetation in arid ecosystems. *Am. Naturalist* **160**: 524–530.

Rietkerk, M., Dekker, S. C., de Ruiter, P. C., and van de Koppel, J. (2004). Self-organized patchiness and catastrophic shifts in ecosystems. *Science* **305**: 1926–1929.

Rietkerk, M. and van de Koppel, J. (1997). Alternate stable states and threshold effects in semi-arid grazing systems. *Oikos* **79**: 69–76.

Risken, H. (1984). *The Fokker–Planck Equation.* Springer, Berlin.

Rodriguez-Iturbe, I., D'Odorico, P., Porporato, A., and Ridolfi, L. (1999a). Tree–grass coexistence in savannas: The role of spatial dynamics and climate fluctuations. *Geophys. Res. Lett.* **26**: 247–250.

Rodriguez-Iturbe, I., Entekhabi, D., and Bras, R. L. (1991). Nonlinear dynamics of soil moisture at climate scales. i. Stochastic analysis. *Water Resource Res.* **27**: 1899–1906.

Rodriguez-Iturbe, I. and Porporato, A. (2005). *Ecohydrology of Water-Controlled Ecosystems.* Cambridge University Press, Cambridge.

Rodriguez-Iturbe, I., Porporato, A., Ridolfi, L., V., I., and Cox, D. R. (1999b). Probabilistic modelling of water balance at a point: The role of climate, soil and vegetation. *Proc. R. Soc. London Ser. A* **455**: 3789–3805.

Rodriguez-Iturbe, I. and Rinaldo, A. (2001). *Fractal River Basins: Chance and Self-Organization*. Cambridge University Press, Cambridge.

Roering, J. J., Kirchner, J. W., and Dietrich, W. E. (1999). Evidence for nonlinear, diffusive sediment transport on hillslopes and implications for landscape morphology. *Water Resource Res.* **35**: 853–870.

Rohani, P., Lewis, T. J., Grunbaum, D., and Ruxton, G. D. (1997). Spatial self-organization in ecology: Pretty patterns or robust reality? *Trends Ecol. Evol.* **12**(2): 70–74.

Ross, S. M. (1996). *Stochastic Processes*. Wiley, New York.

Rovinsky, A. B. and Menzinger, M. (1992). Chemical-instability induced by a differential flow. *Phys. Rev. Lett.* **69**: 1193–1196.

Rovinsky, A. B. and Zhabotinsky, A. M. (1984). Mechanism and mathematical model of the oscillating bromate-ferroin-bromomalonic acid reaction. *J. Phys. Chem.* **88**: 6081–6084.

Sagues, F., Sancho, J. M., and García-Ojalvo, J. (2007). Spatio-temporal order out of noise. *Rev. Mod. Phys.* **79**: 829–882.

Salati, E., Dallolio, A., Matsui, E., and Gat, J. R. (1979). Recycling of water in the Amazon basin: An isotopic study. *Water Resource Res.* **15**: 1250–1255.

Sancho, J. and San Miguel, M. (1989). Langevin equations with colored noise. In Moss, F. and P.V.E., McClintock editors, *Noise in Nonlinear Dynamical Systems*. Cambridge University Press, Cambridge, pp. 72–109.

Sancho, J., San Miguel, M., Katz, S., and J.D., G. (1982). Analytical and numerical studies of multiplicative noise. *Phys. Rev. A* **26**: 1589–1609.

Sankaran, M., Ratnam, J., and Hanan, N. P. (2004). Tree–grass coexistence in savannas revisited – Insights from an examination of assumptions and mechanisms invoked in existing models. *Ecol. Lett.* **7**: 480–490.

Sanz-Anchelergues, A., Zhabotinsky, N., Epstein, I., and Munuzuri, A. (2001). Turing pattern formation induced by spatially correlated noise. *Phys. Rev. E* **63**: 056124.

Sauga, A. and Mankin, R. (2005). Addendum to "colored-noise-induced discontinuous transitions in symbiotic ecosystems." *Phys. Rev. E.* **71**: 062103.

Scheffer, M., Carpenter, S., Foley, J. A., Folke, C., and Walker, B. (2001). Catastrophic shifts in ecosystems. *Nature (London)* **413**: 591–596.

Schlesinger, W., Reynolds, J., Cunningham, G. L., Huenneke, L., Jarrell, W., Virginia, R., and Whitford, W. (1990). Biological feedbacks in global desertification. *Science* **247**: 1043–1048.

Scholes, R. J. and Archer, S. R. (1997). Tree–grass interactions in savannas. *Annu. Rev. Ecol. Syst.* **28**: 517–544.

Schwager, M., Johst, K., and Jeltsch, F. (2006). Does red noise increase or decrease extinction risk? Single extreme events versus series of unfavorable conditions. *Am. Naturalist* **167**: 879–888.

Scott, A. (1975). The electrophysics of a nerve fiber. *Rev. Mod. Phys.* **47**: 487–533.

Sellers, W. D. (1969). A global climate model based on the energy balance of the earth-atmosphere system. *J. Appl. Meteorol.* **8**: 396–400.

Seminara, G. (2010). Fluvial sedimentary patterns. *Annu. Rev. Fluid Mech.* **42**: 43–66.

Sherrat, J. A. (2005). An analysis of vegetation stripe formation in semi-arid landscapes. *J. Math. Biol.* **51**: 183–197.

Shnerb, N. M., Sarah, P., Lavee, H., and Solomon, S. (2003). Reactive glass and vegetation patterns. *Phys. Rev. Lett.* **90**: 038101-1–038101-4.

Sidle, R. C. and Swanston, D. N. (1982). Analysis of a small debris slide in coastal Alaska. *Can. Geotech. J.* **19**: 167–174.

Sieber, M., Malchow, H., and Schimansky-Geier, L. (2007). Constructive effects of environmental noise in an excitable prey-predator plankton system with infected prey. *Ecol. Complex.* **4**: 223–233.

Smit, G. N. and Rethman, N. F. G. (2000). The influence of tree thinning on the soil water in a semi-arid savanna of southern Africa. *J. Arid Environ.* **44**: 41–59.

Smoluchowski, M. V. (1912). Experimentell nachweisbare, der üblichen thermodynamik widersprechende molekularphänomene. *Phys. Zeitschr.* **13**: 1069.

Spagnolo, B., Valenti, D., and Fiasconaro, A. (2004). Noise in ecosystems: A short review. *Math. Biosci.* **1**: 185–211.

Staliunas, K. and Sanchez-Morchillo, V. (2000). Turing patterns in non-linear optics. *Opt. Commun.* **117**: 389–395.

Steneck, R. S., Graham, M. H., Bourque, B. J., Corbett, D., Erlandson, J. M., Estes, J. A., and Tegner, M. J. (2002). Kelp forest ecosystems: Biodiversity, stability, resilience and future. *Environ. Conserv.* **29**: 436–459.

Strikwerda, J. (2004). *Finite Difference Schemes and Partial Differential Equations*. Society of Industrial and Applied Mathematics, Philadelphia.

Swift, J. and Hohenberg, P. (1977). Hydrodynamics fluctuations at the convective instability. *Phys. Rev. A* **15**: 319.

Thiery, J. M., D'Herbes, J.-M., and Valentin, C. (1995). A model simulating the genesis of banded vegetation patterns in Niger. *J. Ecol.* **83**: 497–507.

Tilman, D. (1994). Competition and biodiversity in spatially structured habitats. *Ecology* **75**: 2–16.

Timmermann, A., Gildor, H., Schulz, M., and Tziperman, E. (2003). Coherent resonant millennial-scale oscillations triggered by massive meltwater pulses. *J. Clim.* **16**: 2569–2585.

Todorovic, P. and Woolhiser, D. A. (1975). A stochastic model of *n*-day precipitation. *J. Appl. Meteorol.* **14**: 17–24.

Trefethen, L. N., Trefethen, A. I., Reddy, S., and Driscoll, T. (1993). Hydrodynamic stability without eigenvalues. *Science* **261**: 578–584.

Trenberth, K. E. (1998). Atmospheric moisture recycling: Role of advection and local evaporation. *J. Clim.* **12**: 1368–1381.

Truscott, J. and Brindley, J. (1994). Ocean plankton populations as excitable media. *Bull. Math. Biol.* **56**: 981–998.

Trustrum, N. A. and De Rose, R. C. (1988). Soil depth-age relationship of landslides on deforested hillslopes, Taranaki, New Zealand. *Geomorphology* **1**: 143–160.

Tuckwell, H. C. (1988). *Elementary Applications of Probability Theory*. Chapman & Hall, London.

Turelli, M. (1978). Does environmental variability limit niche overlap? *Proc. Natl. Acad. Sci. USA* **75**: 5085–5089.

Turing, A. (1952). The chemical basis of morphogenesis. *Philos. Trans. R. Soc. B* **237**: 37–72.

Valentin, C., D'Herbes, J. M., and Poesen, J. (1999). Soil and water components of banded vegetation patterns. *Catena* **37**(1-2): 1–24.

van de Koppel, J., Altieri, A. H., Silliman, B. R., Bruno, J. F., and Bertness, M. D. (2006). Scale-dependent interactions and community structure on cobble beaches. *Ecol. Lett.* **9**: 45–50.

van de Koppel, J. and Rietkerk, M. (2004). Spatial interactions and resilience in arid ecosystems. *Am. Naturalist* **163**: 113–121.

van de Koppel, J., Rietkerk, M., Dankers, N., and Herman, P. M. J. (2005). Scale-dependent feedback and regular spatial patterns in young mussel beds. *Am. Naturalist* **165**: E66–E77.

van den Broeck, C. (1983). On the relation between white shot noise, Gaussian white noise, and the dichotomic Markov process. *J. Stat. Phys.* **31**: 467–483.

van den Broeck, C. (1990). Taylor dispersion revised. *Physica A* **168**: 677–696.

van den Broeck, C. (1997). From Stratonovich calculus to noise-induced phase transitions. In Schimansky-Geier, L. and Poschel, T., editors, *Stochastic Dynamics*. Springer, Berlin.

van den Broeck, C., Parrondo, J. M. R., and Toral, R. (1994). Noise-induced nonequilibrium phase transition. *Phys. Rev. Lett.* **73**: 3395–3398.

van Kampen, N. G. (1981). Ito versus Stratonovich. *J. Stat. Phys.* **24**: 175–187.

van Kampen, N. G. (1992). *Stochastic Processes in Physics and Chemistry*. North-Holland, Amsterdam.

van Langevelde, F., van de Vijver, C. A. D. M., Kumar, L., van de Koppel, J., de Ridder, N., van Andel, J., Skidmore, A. K., Hearne, J. W., Stroosnijder, L., Bond, W. J., Prins, H. H. T., and Rietkerk, M. (2003). Effects of fire and herbivory on the stability of savanna ecosystems. *Ecology* **84**: 337–350.

van Wijk, M. T. and Rodriguez-Iturbe, I. (2002). Tree-grass competition in space and time: Insights from a simple cellular automata model based on ecohydrological dynamics. *Water Resource Res.* **38**: 18.11–18.15.

van Wilgen, B. W., Trollope, W. S. W., Biggs, H. C., Potgieter, A. L. F., and Brockett, B. H. (2003). Fire as a driver of ecosystem variability. In Du Toit, J. T., Rogers, K. H., and Biggs, H. C., editors, *The Kruger Experience: Ecology and Management of Savanna Heterogeneity*. Island Press, Washington, D.C., pp. 149–170.

VanMarcke, E. (1983). *Random Fields: Analysis and Synthesis*. MIT Press, Cambridge, MA.

Vasseur, D. A. (2007). Populations embedded in trophic communities respond differently to coloured environmental noise. *Theor. Popul. Biol.* **72**: 186–196.

Vasseur, D. A. and Fox, J. W. (2007). Environmental fluctuations can stabilize food web dynamics by increasing synchrony. *Ecol. Lett.* **10**: 1066–1074.

Vélez-Belchi, P., Alvarez, A., Colet, P., Tintoré, J., and Haney, R. L. (2001). Stochastic resonance in the thermohaline circulation. *Geophys. Res. Lett.* **28**: 2053–2056.

Vetaas, O. R. (1992). Micro-site effects of threes and shrubs in dry savannas. *J. Veg. Sci.* **3**: 337–344.

Vilar, J. and Rubi, J. (1997). Spatiotemporal stochastic resonance in the Swift–Hohenberg equation. *Phys. Rev. Lett.* **78**: 2886–2889.

Vilar, J. and Rubi, J. (2000). Ordering periodic spatial structures by non-equilibrium fluctuations. *Physica A* **277**: 327–334.

von Hardenberg, J., Kletter, A., Yizhaq, H., Nathan, J., and Meron, E. (2010). Periodic versus scale-free patterns in dryland vegetation. *Proc. R. Soc. London Ser. B* **277**: 1771–1776.

von Hardenberg, J., Meron, E., Shachak, M., and Zarmi, Y. (2001). Diversity of vegetation patterns and desertification. *Phys. Rev. Lett.* **87**: 198101.

Walker, B. H., Ludwig, D., Holling, C. S., and Peterman, R. M. (1981). Stability of semi-arid savanna grazing systems. *J. Ecol.* **69**: 473–498.

Walker, B. H. and Noy-Meir, I. (1982). Aspects of stability and resilience of savanna ecosystems. In Walker, B. H. and Huntley, B. H., editors, *Ecology of Subtropical Savannas*. Springer-Verlag, Berlin, pp. 556–590.

Walker, B. H. and Salt, D. (2006). *Resilience Thinking: Sustaining Ecosystems and People in a Changing World*. Island Press, Washington, D.C.

Walter, H. (1971). *Ecology of Tropical and Subtropical Vegetation*. Oliver and Boyd, Edinburgh.

Wang, G. and Eltahir, E. A. B. (2000). Ecosystem dynamics and the Sahel drought. *Geophys. Res. Lett.* **27**: 795–798.

Watt, A. S. (1947). Pattern and process in plant community. *J. Ecol.* **35**: 1–22.

Weiss, G. H., Masoliver, J., Lindenberg, K., and West, B. J. (1987). First passage times for non-Markovian processes – Multivalued noise. *Phys. Rev. A* **36**: 1435–1439.

Wellens, T., Shatokhin, V., and Buchleitner, A. (2004). Stochastic resonance. *Rep. Progr. Phys.* **67**: 45–105.

White, L. P. (1969). Vegetation arcs in Jordan. *J. Ecol.* **57**: 461–464.

White, L. P. (1971). Vegetation stripes on sheet wash surfaces. *J. Ecol.* **59**: 615–622.

Wilde, S. A., Steinbrenner, R. S., Dosen, R. C., and Pronin, D. T. (1953). Influence of forest cover on the state of the ground water table. *Proc. Soil Sci. Soc. Am.* **17**: 65–67.

Wilson, J. B. and Agnew, A. D. Q. (1992). Positive-feedback switches in plant communities. *Adv. Ecol. Res.* **23**: 263–336.

Worral, G. A. (1959). The Butana grass patterns. *J. Soil Sci.* **10**: 34–55.

Worral, G. A. (1960). Patchiness in vegetation in the northern Sudan. *J. Ecol.* **48**: 107–115.

Wu, T. H. and Swanston, D. N. (1980). Risk of landslides in shallow soils and its relation to clearcutting in southeastern Alaska. *Forest. Sci.* **26**: 495–510.

Xu, C. and Li., Z. (2003). Population dynamics and the color of environmental noise: A study on a three-species food chain system. *Ecol. Res.* **18**: 145–154.

Xue, Y. (1997). Biosphere feedback on regional climate in tropical north Africa. *Q. J. R. Meteorol. Soc.* **123**: 1483–1515.

Xue, Y. and Shukla, J. (1993). The influence of land surface properties on Sahel climate. Part i. Desertification. *J. Clim.* **6**: 2232–2245.

Yachi, S. and Loreau, M. (1999). Biodiversity and ecosystem productivity in a fluctuating environment: The insurance hypothesis. *Proc. Natl. Acad. Sci USA* **96**: 1463–1468.

Yevjevich, V. (1970). *Stochastic Processes in Hydrology*. Water Resources Publications, LLC, Edinburgh.

Yizhaq, H., Gilad, E., and Meron, E. (2005). Banded vegetation: Biological productivity and resilience. *Physica A* **356**: 139–144.

Yokozawa, M., Kubota, Y., and Hara, T. (1998). Effects of competition mode on spatial pattern dynamics in plant communities. *Ecol. Model.* **106**: 1–16.

Yokozawa, M., Kubota, Y., and Hara, T. (1999). Effects of competition mode on the spatial pattern dynamics of wave regeneration in subalpine tree stands. *Ecol. Model.* **118**: 73–86.

Zaikin, A., Garcia-Ojalvo, J., and Schimansky-Geier, L. (1999). Nonequilibrium first-order phase transition induced by additive noise. *Phys. Rev. E* **60**: R6275–R6278.

Zaikin, A. and Schimansky-Geier, L. (1998). Spatial patterns induced by additive noise. *Phys. Rev. E* **58**: 4355–4360.

Zeng, N. and Neelin, J. D. (2000). The role of vegetation-climate interaction and interannual variability in shaping the african savanna. *J. Clim.* **13**: 2665–2670.

Zeng, N., Neelin, J. D., Lau, K. M., and Tucker, C. J. (1999). Enhancement of interdecadal climate variability in the sahel by vegetation interaction. *Science* **286**: 1537–1540.

Zeng, Q.-C. and Zeng, X. D. (1996). An analytical dynamic model of grass field ecosystem with two variables. *Ecol. Model.* **85**: 187–196.

Zeng, X., Shen, S. S. P., Zeng, X., and Dickinson, R. E. (2004). Multiple equilibrium states and the abrupt transitions in a dynamical system of soil water interacting with vegetation. *Geophys. Res. Lett.* **31**:10.129/2003GL018910.

Zhonghuai, H., Lingfa, Y., Zuo, X., and Houwen, X. (1998). Noise induced pattern transition and spatiotemporal stochastic resonance. *Phys. Rev. Lett.* **81**: 2854–2857.

Zhou, Y. R., Guo, F., Jiang, S. Q., and Pang, X. F. (2008). Stochastic resonance of a linear system induced by dichotomous noise. *Int. J. Mod. Phys. B* **22**: 697–708.

Index

Index

Printed in the United States
By Bookmasters